RATING OF ELECTRIC POWER CABLES IN UNFAVORABLE THERMAL ENVIRONMENT

RATING OF ELECTRIC POWER CABLES IN UNFAVORABLE THERMAL ENVIRONMENT

GEORGE J. ANDERS

IEEE PRESS

A JOHN WILEY & SONS, INC., PUBLICATION

Library of Congress Cataloging-in-Publication Data is available.

ISBN 0-471-67909-7

Printed in the United States of America.

10 9 8 7 6 5 4 3 2 1

Contents

Preface   xi

1 Review of Power Cable Standard Rating Methods   1

1.1 Introduction   1
1.2 Energy Conservation Equations   2
    1.2.1 Heat Transfer Mechanism in Power Cable Systems   2
        1.2.1.1 Conduction   2
        1.2.1.2 Convection   3
        1.2.1.3 Radiation   3
        1.2.1.4 Energy Balance Equations   4
    1.2.2 Heat Transfer Equations   5
        1.2.2.1 Underground Directly Buried Cables   5
        1.2.2.2 Cables in Air   6
    1.2.3 Analytical Versus Numerical Methods of Solving Heat   7
        Transfer Equations
1.3 Thermal Network Analogs   7
    1.3.1 Thermal Resistance   8
    1.3.1 Thermal Capacitance   10
    1.3.3 Construction of a Ladder Network of a Cable   10
        1.3.3.1 Representation of Capacitances of the Dielectric   11
        1.3.3.2 Reduction of a Ladder Network to a Two-Loop   15
        Circuit
1.4 Rating Equations—Steady-State Conditions   19
    1.4.1 Buried Cables   19
        1.4.1.1 Steady-State Rating Equation without Moisture   19
        Migration
        1.4.1.2 Steady-State Rating Equation with Moisture Migration   21
    1.4.2 Cables in Air   24
1.5 Rating Equations—Transient Conditions   24
    1.5.1 Response to a Step Function   25
        1.5.1.1 Preliminaries   25
        1.5.1.2 Temperature Rise in the Internal Parts of the Cable   26
        1.5.1.3 Second Part of the Thermal Circuit—Influence of   29
        the Soil

	1.5.1.4 Groups of Equally or Unequally Loaded Cables	30
	1.5.1.5 Total Temperature Rise	31
	1.5.2 Transient Temperature Rise under Variable Loading	32
	1.5.3 Conductor Resistance Variations during Transients	32
	1.5.4 Cyclic Rating Factor	32
1.6	Evaluation of Parameters	36
	1.6.1 List of Symbols	36
	1.6.1.1 General Data	36
	1.6.1.2 Cable Parameters	36
	1.6.1.3 Installation Conditions	38
	1.6.2 Conductor ac Resistance	38
	1.6.3 Dielectric Losses	40
	1.6.4 Sheath Loss Factor	41
	1.6.4.1 Sheath Bonding Arrangements	42
	1.6.4.2 Loss Factors for Single-Conductor Cables	45
	1.6.4.3 Three-Conductor Cables	48
	1.6.4.4 Pipe-Type Cables	49
	1.6.5 Armor Loss Factor	49
	1.6.5.1 Single-Conductor Cables	50
	1.6.5.2 Three-Conductor Cables—Steel Wire Armor	51
	1.6.5.3 Three-Conductor Cables—Steel Tape Armor or Reinforcement	52
	1.6.5.4 Pipe-Type Cables in Steel Pipe—Pipe Loss Factor	52
	1.6.6 Thermal Resistances	52
	1.6.6.1 Thermal Resistance of the Insulation	54
	1.6.6.2 Thermal Resistance between Sheath and Armor, T_2	58
	1.6.6.3 Thermal Resistance of Outer Covering (Serving), T_3	58
	1.6.6.4 Pipe-Type Cables	59
	1.6.6.5 External Thermal Resistance	60
	1.6.7 Thermal Capacitances	69
	1.6.7.1 Oil in the Conductor	69
	1.6.7.2 Conductor	71
	1.6.7.3 Insulation	71
	1.6.7.4 Metallic Sheath or Any Other Concentric Layer	72
	1.6.7.5 Reinforcing Tapes	72
	1.6.7.6 Armor	72
	1.6.7.7 Pipe-Type Cables	72
1.7	Concluding Remarks	72
	References	73
2	**Ampacity Reduction Factors for Cables Crossing Thermally Unfavorable Regions**	**77**
2.1	Cables Crossing Thermally Unfavorable Regions	77
	2.1.1 Introduction	77
	2.1.2 Ampacity Reduction Modeling	78

2.1.3 Temperature Distribution Along the Rated Cable 78
2.1.4 Derating Factor 84
2.1.5 Cyclic Rating for Cable Crossing Unfavorable Region 90
2.1.6 Soil Dryout Caused by Moisture Migration 99
2.2 Ventilated Routes 101
  2.2.1 Cable Laid in a Pipe with Air Convection 102
    2.2.1.1 Self-Supporting Air Convection 102
    2.2.1.2 Forced Air Convection 108
  2.2.2 Numerical Example Illustrating the Cooling Pipe Concepts 108
  2.2.3 Reduction of a Magnetic Field 112
2.3 Concluding Remarks 117
2.4 Chapter Summary 117
  References 119

3 Cable Crossings—Derating Considerations **121**

3.1 Introduction 121
3.2 Utility Practices 122
3.3 Derating Factor 124
3.4 Temperature Distribution Along the Rated Cable and the Mutual 125
  Thermal Resistance
  3.4.1 Single External Heat Source 125
  3.4.2 Heating by a Steam Pipe 131
  3.4.3 Multiple Crossing Heat Sources 134
3.5 Consideration of a Screen Longitudinal Heat Flow 140
3.6 Transient Temperature Rise of Cable Crossings 145
  3.6.1 Transient External Thermal Resistance 145
  3.6.2 Cyclic Rating 147
    3.6.2.1 Cyclic Loading of the External Heat Source 147
    3.6.2.2 Rated Cable with Cyclic Load 149
3.7 Soil Dryout Caused by Moisture Migration 156
3.8 Concluding Remarks 161
3.9 Chapter Summary 161
  References 164

4 Application of Thermal Backfills for Cables Crossing Unfavorable **165**
Thermal Environments

4.1 A Brief History of Soil Thermal Resistivity Measurements 166
4.2 Optimization of Power Cable and Thermal Backfill Configurations 168
  4.2.1 Analysis of the Effect of Parameter Variations 169
  4.2.2 Formulation of the Optimization Problem 170
  4.2.3 Assessment of Sensitivities to Ambient Fluctuations 178
4.3 Parameter Uncertainty in Rating Analysis of Cables Crossing 187
  Unfavorable Thermal Environments
  4.3.1 Sample Cable System 188

	4.3.2	Statistical Variations of Cable Circuit Parameters	190
		4.3.2.1 Load Probability Distribution	190
		4.3.2.2 Ambient Temperature Probability Distribution	190
		4.3.2.3 Native Soil and Backfill Probability Distributions	190
	4.3.3	Computation of Temperature Distribution Using Monte Carlo Simulation	194
		4.3.3.1 Steady-State Analysis	194
		4.3.3.2 Transient Analysis	196
	4.3.4	Sample Applications	197
4.4	Stochastic Optimization		202
4.5	Concluding Remarks		206
4.6	Chapter Summary		207
	References		209

5 Special Considerations for Real-Time Rating Analysis and Deeply Buried Cables — **211**

5.1	Introduction	211
5.2	Prediction of Conductor Temperature from the Conductor Current	213
	5.2.1 Introduction	213
	5.2.2 Mathematical Model	213
5.3	Practical Application of the Temperature Prediction Equation	222
5.4	Field Verification of the Temperature Calculations	223
5.5	Loss Factor Calculations in Rating Standards	224
	5.5.1 Daily Load Cycle	224
	5.5.2 Consideration of Weekly Load Variations	230
	5.5.2.1 Characteristic Diameter	233
5.6	Deeply Buried Cables	237
	5.6.1 Characteristic Diameter	238
	5.6.2 Temperature Changes for Deeply Buried Cables	242
5.7	Concluding Remarks	242
5.8	Chapter Summary	243
	References	245

6 Installations Involving Multiple Cables in Air — **247**

6.1	Introduction	247
6.2	Current Rating of Multicore Cables	247
	6.2.1 Introduction	248
	6.2.2 Background	249
	6.2.2.1 Evaluation of the Jacket Thermal Resistance	249
	6.2.2.2 Rating Equations	250
	6.2.2.3 Evaluation of the External Thermal Resistance	251
	6.2.3 Heat Conduction Inside the Cable Bundle	254
	6.2.3.1 Uniform Loss Density	254

6.2.3.2 Unequally Loaded Bundle — 262

6.3 Examples of Derating Factors — 266

6.3.1 Rating Factors — 267

6.3.1.1 Laying in a Single Layer on Wall, Floors, or in Cable Trays — 267

6.3.1.2 Laying in a Single Layer under Ceilings — 268

6.3.1.3 Laying in a Single Layer on Ventilated Cable Trays — 268

6.3.1.4 Laying in a Single Layer on Cable Ladders, Brackets, or Wire Mesh — 269

6.3.1.5 Laying in Several Layers — 270

6.3.1.6 Several Cables Connected in Parallel — 270

6.4 Concluding Remarks — 270

6.5 Chapter Summary — 273

6.5.1 Cables in Free Air — 273

6.5.2 Cables in Moving Air — 274

6.5.3 Unequally Loaded Bundle — 275

6.5.3.1 Central Part Loaded — 275

6.5.3.2 Outer Part Loaded — 275

References — 276

7 Rating of Pipe-Type Cables with Slow Circulation of Dielectric Fluid — **279**

7.1 Nomenclature — 279

7.2 Introduction — 280

7.3 Thermal Effects of Dielectric Fluid Circulation — 283

7.3.1 Background Information — 284

7.3.1.1 Calculation of Coolant Parameters — 284

7.3.1.2 Thermal Calculations — 285

7.3.2 Buller's Model — 285

7.3.2.1 Effective Cooling Distance — 287

7.3.3 Model for Real-Time Rating Computations — 291

7.3.3.1 Oil Velocity Profile — 292

7.3.3.2 Oil Temperature Distribution — 293

7.3.3.3 Thermal Resistances of the Oil Film — 294

7.4 Concluding Remarks — 299

7.5 Chapter Summary — 300

References — 302

Appendix A: Model Cables — **303**

Model Cable No. 1 — 303

Model Cable No. 2 — 303

Model Cable No. 3 — 305

Model Cable No. 4 — 305

Model Cable No. 5 — 305

Model Cable No. 6   308
References   312

**Appendix B: Computations of the Mean Moisture Content in   313
Media Surrounding Underground Power Cables**

Introduction   313
Soil–Water Balance   314
Determination of Potential Evapotranspiration   315
Weekly Soil–Water Balance Calculations   315
References   315

Appendix C: Estimation of Backfill Thermal Resistivity   317
Reference   319

Appendix D: Equations for Dielectric Fluid Parameters   321

Index   323

▬▬▬ Preface

The focus of this book is the calculation of the current-carrying capabilities[1] of the cables crossing unfavorable thermal environments. However, in order to make this book self-contained and more accessible to a wide group of interested readers, a comprehensive review of the rating methods for standard installation conditions is also included. Cable rating standards deal with uniform laying conditions only; however, in modern cable installations, such conditions are encountered very seldom. Cable routes often cross heat sources, including other cables, or pass through regions of high soil thermal resistivity, for example, in the vicinity of trees or shrubs. Air, walls, ceilings, or floors form an impediment to heat flowing away from the cables. In all such situations, the rating of the cables should be reduced to avoid overheating.

There has been little attention devoted in the past to the requirement of ampacity derating in such cases. The fact that there have been relatively few failures attributed to cable overheating is a result of the conservative design procedures used by cable engineers. With the economic pressure arising from the restructuring of the electric power industry around the world, the transmission circuits are becoming more heavily loaded and hence more prone to thermal overloads.

Better understanding of the heat transfer phenomena around loaded electric power cables will not only help to establish correct transmission line limits, but also may help the circuit owner in the implementation of corrective measures needed to increase cable ratings. Thus, in addition to improved computational procedures, including probabilistic analysis, optimization of thermal backfill design will become more common. This book is aimed at providing cable design engineers and power network analysts and operators with the computational tools and techniques to address challenges arising from installation of power cables in a complex thermal environment.

Different readers may read the book in different ways. The readers not very familiar with power cable heat transfer theory may wish to consult my earlier book,

[1]We will use the term cable rating or ampacity to denote current-carrying capability.

Rating of Electric Power Cables: Ampacity Computations for Transmission, Distribution and Industrial Applications, published by IEEE Press in 1997 and McGraw Hill in 1998. The present book can be viewed as a continuation of the earlier work and it makes many references to it. However, in order to make the present book self-contained, the most important rating equations developed in the earlier book are quoted here but without the background mathematical developments. The new formulae presented in this book are, however, always supported by a rigorous mathematical thought. The readers familiar with the theory can follow all the developments presented in this text, whereas some engineers may wish to use only the final results. For those readers, I added at the end of each chapter a summary of the salient findings developed in the text.

The book is organized as follows. Chapter 1 summarizes the standard methods of rating calculations. Some background information on heat transfer theory is offered, followed by a brief description of the lump parameter method used in most standard applications. Continuous and time-dependent rating equations for underground and aerial cables are introduced in this chapter, together with a summary of the methods for evaluation of the parameters appearing in these equations. After reading Chapter 1, the reader should be able to perform ampacity calculations for the majority of cable constructions and installation conditions covered in the international and national standards.

Chapters 2 and 3 address the main topics of this book, namely, cables crossing an unfavorable thermal environment. Chapter 2 looks at short sections of the cable right-of-way where either the thermal resistivity of the surrounding medium or the soil ambient temperature are higher than in the rest of the route. Chapter 3 addresses the issue of cables crossing other heat sources. In both chapters, the possibility of a soil dryout is considered and the issue of transient rating calculations is also addressed.

Chapter 4 is devoted to the topics related to the application of the thermal backfill around power cables. Advanced computational techniques presented in this chapter include nonlinear optimization of backfill design and the probabilistic analysis of cable ampacity.

The daily variations of cable loading are addressed in transient rating calculations. These are examined in several chapters. The simplest way to include load variability is to consider the load-loss factor in thermal calculations. A fresh look at this issue is offered in Chapter 5. Especially, analysis of weekly and monthly load cycles, not considered generally in practice, sheds new light on the cable rating capabilities. An application of the theory presented in this chapter to cables buried in deep tunnels or using a horizontal direct drilling technique might be of particular importance in the future since more and more installations follow this method of laying.

Chapter 6 deals with cables in air. In particular, rating issues for bundled cables found mostly in the telecommunication industry are discussed. However, the mathematical theory applied to rating these cables is also applicable to any other group of cables in air bundled together. In addition, derating factors for some medium voltage cable installations in an unfavorable thermal environment are given. These are mostly based on the work reported in the IEC standards.

The final chapter presents mathematical models for rating calculation of pipe-type cables with slow oil circulation. The classical model by Buller is thoroughly reviewed and compared with a new approach proposed in this chapter.

The book contains a large number of numerical examples that explain the various concepts discussed in the text. Each new concept is illustrated through examples based on practical cable constructions and installations. To facilitate the computational tasks, I have selected six model cables that will be used throughout the book. Five are transmission-class, high-voltage cables and one is a distribution cable. The model cables were selected to represent major constructions encountered in practice and are described in Appendix A. Chapter 6 considers a special type of cable, found mostly in the telecommunication and aircraft industry, namely a cable composed of many cores bundled together.

Appendix B contains a summary of a method, due to Thorntwaite, to calculate moisture content in the soil. Moisture content is a critical parameter determining soil thermal resistivity. An algorithm to calculate the probability distribution of this resistivity based on the distribution of the moisture content in the soil is discussed in Appendix C.

There are literally hundreds of symbols used in the book in the derivations of the mathematical expressions. Even though it might be difficult to create a completely consistent set of variables, the author's aim was to apply as much as possible the notation used by the International Electrotechnical Commission (IEC), which is a publisher of the international cable rating standards. Several topics covered in this book go beyond the material discussed in the standards. In such cases, generally accepted terminology was applied.

Several of the topics addressed in this book originated with the work of my colleague, Prof. Heinrich Brakelmann of the University of Duisburg in Germany. I am particularly indebted for the encouragement and constant support that I received from him while working on this book. Several complex mathematical derivations presented in the text were developed jointly with another colleague of mine, Mr. Eric Dorison of Électricté de France. His brilliant insight into the complex issue of thermal rating calculations is reflected in several parts of this work. A part of the material covered in this book was derived from various IEC reports dealing with the thermal rating of power cables. These publications are being prepared by Working Group 19 of Study Committee 20A (High Voltage Cables) of the IEC as an ongoing activity. The author is indebted to Mr. Mark Coates from ERA Technology in Britain, the Convener of Working Group 19, who reviewed several chapters of the book and offered his valuable comments. In addition, I could have not written this book without the involvement and close association of Dr. John Endrenyi from Kinectrics Inc., who reviewed the entire manuscript and provided several helpful comments. I would also like to acknowledge the financial assistance of the Standards Council of Canada and Kinectrics Inc. in supporting my participation in the activities of Working Group 19 of the IEC over many years.

The author thanks the International Electrotechnical Commission (IEC) for permission to reproduce information from its International Standards IEC 60287 and 60364-5-52. All such extracts are copyright of IEC, Geneva, Switzerland. All rights

Finally, but by no means last, I would like to thank my wife Justyna and my son Adam who supported wholeheartedly this difficult endeavor.

CHAPTER 1

Review of Power Cable Standard Rating Methods

1.1 INTRODUCTION

Calculation of the current-carrying capability, or ampacity, of electric power cables has been extensively discussed in the literature and is the subject of several international and national standards. The main international standards are those issued by the International Electrotechnical Commission (IEC) and the Institute of Electrical and Electronic Engineers (IEEE). References at the end of this chapter list the documents either issued or sponsored by these two organizations. The calculation procedures in both standards are, in principle, the same, with the IEC method incorporating several new developments that took place after the publication of the Neher and McGrath (NM) paper (1957). Similarities in the approaches are not surprising, since during the preparation of the standard, Mr. McGrath was in touch with the Chairman of Working Group 10 of IEC Subcommittee 20A (responsible for the preparation of ampacity calculation standards). The major difference between the two approaches is the use of metric units in IEC 60287 and imperial units in NM paper (the same equations look completely different because of this). Even though the methods are similar in principle, the IEC document is more comprehensive than the NM paper. IEC 60287 not only contains all the formulas (with minor exceptions listed in Appendix F of Anders, 1997) of the NM paper, but, in several cases, it makes a distinction between different cable types and installation conditions where the NM paper does not make such a distinction. Also, the constants used in the IEC document are more up to date.

Nevertheless, the principles of heat transfer for buried cables applied in both standards are the same, and the resulting equations will be reviewed in this chapter. The ampacity calculations are usually carried out in two different ways. On the one hand, steady-state or continuous ratings are sought; and, on the other, time-dependent or transient calculations are performed. In both cases, for cables buried underground, soil dry out caused by the heat generated in the cable might be considered. The calculation methods for cables in air are slightly different in both standards and, where applicable, the differences will be brought up in this book. For cables installed in air, the presence of solar radiation and wind may have profound effects on

Rating of Electric Power Cables in Unfavorable Thermal Environment. By George J. Anders
ISBN 0-471-67909-7 © 2004 the Institute of Electrical and Electronics Engineers.

the cable rating. Again, where applicable, these influences will be discussed in the subsequent chapters.

The focus of this book is on the installations that are not covered in the international standards mentioned above. However, the starting point of the analysis will be the standard methods and they will be reviewed in this chapter based on the presentation in Anders (1997).[1] This approach will make this book self-contained, thus allowing analysis of both standard and nonstandard calculation methods. The developments leading to the standard rating equations will not, however, be repeated here and the interested reader is referred to Anders (1997) for the relevant background information.

1.2 ENERGY CONSERVATION EQUATIONS

Ampacity computations of power cables require solution of the heat transfer equations, which define a functional relationship between the conductor current and the temperature within the cable and in its surroundings. In this section, we will analyze how the heat generated in the cable is dissipated to the environment. We will also show the basic heat transfer equations, and discuss how these equations are solved, thus laying the groundwork for cable rating calculations.

1.2.1 Heat Transfer Mechanism in Power Cable Systems

The two most important tasks in cable ampacity calculations are the determination of the conductor temperature for a given current loading, or conversely, determination of the tolerable load current for a given conductor temperature. In order to perform these tasks, the heat generated within the cable and the rate of its dissipation away from the conductor for a given conductor material and given load must be calculated. The ability of the surrounding medium to dissipate heat plays a very important role in these determinations, and varies widely because of several factors such as soil composition and moisture content, ambient temperature, and wind conditions. The heat is transferred through the cable and its surroundings in several ways and these are described in the following sections.

1.2.1.1 Conduction. For underground installations, the heat is transferred by conduction from the conductor and other metallic parts as well as from the insulation. It is possible to quantify heat transfer processes in terms of appropriate rate equations. These equations may be used to compute the amount of energy being transferred per unit time. For heat conduction, the rate equation is known as Fourier's law. For a wall having a temperature distribution $\theta(x)$, the rate equation is expressed as

$$q = -\frac{1}{\rho}\frac{d\theta}{dx} \tag{1.1}$$

[1]The author would like to acknowledge the permission received from the IEEE Press and the McGraw-Hill Company for extracting information from my first book for the purposes of this chapter.

The heat flux q (W/m^2) is the heat transfer rate in the x direction per unit area perpendicular to the direction of transfer, and is proportional to the temperature gradient $d\theta/dx$ in this direction. The proportionality constant ρ is a transport property known as thermal resistivity (K · m/W) and is a characteristic of the material. The minus sign is a consequence of the fact that heat is transferred in the direction of decreasing temperature.

1.2.1.2 Convection.

For cables installed in air, convection and radiation are important heat transfer mechanisms from the surface of the cable to the surrounding air. Convection heat transfer may be classified according to the nature of the flow. We speak of forced convection when the flow is caused by external means, such as by wind, pump, or fan. In contrast, for free (or natural) convection, the flow is induced by buoyancy forces, which arise from density differences caused by temperature variations in the air. In order to be somewhat conservative in cable rating computations, we usually assume that only natural convection takes place at the outside surface of the cable. However, both convection modes will be considered in Chapter 6.

Regardless of the particular nature of the convection heat transfer process, the appropriate rate equation is of the form

$$q = h(\theta_s - \theta_{amb}) \tag{1.2}$$

where q, the convective heat flux (W/m^2), is proportional to the difference between the surface temperature and the ambient air temperature, θ_s and θ_{amb}, respectively. This expression is known as Newton's law of cooling, and the proportionality constant h (W/m^2 · K) is referred to as the convection heat transfer coefficient. Determination of the heat convection coefficient is perhaps the most important task in computation of ratings of cables in air. The value of this coefficient varies between 2 and 25 W/m^2 · K for free convection and between 25 and 250 W/m^2 · K for forced convection.

1.2.1.3 Radiation.

Thermal radiation is energy emitted by cable or duct surface. The heat flux emitted by a cable surface is given by the Stefan–Boltzmann law:

$$q = \varepsilon\sigma_B\theta_s^{*4} \tag{1.3}$$

where θ_s^* is the absolute temperature (K) of the surface,[2] σ_B is the Stefan–Boltzmann constant ($\sigma_B = 5.67 \cdot 10^{-8}$ W/m^2 · K^4), and ε is a radiative property of the surface called the emissivity. This property, whose value is in the range $0 \leq \varepsilon \leq 1$, indicates how efficiently the surface emits compared to an ideal radiator. Conversely,

---

[2]Throughout this book, the temperature with an asterisk will denote absolute value in degrees Kelvin. Similarly, dimensions with an asterisk will denote the measurements in meters rather than in millimeters, as is usually the case.

if radiation is incident upon a surface, a portion will be absorbed, and the rate at which energy is absorbed per unit surface area may be evaluated from knowledge of the surface radiative property known as absorptivity, α. That is,

$$q_{abs} = \alpha q_{inc} \tag{1.4}$$

where $0 \leq \alpha \leq 1$. Equations (1.3) and (1.4) determine the rate at which radiant energy is emitted and absorbed, respectively, at a surface. Since the cable both emits and absorbs radiation, radiative heat exchange can be modeled as an interaction between two surfaces. Determination of the net rate at which radiation is exchanged between two surfaces is generally quite complicated. However, for cable rating computations, we may assume that a cable surface is small and the other surface is remote and much larger. Assuming this surface is one for which $\alpha = \varepsilon$ (a gray surface), the net rate of radiation exchange between the cable and its surroundings, expressed per unit area of the cable surface, is

$$q = \varepsilon \sigma_B (\theta_s^{*4} - \theta_{amb}^{*4}) \tag{1.5}$$

Throughout this book, we will use a notion of heat rate rather than heat flux. The heat transfer rate is obtained by multiplying heat flux by the area. Thus, the heat rate for radiative heat transfer will be given by the following equation[3]:

$$W_{rad} = \varepsilon \sigma_B A_{sr} (\theta_s^{*4} - \theta_{amb}^{*4}) \tag{1.6}$$

where A_{sr} (m^2) is the effective radiation area per meter length.

In power cables installed in air, the cable surface within the surroundings will simultaneously transfer heat by convection and radiation to the adjoining air. The total rate of heat transfer from the cable surface is the sum of the heat rates due to the two modes.[4] That is,

$$W = hA_s(\theta_s - \theta_{amb}) + \varepsilon A_{sr}\sigma_B(\theta_s^{*4} - \theta_{amb}^{*4}) \tag{1.7}$$

where A_s (m^2) is the convective area per meter length.

For some special cable installations, the ambient temperature used for heat convection can be different from the one used for heat transfer by radiation. The appropriate temperatures to be used are described in Chapter 10 of Anders (1997).

1.2.1.4 *Energy Balance Equations.*

In the analysis of heat transfer in a cable system, the law of conservation of energy plays an important role. We will formulate this law on a rate basis; that is, at any instant, there must be a balance between all energy rates, as measured in joules per second (W). The energy conservation law can be expressed by the following equation:

---

[3]Throughout the book, the symbol W will be used for heat transfer rate.
[4]The heat conduction in air is often neglected in cable rating computations.

$$W_{ent} + W_{int} = W_{out} + \Delta W_{st} \tag{1.8}$$

Where W_{ent} is the rate of energy entering the cable. This energy may be generated by other cables located in the vicinity of the given cable or by solar radiation. W_{int} is the rate of heat generated internally in the cable by joule or dielectric losses and ΔW_{st} is the rate of change of energy stored within the cable. The value of W_{out} corresponds to the rate at which energy is dissipated by conduction, convection, and radiation. For underground installations, the cable system will also include the surrounding soil.

We will use the fundamental equations described in this section to develop rating equations throughout the reminder of the book.

1.2.2   Heat Transfer Equations

As we mentioned earlier, current flowing in the cable conductor generates heat, which is dissipated through the insulation, metal sheath, and cable servings into the surrounding medium. The cable ampacity depends mainly upon the efficiency of this dissipation process and the limits imposed on the insulation temperature. To understand the nature of the heat dissipation process, we need to use the relevant heat transfer equations.

1.2.2.1   *Underground, Directly Buried Cables.*   Let us consider an underground cable located in a homogeneous soil. In such a cable, the heat is transferred by conduction through cable components and the soil. Since the length of the cable is much greater than its diameter, end effects can be disregarded and the heat transfer problem can be formulated in two dimensions only.[5]

The differential equation describing heat conduction in the soil has the following form:

$$\frac{\partial}{\partial x}\left(\frac{1}{\rho}\frac{\partial \theta}{\partial x}\right) + \frac{\partial}{\partial y}\left(\frac{1}{\rho}\frac{\partial \theta}{\partial y}\right) + W_{int} = c\frac{\partial \theta}{\partial t} \tag{1.9}$$

where
ρ = thermal resistivity, K · m/W
S = surface area perpendicular to heat flow, m^2
$\dfrac{\partial \theta}{\partial x}$ = temperature gradient in x direction

c = the volumetric thermal capacity of the material

For a cable buried in soil, Equation (1.9) is solved with the boundary conditions usually specified at the soil surface. These boundary conditions can be expressed in two different forms. If the temperature is known along a portion of the boundary, then

---

[5]Because end effects are neglected, all thermal parameters will be expressed in this book on a per-unit-length basis.

$$\theta = \theta_B(s) \tag{1.10}$$

where θ_B is the boundary temperature that may be a function of the surface length s. If heat is gained or lost at the boundary due to convection $h(\theta - \theta_{amb})$ or a heat flux q, then

$$\frac{1}{\rho}\frac{\partial\theta}{\partial n} + q + h(\theta - \theta_{amb}) = 0 \tag{1.11}$$

where n is the direction of the normal to the boundary surface, h is a convection coefficient, and θ is an unknown boundary temperature.

In cable rating computation, the temperature of the conductor is usually given and the maximum current flowing in the conductor is sought. Thus, when the conductor heat loss is the only energy source in the cable, we have $W_{int} = I^2R$, and Equation (1.9) is used to solve for I with the specified boundary conditions.

The challenge in solving Equation (1.9) analytically stems mostly from the difficulty of computing the temperature distribution in the soil surrounding the cable. An analytical solution can be obtained when a cable is represented as a line source placed in an infinite homogenous surrounding. Since this is not a practical assumption for cable installations, another assumption is often used; namely, that the earth surface is an isotherm. In practical cases, the depth of burial of the cables is on the order of ten times their external diameter, and for the usual temperature range reached by such cables, the assumption of an isothermal earth surface is a reasonable one. In cases where this hypothesis does not hold, namely, for large cable diameters and cables located close to the earth surface, a correction to the solution equation has to be used or numerical methods applied. Both are discussed in Anders (1997).

1.2.2.2 Cables in Air. For an insulated power cable installed in air, several modes of heat transfer have to be considered. Conduction is the main heat transfer mechanism inside the cable. Suppose that the heat generated inside the cable (due to joule, ferromagnetic and dielectric losses) is W_t (W/m). Another source of heat energy can be provided by the sun if the cable surface is exposed to solar radiation. Energy outflow is caused by convection and net radiation from the cable surface. Therefore, the energy balance Equation (1.83) at the surface of the cable can be written as

$$W_t + W_{sol} - W_{conv} - W_{rad} = 0 \tag{1.12}$$

where W_{sol} is the heat gain per unit length caused by solar heating, and W_{conv} and W_{rad} are the heat losses due to convection and radiation, respectively. Substituting appropriate formulas for the heat gains and losses at the surface of the cable, the following form of the heat balance equation is obtained:

$$W_t + \sigma D_e^* H - \pi D_e^* h(\theta_e^* - \theta_{amb}^*) - \pi D_e^* \varepsilon \sigma_B(\theta_e^{*4} - \theta_{amb}^{*4}) = 0 \tag{1.13}$$

where
θ_e^* = cable surface temperature, K
σ = solar absorption coefficient
H = intensity of solar radiation, W/m^2
σ_B = Stefan–Boltzmann constant, equal to $5.67 \cdot 10^{-8}$ W/m^{2}K^4
ε = emissivity of the cable outer covering
D_e^* = cable external diameter,[6] m
θ_{amb}^* = ambient temperature, K

This equation is usually solved iteratively. In steady-state rating computations, the effect of heat gain by solar radiation and heat loss caused by convection are taken into account by suitably modifying the value of the external thermal resistance of the cable. Computation of the convection coefficient h can be quite involved. Suitable approximations are summarized in Section 1.6.6.5 and, for special installations discussed in this book, are revisited in Chapter 6.

1.2.3 Analytical Versus Numerical Methods of Solving Heat Transfer Equations

Equations (1.9) and (1.13) can be solved analytically, with some simplifying assumptions, or numerically. Analytical methods have the advantage of producing current rating equations in a closed formulation, whereas numerical methods require iterative approaches to find cable ampacity. However, numerical methods provide much greater flexibility in the analysis of complex cable systems and allow representation of more realistic boundary conditions. In practice, analytical methods have found much wider application than the numerical approaches. There are several reasons for this situation. Probably the most important one is historical: cable engineers have been using analytical solutions based on either Neher/McGrath (1957) formalism or IEC Publication 60287 (1994) for a long time. Computations for a simple cable system can often be performed using pencil and paper or with the help of a hand-held calculator. Numerical approaches, on the other hand, require extensive manipulation of large matrices and have only become popular with an advent of powerful computers. Both approaches will be used in this book; analytical methods are discussed in Chapters 2 and 3, whereas the numerical approaches are dealt with in Chapter 4.

1.3 THERMAL NETWORK ANALOGS

Analytical solutions to the heat transfer equations are available only for simple cable constructions and simple laying conditions. In solving the cable heat dissipation problem, electrical engineers use a fundamental similarity between the heat flow

---

[6]We recall that the dimension symbols with an asterisk refer to the length in meters and without it to the length in millimeters.

due to the temperature difference between the conductor and its surrounding medium and the flow of electrical current caused by a difference of potential. Using their familiarity with the lumped parameter method to solve differential equations representing current flow in a material subjected to potential difference, they adopt the same method to tackle the heat conduction problem. The method begins by dividing the physical object into a number of volumes, each of which is represented by a thermal resistance and a capacitance. The thermal resistance is defined as the material's ability to impede heat flow. Similarly, the thermal capacitance is defined as the material's ability to store heat. The thermal circuit is then modeled by an analogous electrical circuit in which voltages are equivalent to temperatures and currents to heat flows. If the thermal characteristics do not change with temperature, the equivalent circuit is linear and the superposition principle is applicable for solving any form of heat flow problem.

In a thermal circuit, charge corresponds to heat; thus, Ohm's law is analogous to Fourier's law. The thermal analogy uses the same formulation for thermal resistances and capacitances as in electrical networks for electrical resistances and capacitances. Note that there is no thermal analogy to inductance or in steady-state analysis; only resistance will appear in the network.

Since the lumped parameter representation of the thermal network offers a simple means for analyzing even complex cable constructions, it has been widely used in thermal analysis of cable systems. A full thermal network of a cable for transient analysis may consist of several loops. Before the advent of digital computers, the solution of the network equations was a formidable numerical task. Therefore, simplified cable representations were adopted and methods to reduce a multiloop network to a two-loop circuit were developed. A two-loop representation of a cable circuit turned out to be quite accurate for most practical applications and, consequently, was adopted in international standards. In this section, we will explain how the thermal circuit of a cable is constructed, and we will show how the required parameters are computed. We will also explain how full network equations are solved.

1.3.1 Thermal Resistance

All nonconducting materials in the cable will impede heat flow away from the cables (the thermal resistance of the metallic parts in the cable, even though not equal to zero, is so small that it is usually neglected in rating computations). Thus, we can talk about material resistance to heat flow. Of particular interest is an expression for the thermal resistance of a cylindrical layer, for example, cable insulation, with constant thermal resistivity ρ_{th}. If the internal and external radii of this layer are r_1 and r_2, respectively, then the thermal resistance for conduction of a cylindrical layer per unit length is

$$T = \frac{\rho_{th}}{2\pi} \ln \frac{r_2}{r_1} \qquad (1.14)$$

For a rectangular wall, we have

$$T = \rho_{th} \frac{l}{S}$$

(1.15)

where
ρ_{th} = thermal resistivity of a material, K · m/W
S = cross-section area of the body, m^2
l = thickness of the body, m

In analogy to electrical and thermal networks, we also can write that

$$W = \frac{\Delta \theta}{T}$$

(1.16)

which is the thermal equivalent of Ohm's law.

A thermal resistance may also be associated with heat transfer by convection at a surface. From Newton's law of cooling [Equation (1.2)],

$$W = h_{conv} A_s (\theta_e - \theta_{amb})$$

(1.17)

where A_s is the area of the outside surface of the cable for unit length, h_{conv} is the cable surface convection coefficient, and θ_e is the cable surface temperature.

The thermal resistance for convection is then

$$T_{conv} = \frac{\theta_e - \theta_{amb}}{W} = \frac{1}{h_{conv} A_s}$$

(1.18)

Yet another resistance may be pertinent for a cable installed in air. In particular, radiation exchange between the cable surface and its surroundings may be important. It follows that a thermal resistance for radiation may be defined as

$$T_{rad} = \frac{\theta_e^* - \theta_{gas}^*}{W_{rad}} = \frac{1}{h_r A_{sr}}$$

(1.19)

where A_{sr} is the area of the cable surface effective for heat radiation for unit length of the cable and θ_{gas}^* is the temperature of the air surrounding the cable which, when cable is installed in free air, is equal to the ambient temperature θ_{amb}^*. h_r is the radiation heat transfer coefficient obtained from Equation (1.6) for radiation heat transfer rate:

$$h_r = \varepsilon \sigma_B (\theta_e^* + \theta_{gas}^*)(\theta_e^{*2} + \theta_{gas}^{*2})$$

(1.20)

The total heat transfer coefficient for a cable in air is given by

$$h_t = h_{conv} + h_r$$

(1.21)

1.3.2   Thermal Capacitance

Many cable rating problems are time dependent. To determine the time dependence of the temperature distribution within the cable and its surroundings, we could begin by solving the appropriate form of the heat equation, for example, Equation (1.9) In the majority of practical cases, it is very difficult to obtain analytical solutions of this equation and, where possible, a simpler approach is preferred. One such approach may be used where temperature gradients within the cable components are small. It is termed the lumped capacitance method. In order to satisfy the requirement that the temperature gradient within the body must be small, some components of the cable system, for example, the insulation and surrounding soil, must be subdivided into smaller entities. This is done using a theory developed by Van Wormer (1955), application of which is briefly reviewed in this section.

As mentioned above, an equivalent thermal network will contain only thermal resistances T and thermal capacitances Q. The thermal capacitance Q can be defined as the "ability to store the heat," and is defined by

$$Q = V \cdot c \tag{1.22}$$

where
V = volume of the body, m^3
c = volumetric specific heat of the material, $J/m^3°C$

As an illustration, the formula for the thermal capacitance for a coaxial configuration with internal and external diameters D_1^* and D_2^* (m), respectively, which may represent, for example, a cylindrical insulation, is given by

$$Q = \frac{\pi}{4}(D_2^{*2} - D_1^{*2})c \tag{1.23}$$

Thermal capacitances and resistances are used to construct a thermal ladder network to obtain the temperature distribution within the cable and its surroundings as a function of time. This topic is discussed in the next section.

1.3.3   Construction of a Ladder Network of a Cable

The electrical and thermal analogy discussed in Section 1.3.1 allows the solution of many thermal problems by applying mathematical tools well known to electrical engineers. An ability to construct a ladder network is particularly useful in transient computations. To build a ladder network, the cable is considered to extend as far as the inner surface of the soil for buried cables, and to free air for cables in air.

In constructing ladder networks, dielectric losses require special attention. Although the dielectric losses are distributed throughout the insulation, it may be shown that for a single-conductor cable and also for multicore, shielded cables with round conductors, the correct temperature rise is obtained by considering for transients and steady-state that all of the dielectric loss occurs at the middle of the thermal resistance between the conductor and the sheath. For multicore belted cables, dielectric losses can generally be neglected, but if they are represented, the conductors are taken as the source of dielectric loss (Neher and McGrath, 1957).

Thermal capacitances of the metallic parts are placed as lumped quantities corresponding to their physical position in the cable. The thermal capacitances of materials with high thermal resistivity and possibly large temperature gradients across them (e.g., insulation and coverings) are allocated by the technique described below.

1.3.3.1 Representation of Capacitances of the Dielectric.

To improve the accuracy of the approximate solution using lumped constants, Van Wormer (1955) proposed a simple method for allocating the thermal capacity of the insulation between the conductor and the sheath so that the total heat stored in the insulation is represented. An assumption made in the derivation is that the temperature distribution in the insulation follows a steady-state logarithmic distribution for the period of the transient. The ladder networks for short and long duration transients are somewhat different and are discussed below.

Whether the transient is long or short depends on the cable construction. For the purpose of transient rating computations, long-duration transients are those lasting longer than $\frac{1}{3}\Sigma T \cdot \Sigma Q$, where ΣT and ΣQ are the internal cable thermal resistance and capacitance, respectively. The methods for computing the values of T and Q are summarized in Section 1.6 in this chapter.

Ladder Network for Long-Duration Transients. The dielectric is represented by lumped thermal constants. The total thermal capacity of the dielectric (Q_i) is divided between the conductor and the sheath, as shown in Figure 1-1.

When screening layers are present, metallic tapes are considered to be part of the conductor or sheath, whereas semiconducting layers (including metallized carbon paper tapes) are considered part of the insulation in thermal calculations.

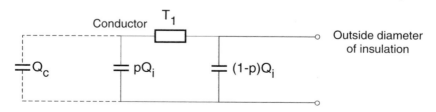

Figure 1-1 Representation of the dielectric for times greater that $\frac{1}{3}\Sigma T \cdot \Sigma Q$. T_1 = total thermal resistance of dielectric per conductor (or equivalent single-core conductor of a three-core cable; see below). Q_i = total thermal capacitance of dielectric per conductor (or equivalent single-core conductor of a three-core cable). Q_c = thermal capacitance of conductor (or equivalent single-core conductor of a three-core cable).

The Van Wormer coefficient p is given by

$$p = \frac{1}{2 \ln\left(\dfrac{D_i}{d_c}\right)} - \frac{1}{\left(\dfrac{D_i}{d_c}\right)^2 - 1} \tag{1.24}$$

Equation (1.24) is also used to allocate the thermal capacitance of the outer covering in a similar manner to that used for the dielectric. In this case, the Van Wormer factor is given by

$$p' = \frac{1}{2 \ln\left(\dfrac{D_e}{D_s}\right)} - \frac{1}{\left(\dfrac{D_e}{D_s}\right)^2 - 1} \tag{1.25}$$

where D_e and D_s are the outer and inner diameters of the covering.

For long-duration transients and cyclic factor computations, the three-core cable is replaced by an equivalent single-core construction dissipating the same total conductor losses (Wollaston, 1949). The diameter d_c^* of the equivalent single-core conductor is obtained on the assumption that new cable will have the same thermal resistance of the insulation as the thermal resistance of a single core of the three-core cable; that is,

$$\frac{T_1}{3} = \frac{\rho_i}{2\pi} \ln \frac{D_i^*}{d_c^*} \tag{1.26}$$

where D_i^* is the same value of diameter over dielectric (under the sheath) as for the three-core cable, and T_1 is the thermal resistance of the three-conductor cable as given in Section 1.6.6.1; ρ_i is the thermal resistivity of the dielectric.

Hence, we have

$$d_c^* = D_i^* e^{-2\pi T_1/3\rho_i} \tag{1.27}$$

Thermal capacitances are calculated on the following assumptions:

1. The actual conductors are considered to be completely inside the diameter of the equivalent single conductor, the remainder of the equivalent conductor being occupied by insulation.
2. The space between the equivalent conductor and the sheath is considered to be completely occupied by insulation (for fluid-filled cable, this space is filled partly by the total volume of oil in the ducts and the remainder is oil-impregnated paper).

Factor p is then calculated using the dimensions of the equivalent single-core cable, and is applied to the thermal capacitance of the insulation based on assumption (2) above.

Ladder Network for Short-Duration Transients. Short-duration transients last usually between 10 min and about 1 h. In general, for a given cable construction, the formula for the Van Wormer coefficient shown in this section applies when the duration of the transient is not greater than $\frac{1}{3}\Sigma T \cdot \Sigma Q$. The heating process for short-duration transients can be assumed to be the same as if the insulation were thick.

The method is the same as for long-duration transients except that the cable insulation is divided at diameter $d_x = \sqrt{D_i \cdot d_c}$, giving two portions having equal thermal resistances, as shown in Figure 1-2. The thermal capacitances Q_{i1} and Q_{i2} are defined in Section 1.6.7.

The Van Wormer coefficient is given by

$$p^* = \frac{1}{\ln\left(\dfrac{D_i}{d_c}\right)} - \frac{1}{\left(\dfrac{D_i}{d_c}\right) - 1} \tag{1.28}$$

Example 1.1

Construct a ladder network for model cable No. 4 in Appendix A for a short-duration transient.

This network is shown in Figure 1-3, where it is shown that the insulation thermal resistance is divided into two equal parts, the insulation capacitance into four parts, and the capacitance of the cable serving into two parts. ∎

A three-core cable is represented as an equivalent single-core cable as described above for durations of about $\frac{1}{2}\Sigma T \cdot \Sigma Q$ or longer (the quantities ΣT and ΣQ refer to the whole cable). However, for very short transients (i.e., for durations up to the value of the product $\Sigma T \cdot \Sigma Q$, where ΣT and ΣQ now refer to the single core), the mutual heating of the cores is neglected, and a three-core cable is treated as a single-core cable with the dimensions corresponding to the one core. For durations between these two limits, $\Sigma T \cdot \Sigma Q$ for one core and $\frac{1}{2}\Sigma T \cdot \Sigma Q$ for the whole cable, the transient is assumed to be given by a straight-line interpolation in a diagram with axes of linear temperature rise and logarithmic times.

Van Wormer Coefficient for Transients Due to Dielectric Loss. In the preceding sections, it has been assumed that the temperature rise of the conductor due to

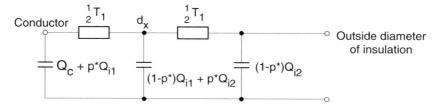

Figure 1-2   Representation of the dielectric for times less than or equal to $\frac{1}{3}\Sigma T \cdot \Sigma Q$.

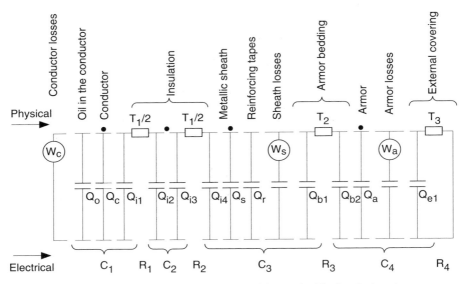

Figure 1-3   Thermal network for model cable No. 4 with electrical analogy.

dielectric loss has reached its steady state, and that the total temperature at any time during the transient can be obtained simply by adding the constant temperature value due to the dielectric loss to the transient value caused by the load current.

If changes in load current and system voltage occur at the same time, then an additional transient temperature rise due to the dielectric loss has to be calculated (Morello, 1958). For cables at voltages up to and including 275 kV, it is sufficient to assume that half of the dielectric loss is produced at the conductor and the other half at the insulation screen or sheath. The cable thermal circuit is derived by the method given above with the Van Wormer coefficient computed from equations and for long- and short-duration transients, respectively.

For paper-insulated cables operating at voltages higher than 275 kV, the dielectric loss is an important fraction of the total loss and the Van Wormer coefficient is calculated by (IEC 853-2, 1989)

$$p_d = \frac{\left[\left(\dfrac{D_i}{d_c}\right)^2 \ln\left(\dfrac{D_i}{d_c}\right)\right] - \left[\ln\left(\dfrac{D_i}{d_c}\right)\right]^2 - \dfrac{1}{2}\left[\left(\dfrac{D_i}{d_c}\right)^2 - 1\right]}{\left[\left(\dfrac{D_i^2}{d_c}\right) - 1\right]\left[\ln\left(\dfrac{D_i}{d_c}\right)\right]^2} \qquad (1.29)$$

In practical calculations for all voltage levels for which dielectric losses are important, half of the dielectric loss is added to the conductor loss and half to the sheath loss; therefore, the loss coefficients $(1 + \lambda_1)$ and $(1 + \lambda_1 + \lambda_2)$ used to evaluate thermal resistances and capacitances are set equal to 2.

Example 1.2

We will compute the Van Wormer coefficient for dielectric losses for cable No. 3 described in Appendix A. From Table A1, we have $D_j = 67.26$ mm and $d_c = 41.45$ mm. Hence,

$$p_d = \frac{p_d = \left[\left(\frac{67.26}{41.45}\right)^2 \ln\left(\frac{67.26}{41.45}\right)\right] - \left[\ln\left(\frac{67.26}{41.45}\right)\right]^2 - \frac{1}{2}\left[\left(\frac{67.26}{41.45}\right)^2 - 1\right]}{\left[\left(\frac{67.26}{41.45}\right)^2 - 1\right]\left[\ln\left(\frac{67.26}{41.45}\right)\right]^2} = 0.585$$

An example of the transient analysis with the voltage applied simultaneously with the current is given in Example 5.4 in (Anders, 1997) ∎

1.3.3.2 *Reduction of a Ladder Network to a Two-Loop Circuit.* CIGRE (1972, 1976) and later IEC (1985, 1989) introduced computational procedures for transient rating calculations employing a two-loop network with the intention of simplifying calculations and with the objective of standardizing the procedure for basic cable types. Even though with the advent and wide availability of fast desktop computers the advantage of simple computations is no longer so pronounced, there is some merit in performing some computations by hand, if only for the purpose of checking sophisticated computer programs. To perform hand computations for the transient response of a cable to a variable load, the cable ladder network has to be reduced to two sections. The procedure to perform this reduction is described below.

Consider a ladder network composed of v resistances and $(v + 1)$ capacitances, as shown in Figure 1-4. If the last component of the network is a capacitance, the last capacitance Q_{v+1} is short-circuited. An equivalent network, which represents the cable with sufficient accuracy, is derived with two sections $T_A Q_A$ and $T_\beta Q_B$, as shown in Figure 1-5.

The first section of the derived network is made up of $T_A = T_\alpha$ and $Q_A = Q_\alpha$ without modification, in order to maintain the correct response for relatively short durations.

The second section $T_\beta Q_B$ of the derived network is made up from the remaining sections of the original circuit by equating the thermal impedance of the second de-

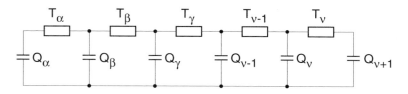

Figure 1-4   General ladder network representing a cable.

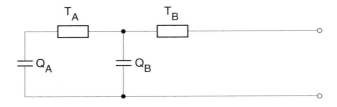

Figure 1-5   Two-loop equivalent network.

rived section to the total impedance of the multiple sections. The resulting expressions are then equal to (Anders, 1997)

$$T_B = T_\beta + T_\gamma + \ldots + T_v \tag{1.30}$$

$$Q_B = Q_\beta + \left(\frac{T_\gamma + T_\delta + \ldots + T_v}{T_\beta + T_\gamma + \ldots + T_v}\right)^2 Q_\gamma + \left(\frac{T_\delta + T_\varepsilon + \ldots + T_v}{T_\beta + T_\gamma + \ldots + T_v}\right)^2 Q_\delta + \ldots$$

$$+ \left(\frac{T_v}{T_\beta + T_\gamma + \ldots + T_v}\right)^2 Q_v \tag{1.31}$$

Even though formulas and are straightforward, a great deal of care is required when the equivalent thermal resistances and capacitances are computed in the case when sheath, armor, and pipe losses are present (IEC, 1985). This is because the location of these losses inside the original network has to be carefully taken into account. The following example illustrates this point.

Example 1.3
We will construct a two-loop equivalent network for model cable No. 1 assuming (1) a short-duration transient and (2) a long-duration transient.

   1. *Short-Duration Transient*
   From Table A1 we observe that for this cable, short-duration transients are those lasting half an hour or less. The diagram of the full network for a short-duration transient is shown in Figure 1-6.

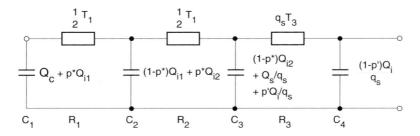

Figure 1-6   Network diagram for cable No. 1 for short-duration transients.

The method of dividing insulation and jacket capacitances into parts is discussed in Section 1.6.7. Before we apply the reduction procedure, we combine parallel capacitances into four equivalent capacitances. In the equivalent network, only conductor losses are represented. Therefore, to account for the presence of sheath losses, the thermal resistances beyond the sheath must be multiplied, and the thermal capacitances divided by the ratio of the losses in the conductor and the sheath to the conductor losses.[7] By performing these multiplications and divisions, the time constants of the thermal circuits involved are not changed. Thus,

$$Q_1 = Q_c + p^*Q_{i1} \qquad Q_2 = (1 - p^*)Q_{i1} + p^*Q_{i2}$$

$$Q_3 = (1 - p^*)Q_{i2} \qquad Q_4 = \frac{Q_s + p'Q_{j1}}{1 + \lambda_1} \qquad Q_5 = \frac{(1 - p')Q_{j2}}{1 + \lambda_1} \tag{1.32}$$

To compute numerical values, we will require expressions for Q_{i1} and Q_{i2}. These expressions are given in Section 1.6.7. The numerical values are as follows: $Q_{i1} = 763$ J/K · m, $Q_{i2} = 453.9$ J/K · m, and $Q_j = 394.8$ J/K · m. With these values and with the additional numerical values in Table A1, $D_i = 30.1$ mm, $d_c = 20.5$ mm, $D_e = 35.8$ mm, and $D_s = 31.4$ mm, we have

$$p^* = \frac{1}{\ln\left(\dfrac{D_i}{d_c}\right)} - \frac{1}{\left(\dfrac{D_i}{d_c}\right) - 1} = \frac{1}{\ln\dfrac{30.1}{20.5}} - \frac{1}{\dfrac{30.1}{20.5} - 1} = 0.468$$

$$p' = \frac{1}{2\ln\left(\dfrac{D_e}{D_s}\right)} - \frac{1}{\left(\dfrac{D_e}{D_s}\right)^2 - 1} = \frac{1}{2\ln\left(\dfrac{35.8}{31.4}\right)} - \frac{1}{\left(\dfrac{35.8}{31.4}\right)^2 - 1} = 0.478$$

$$Q_1 = 1035 + 0.468 \cdot 763 = 1392.1 \text{ J/K} \cdot \text{m}$$

$$Q_2 = (1 - 0.468)763 + 0.468 \cdot 453.9 = 618.3 \text{ J/K} \cdot \text{m}$$

$$Q_3 = (1 - 0.468)453.9 = 241.5 \text{ J/K} \cdot \text{m} \qquad Q_4 = \frac{4 + 0.478 \cdot 394.8}{1.09} = 176.8 \text{ J/K} \cdot \text{m}$$

$$Q_5 = \frac{(1 - 0.478)394.8}{1.09} = 189.1 \text{ J/K} \cdot \text{m}$$

The final capacitance Q_5 is omitted in further analysis because the transient for the cable response is calculated on the assumption that the output terminals on the right-hand side are short-circuited.

Since the first section of the network in Figure 1-6 represents the conductor, and in rating computations the conductor temperature is of interest, the equivalent

---

[7]These ratios are called sheath and armor loss factors and are defined in Sections 1.6.4 and 1.6.5, respectively.

network will have the first section equal to the first section of the full network; that is,

$$T_A = \tfrac{1}{2}T_1 \quad \text{and} \quad Q_A = Q_1 \tag{1.33}$$

From Equation (1.30), we have

$$T_B = \tfrac{1}{2}T_1 + (1 + \lambda_1)T_3 \tag{1.34}$$

Thermal capacitance of the second part is obtained by applying Equation (1.31):

$$Q_B = Q_2 + \left[\frac{(1 + \lambda_1)T_3}{\tfrac{1}{2}T_1 + (1 + \lambda_1)T_3} \right]^2 (Q_3 + Q_4) \tag{1.35}$$

The sheath loss factor and thermal resistances for this cable are given in Table A1 as $\lambda_1 = 0.09$, $T_1 = 0.214$ K · m/W, and $T_3 = 0.104$ K · m/W. Substituting numerical values in Equations (1.33) to (1.35), we obtain

$$T_A = 0.107 \text{ K · m/W} \qquad Q_A = 1392.1 \text{ J/K · m}$$

$$T_B = 0.107 + 1.09 \cdot 0.104 = 0.220 \text{ K · m/W}$$

$$Q_B = 618.3 + \left(\frac{1.09 \cdot 0.104}{0.107 + 1.09 \cdot 0.104} \right)^2 (241.5 + 176.8) = 729.4 \text{ J/K · m}$$

2. Long-Duration Transients

Long-duration transient for this cable are those lasting longer than 0.5 h. The appropriate diagram is shown in Figure 1-7. In this case, we have

$$T_A = T_1 \qquad T_B = (1 + \lambda_1)T_3 \tag{1.36}$$

The insulation and jacket are split into two parts with the Van Wormer coefficients given by Equations (1.24) and (1.25), respectively. Since the last part of the jacket capacitance is short-circuited, Q_A and Q_b are simply obtained as the sums of relevant capacitances:

$$Q_A = Q_c + pQ_i \qquad Q_B = (1-p)Q_i + \frac{Q_s + p'Q_j}{1 + \lambda_1} \tag{1.37}$$

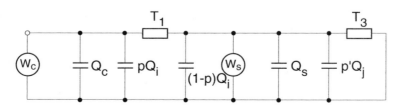

Figure 1-7 Network diagram for a long-duration transient for model cable No. 1.

Substituting numerical values, we obtain

$$p = \frac{1}{2\ln\left(\dfrac{D_i}{d_c}\right)} - \frac{1}{\left(\dfrac{D_i}{d_c}\right)^2 - 1} = \frac{1}{2\ln\left(\dfrac{30.1}{20.5}\right)} - \frac{1}{\left(\dfrac{30.1}{20.5}\right)^2 - 1} = 0.437$$

$$T_A = 0.214 \text{ K} \cdot \text{m/W} \qquad Q_A = 1035 + 0.437 \cdot 915.6 = 1435.1 \text{ J/K} \cdot \text{m}$$

$$T_B = 1.09 \cdot 0.104 = 0.113 \text{ K} \cdot \text{m/W}$$

$$Q_B = (1 - 0.437)915.6 + \frac{4 + 0.478 \cdot 394.8}{1.09} = 692.3 \text{ J/K} \cdot \text{m} \qquad \blacksquare$$

1.4 RATING EQUATIONS—STEADY-STATE CONDITIONS

The current-carrying capability of a cable system will depend on several parameters. The most important of these are:

1. The number of cables and the different cable types in the installation under study
2. The cable construction and materials used for the different cable types
3. The medium in which the cables are installed
4. Cable locations with respect to each other and with respect to the earth surface
5. The cable bonding arrangement

For some cable constructions, the operating voltage may also be of significant importance. All of the above issues are taken into account, some of them explicitly, the others implicitly, in the rating equations summarized in this chapter. The lumped parameter network representation of the cable system is used for the development of steady-state and transient rating equations. These equations are developed for a single cable, either with one core or with multiple cores. However, they can be applied to multicable installations, for both equally and unequally loaded cables, by suitably selecting the value of the external thermal resistance, as discussed in Section 1.6.6.5.

The development of cable rating equations is quite different for steady-state and transient conditions. We will start with analysis of the steady-state conditions, which could be a result of either constant or cyclic loading. We will present only the very fundamental equations that will form the basis of the developments presented in the subsequent chapters. The parameters appearing in these equations can occasionally involve very complex calculations. We will review these calculations when a need arises before introducing modifications that are the subject of this book.

1.4.1 Buried Cables

1.4.1.1 Steady-State Rating Equation without Moisture Migration.
Steady-state rating computations involve solving the equation for the ladder net-

work shown in Figure 1-8. With reference to Figure 1-8, W_c, W_d, W_s, and W_a (W/m) represent conductor, dielectric, sheath, and armor losses, respectively, and n denotes the number of conductors in the cable. T_1, T_2, T_3, and T_4 (K · m/W) are the thermal resistances, where T_1 is the thermal resistance per unit length between one conductor and the sheath, T_2 is the thermal resistance per unit length of the bedding between sheath and armor, T_3 is the thermal resistance per unit length of the external serving of the cable, and T_4 is the thermal resistance per unit length between the cable surface and the surrounding medium.

Since losses occur at several positions in the cable system (for this lumped parameter network), the heat flow in the thermal circuit shown in Figure 1-8 will increase in steps. Thus, the total joule loss W_I in a cable can be expressed as

$$W_I = W_c + W_s + W_a = W_c(1 + \lambda_1 + \lambda_2) \tag{1.38}$$

The quantity λ_1 is called the sheath loss factor and is equal to the ratio of the total losses in the metallic sheath to the total conductor losses. Similarly, λ_2 is called the armor loss factor and is equal to the ratio of the total losses in the metallic armor to the total conductor losses. Incidentally, it is convenient to express all heat flows caused by the joule losses in the cable in terms of the loss per meter of the conductor.

Referring now to the diagram in Figure 1-8, and remembering the analogy between the electrical and thermal circuits, we can write the following expression for $\Delta\theta$, the conductor temperature rise above the ambient temperature:

$$\Delta\theta = (W_c + \tfrac{1}{2}W_d)T_1 + [W_c(1 + \lambda_1) + W_d]nT_2 + [W_c(1 + \lambda_1 + \lambda_2) + W_d]n(T_3 + T_4) \tag{1.39}$$

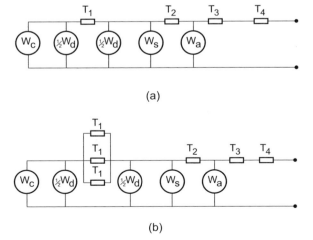

(a)

(b)

Figure 1-8   The ladder diagram for steady-state rating computations. (a) Single-core cable. (b) Three-core cable.

where W_d, W_c, λ_1, and λ_2 are defined above, and n is the number of load-carrying conductors in the cable (conductors of equal size and carrying the same load). The ambient temperature is the temperature of the surrounding medium under normal conditions at the location where the cables are installed, or are to be installed, including any local sources of heat, but not the increase of temperature in the immediate neighborhood of the cable due to the heat arising therefrom.

The unknown quantity is either the conductor current I or its operating temperature θ_c (°C). In the first case, the maximum operating conductor temperature is given, and in the second case, the conductor current is specified. The permissible current rating is obtained from Equation (1.39). Remembering that $W_c = I^2R$, we have

$$I = \left[\frac{\Delta\theta - W_d[0.5T_1 + n(T_2 + T_3 + T_4)]}{RT_1 + nR(1 + \lambda_1)T_2 + nR(1 + \lambda_1 + \lambda_2)(T_3 + T_4)} \right]^{0.5} \tag{1.40}$$

where R is the ac resistance per unit length of the conductor at maximum operating temperature.

Equation (1.40) is often written in a simpler form that clearly distinguishes between internal and external heat transfers in the cable. Denoting

$$T = \frac{T_1}{n} + (1 + \lambda_1)T_2 + (1 + \lambda_1 + \lambda_2)T_3$$

$$T_d = \frac{T_1}{2n} + T_2 + T_3 \tag{1.41}$$

Equation (1.40) becomes

$$\Delta\theta = n(W_cT + W_tT_4 + W_dT_d) \tag{1.42}$$

where W_t are the total losses generated in the cable defined by:

$$W_t = W_I + W_d = W_c(1 + \lambda_1 + \lambda_2) + W_d \tag{1.43}$$

and T computed from Equation (1.41) is an equivalent cable thermal resistance. This is an internal thermal resistance of the cable, which depends only on the cable construction. The external thermal resistance, on the other hand, will depend on the properties of the surrounding medium as well as on the overall cable diameter, as shown below.

The last term in Equation (1.42) is the temperature rise caused by dielectric losses. Denoting it by $\Delta\theta_d$,

$$\Delta\theta_d = nW_dT_d \tag{1.44}$$

1.4.1.2 Steady-State Rating Equation with Moisture Migration. The

laying conditions examined in this book are particularly conducive to the formation of a dry zone around the cable. Under unfavorable conditions, the heat flux from the cable entering the soil may cause significant migration of moisture away from the cable. A dried-out zone may develop around the cable, in which the thermal conductivity can be reduced by a factor of three or more over the conductivity of the bulk. The drying-out conditions may occur in both regions of the route, but particularly in the region of high thermal resistivity.

The modeling of the dry zone around the cable is discussed in Anders (1997). For completeness, we will start by recalling the basic developments presented there. This will be followed by the modification of the expression for the conductor temperature, taking into account the drying-out conditions in the unfavorable region in Chapter 2 and in cable crossings in Chapter 3.

The current-carrying capacity of buried power cables depends to a large extent on the thermal conductivity of the surrounding medium. Soil thermal conductivity is not a constant, but is highly dependent on its moisture content. Under unfavorable conditions, the heat flux from the cable entering the soil may cause significant migration of moisture away from the cable. A dried-out zone may develop around the cable, in which the thermal conductivity is reduced by a factor of three or more over the conductivity of the bulk. This, in turn, may cause an abrupt rise in temperature of the cable sheath, which may lead to damage to the cable insulation. The likelihood of soil drying out is even greater when the route of the rated cable is crossed by another heat source.

In order to give some guidance on the effect of moisture migration on cable ratings, CIGRE (1986) has proposed a simple two-zone model for the soil surrounding loaded power cables, resulting in a minor modification of the steady-state rating equation (Anders, 1997). Subsequently, this model has been adopted by the IEC as an international standard (IEC, 1994). We will further extend this model to account for heat sources crossing the rated cable.

The concept on which the method proposed by CIGRE relies can be summarized as follows. Moist soil is assumed to have a uniform thermal resistivity, but if the heat dissipated from a cable and its surface temperature are raised above certain critical limits, the soil will dry out, resulting in a zone that is assumed to have a uniform thermal resistivity higher than the original one. The critical conditions, that is, the conditions for the onset of drying, are dependent on the type of soil, its original moisture content, and temperature.

Given the appropriate conditions, it is assumed that when the surface of a cable exceeds the critical temperature rise above ambient, a dry zone will form around it. The outer boundary of the zone is on the isotherm related to that particular temperature rise. An additional assumption states that the development of such a dry zone does not change the shape of the isothermal pattern from what it was when all the soil was moist, only that the numerical values of some isotherms change. Within the dry zone, the soil has a uniformly high value of thermal resistivity, corresponding to its value when the soil is "oven dried" at not more than 105°C. Outside the dry zone, the soil has uniform thermal resistivity corresponding to the site moisture content. The essential advantages of these assumptions are that the resistivity is uniform over each zone, and that the values are both convenient and sufficiently accurate for practical purposes.

The method presented below assumes that the entire region surrounding a cable or cables has uniform thermal characteristics prior to drying out; the only nonuniformity being that caused by drying. As a consequence, the method should not be applied without further consideration to installations where special backfills with properties different from the site soil are used.

Let θ_e be the cable surface temperature corresponding to the moist soil thermal resistivity ρ_1. Within the area between the cable surface (assumed to be isothermal) and the critical isotherm, the heat transfer equation remains the same, the only change from the uniform soil condition being the thermal resistivity of the dry zone.

Without moisture migration, we obtain the following relations, remembering that the soil thermal resistance is directly proportional to the value of resistivity:

$$nW_t = \frac{\theta_e - \theta_{amb}}{T_4} = \frac{(\theta_e - \theta_x) + (\theta_x - \theta_{amb})}{T_4} \tag{1.45}$$

and

$$nW_t = \frac{\theta_e - \theta_x}{C\rho_1} \tag{1.46}$$

where C is a constant, n is the number of cores in the cable, T_4 is the cable external thermal resistance when the soil is moist, and W_t is the total losses in a single core. θ_{amb} and θ_x are ambient temperature and the temperature of an isotherm at distance x, respectively.

If we now assume that the region between the cable and the θ_x isotherm dries out so that its resistivity becomes ρ_2, and that the power losses W_t remain unchanged, we have

$$nW_t = \frac{\theta'_e - \theta_x}{C\rho_2} \tag{1.47}$$

where θ'_e is the cable surface temperature after moisture migration has taken place.

Combining Equations (1.46) and (1.47), we obtain the following form of Equation (1.45):

$$\theta'_e - \theta_x = \frac{\rho_2}{\rho_1}(\theta_e - \theta_x) = \frac{\rho_2}{\rho_1}[(\theta_e - \theta_{amb}) - (\theta_x - \theta_{amb})] \tag{1.48}$$

After rearranging the last equation, we obtain

$$\theta'_e - \theta_{amb} = v(W_t \cdot T_4) - (v - 1)\,\Delta\theta_x \tag{1.49}$$

where

$$v = \frac{\rho_2}{\rho_1} \qquad \text{and} \qquad \Delta\theta_x = \theta_x - \theta_{amb} \tag{1.50}$$

The rating Equation (1.40) takes the form

$$I = \left[\frac{\theta_c - \theta_{amb} - W_d[0.5T_1 + n(T_2 + T_3 + vT_4)] + (v-1)\Delta\theta_x}{RT_1 + nR(1 + \lambda_1)T_2 + nR(1 + \lambda_1 + \lambda_2)(T_3 + vT_4)} \right]^{0.5} \quad (1.51)$$

We can observe that Equation (1.40) has been modified by the addition of the term $(v-1)\Delta\theta_x$ in the numerator, and the substitution of vT_4 for T_4 in both the numerator and the denominator.

1.4.2 Cables in Air

When cables are installed in free air, the external thermal resistance now accounts for the radiative and convective heat loss. For cables exposed to solar radiation, there is an additional temperature rise caused by the heat absorbed by the external covering of the cable. The heat gain by solar absorption is equal to $\sigma D_e H$, with the meaning of the variables defined below. In this case, the external thermal resistance is different than for shaded cables in air, and the current rating is computed from the following modification of Equation (1.40) (IEC 60287, 1994).

$$I = \left[\frac{\Delta\theta - W_d[0.5T_1 + n(T_2 + T_3 + T_4^*)] + \sigma D_e^* H T_4^*}{RT_1 + nR(1 + \lambda_1)T_2 + nR(1 + \lambda_1 + \lambda_2)(T_3 + T_4^*)} \right]^{0.5} \quad (1.52)$$

where
D_e^* = external diameter of the cable, m
σ = absorption coefficient of solar radiation for the cable surface
H = intensity of solar radiation, W/m^2
T_4^* = external thermal resistance of the cable in free air, adjusted to take account of solar radiation, K · m/W

1.5 RATING EQUATIONS—TRANSIENT CONDITIONS

The procedure to evaluate temperatures is the main computational block in transient rating calculations. This block requires a fairly complex programming procedure to take into account self- and mutual heating, and to make suitable adjustments in the loss calculations to reflect changes in the conductor resistance with temperature.

Transient rating of power cables requires the solution of the equations for the network in Figure 1-6. The unknown quantity in this case is the variation of the conductor temperature rise with time,[8] $\theta(t)$. Unlike in the steady-state analysis, this temperature is not a simple function of the conductor current $I(t)$. Therefore, the process for determining the maximum value of $I(t)$ so that the maximum operating conductor temperature is not exceeded requires an iterative procedure. An excep-

---

[8]Unless otherwise stated, in this section we will follow the notation in IEC (1989) and we will use the symbol θ to denote temperature rise, and not $\Delta\theta$ as in other parts of the book and in IEC (1994).

tion is the simple case of identical cables carrying equal current located in a uniform medium. Approximations have been proposed for this case, and explicit rating equations developed. We will discuss this case first and then we will extend the discussion to multiple cable types.

1.5.1 Response to a Step Function

1.5.1.1 Preliminaries. Whether we consider the simple cable systems mentioned above, or a more general case of several cable circuits in a backfill or duct bank, the starting point of the analysis is the solution of the equations for the network in Figure 1-6. Our aim is to develop a procedure to evaluate temperature changes with time for the various cable components. As observed by Neher (1964), the transient temperature rise under variable loading may be obtained by dividing the loading curve at the conductor into a sufficient number of time intervals, during any one of which the loading may be assumed to be constant. Therefore, the response of a cable to a step change in current loading will be considered first.

This response depends on the combination of thermal capacitances and resistances formed by the constituent parts of the cable itself and its surroundings. The relative importance of the various parts depends on the duration of the transient being considered. For example, for a cable laid directly in the ground, the thermal capacitances of the cable, and the way in which they are taken into account, are important for short-duration transients, but can be neglected when the response for long times is required. The contribution of the surrounding soil is, on the other hand, negligible for short times, but has to be taken into account for long transients. This follows from the fact that the time constant of the cable itself is much shorter than the time constant of the surrounding soil.

The thermal network considered in this work is a derivation of the lumped parameter ladder network introduced early in the history of transient rating computations (Buller, 1951; Van Wormer, 1955; Neher, 1964; CIGRE, 1972; IEC 1985, 1989). For computational purposes, Baudoux et al. (1962) and then Neher (1964) proposed to represent a cable in just two loops. Baudoux et al. provided procedures for combining several loops to obtain a two-section network, which was latter adopted by CIGRE WG 02 and published in *Electra* (CIGRE, 1972). However, transformation of a multiloop network into a two-loop equivalent not only requires substantial manual work before the actual transient computations can be performed, but also inhibits the computation of temperatures at parts of the cable other than the conductor. A procedure is given below for analytical solution of the entire network. Generally, the network will be somewhat different for short- and long-duration transients, and, usually, the limiting duration to distinguish these two cases can be taken to be 1 h. Short transients are assumed to last at least 10 min. A more detailed time division between short and long transients can be found in Table 1-1 presented later in this Chapter.

The temperature rise of a cable component (e.g., conductor, sheath, jacket, etc.) can be represented by the sum of two components: the temperature rise inside and outside the cable. The method of combining these two components, introduced by

Morello (Morello, 1958; CIGRE, 1972; IEC, 1985, 1989), makes allowance for the heat that accumulates in the first part of the thermal circuit and which results in a corresponding reduction in the heat entering the second part during the transient. The reduction factor, known as the attainment factor, $\alpha(t)$, of the first part of the thermal circuit is computed as a ratio of the temperature rise across the first part at time t during the transient to the temperature rise across the same part in the steady state. Then, the temperature transient of the second part of the thermal circuit is composed of its response to a step function of heat input multiplied by a reduction coefficient (variable in time) equal to the attainment factor of the first part. Evaluation of these temperatures is discussed below.

1.5.1.2 *Temperature Rise in the Internal Parts of the Cable.* The internal parts of the cable encompass the complete cable including its outermost serving or anticorrosion protection. If the cable is located in a duct or pipe, the duct and pipe (including pipe protective covering) are also included. For cables in air, the cable extends as far as the free air.

Analyses of linear networks, such as the one in Figure 1-6, involve the determination of the expression for the response function caused by the application of a forcing function. In our case, the forcing function is the conductor heat loss and the response sought is the temperature rise above the cable surface at node i. This is accomplished by utilizing a mathematical quantity called the transfer function of the network. It turns out that this transfer function is the Fourier transform of the unit-impulse response of the network. The Laplace transform of the network's transfer function is given by a ratio

$$H(s) = \frac{P(s)}{Q(s)} \tag{1.53}$$

$P(s)$ and $Q(s)$ are polynomials, their forms depending on the number of loops in the network. Node i can be the conductor or any other layer of the cable. In terms of time, the response of this network is expressed as (Van Valkenburg, 1964)[9]

$$\theta_i(t) = W_c \sum_{j=1}^{n} T_{ij}(1 - e^{P_j t}) \tag{1.54}$$

where:
$\theta_i(t)$ = temperature rise at node i at time t, °C
W_c = conductor losses including skin and proximity effects, W/m
T_{ij} = coefficient, °Cm/W
P_j = time constant, s^{-1}
t = time from the beginning of the step, s
n = number of loops in the network

---

[9]Unless otherwise stated, in the remainder of this section all the temperature rises are caused by the joule losses in the cable.

i = node index

j = index from 1 to n.

The coefficients T_{ij} and the time constants P_j are obtained from the poles and ze-ros of the equivalent network transfer function given by Equation (1.53). Poles and zeros of the function $H(s)$ are obtained by solving equations $Q(s) = 0$ and $P(s) = 0$, respectively. From the circuit theory, the coefficients T_{ij} are given by

$$T_{ij} = -\frac{a_{(n-i)i}}{b_n} \frac{\prod_{k=1}^{n-i}(Z_{ki} - P_j)}{P_j \prod_{\substack{k=1 \\ k \neq j}}^{n}(P_k - P_j)} \tag{1.55}$$

where

$a_{(n-i)i}$ = coefficient of the numerator equation of the transfer function

b_n = first coefficient of the denominator equation of the transfer function

Z_{ki} = zeros of the transfer function

P_j = poles of the transfer function.

An algorithm for the computation of the coefficients of the transfer function equation is given in Appendix B of Anders (1997).

Example 1.4

A simple expression of Equation (1.54) is obtained for the case of $n = 2$. Construc-tion of such a network is discussed in Section 1.3.3.2 and the network is shown in Figure 1-5.

In this simple case, the time-dependent solution for the conductor temperature can easily be obtained directly. However, to illustrate the procedure outlined above, we will compute this temperature from Equations (1.53)–(1.55).

The transfer function for this network is given by

$$H(s) = \frac{(T_A + T_B)sT_AT_BQ_B}{1 + s(T_AQ_A + T_BQ_B + T_BQ_A) + s^2T_AQ_AT_BQ_B} \tag{1.56}$$

Since we are interested in obtaining conductor temperature, $i = 1$ and $j = 1, 2$. To simplify the notation, we will use the following substitutions:

$$T_a = T_{11} \quad T_b = T_{12} \quad M_0 = 0.5(T_AQ_A + T_BQ_B + T_BQ_A) \quad N_0 = T_AQ_AT_BQ_B \tag{1.57}$$

The zeros and poles of the transfer function are easily obtained as follows:

$$Z_{11} = \frac{T_A + T_B}{T_AT_BQ_B} \quad P_1 = -a \quad P_2 = -b$$

where

$$a = \frac{M_0 + \sqrt{M_0^2 - N_0}}{N_0} \quad b = \frac{M_0 - \sqrt{M_0^2 - N_0}}{N_0} \tag{1.58}$$

From Equation (1.56),

$$a_{(2-1)1} = T_A T_B Q_B \qquad b_2 = T_A T_B Q_A Q_B$$

Thus,

$$\frac{a_{11}}{b_2} = \frac{1}{Q_A}$$

From Equation (1.55), we have

$$T_a = -\frac{1}{Q_A} \frac{\dfrac{T_A + T_B}{T_A T_B Q_B} + a}{-a(-b + a)} = \frac{1}{a - b}\left(\frac{1}{Q_A} - \frac{T_A + T_B}{a T_A T_B Q_A Q_B} \right)$$

but

$$ab = \frac{1}{T_A T_B Q_A Q_B}$$

Hence,

$$T_a = \frac{1}{a - b}\left[\frac{1}{Q_A} - b(T_A + T_B) \right] \qquad \text{and} \qquad T_b = T_A + T_B - T_a \qquad (1.59)$$

Finally, the conductor temperature as a function of time is obtained from Equation (1.54):

$$\theta_c(t) = W_c[T_a(1 - e^{-at}) + T_b(1 - e^{-bt})] \qquad (1.60)$$

where W_c is the power loss per unit length in a conductor based on the maximum conductor temperature attained. The power loss is assumed to be constant during the step of the transient. Further,

$$\alpha(t) = \frac{\theta_c(t)}{W_c(T_A + T_B)} \qquad (1.61)$$

■

Because the solution of network equations for a two-loop network is quite simple, IEC publications 853-1 (1985) and 853-2 (1989) recommend that this form be used in transient analysis. The two-loop computational procedure was published at a time when access to fast computers was very limited (CIGRE, 1972). Today, this limitation is no longer a problem and a full network representation is recommended in transient analysis computations. This recommendation is particularly applicable in the case when temperatures of cable components, other than the conductor, are of interest.

1.5.1.3 Second Part of the Thermal Circuit—Influence of the Soil. The

transient temperature rise $\theta_e(t)$ of the outer surface of the cable can be evaluated exactly in the case when the cable is represented by a line source located in a homogeneous, infinite medium with uniform initial temperature. However, for practical applications, we have to use another hypothesis, namely, the hypothesis of Kennelly, which assumes that the earth surface must be an isotherm. Under this hypothesis, the temperature rise at any point M in the soil is, at any time, the sum of the temperature rises caused by the heat source W_t and by its fictitious image placed symmetrically with the earth surface as the axis of symmetry and emitting heat $-W_t$ (see Figure 1-9).

The temperature rise at the cable outside surface is then given by

$$\theta_e(t) = W_t \frac{\rho_s}{4\pi} \left[-Ei\left(-\frac{D_e^{*2}}{16\delta t} \right) + Ei\left(-\frac{L^{*2}}{\delta t} \right) \right] \tag{1.62}$$

where:
D_e^* = external surface diameter of cable, m
L^* = axial depth of burial of the cable, m
δ = soil diffusivity, m²/s

The expression

$$-Ei(-x) = \int_x^\infty \frac{e^{-v}}{v} dv$$

is called the exponential integral. The value of the exponential integral can be developed in the series

$$-Ei(-x) = -0.577 - \ln x + x - \frac{x^2}{2 \cdot 2!} + \frac{x^3}{3 \cdot 3!} \cdots$$

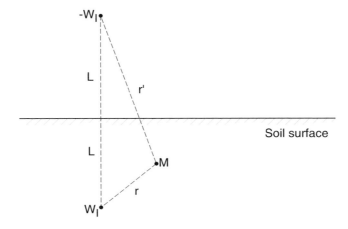

Figure 1-9   Illustration of Kennelly's hypothesis.

When $x < 0.1$,

$$-Ei(-x) = -0.577 - \ln x + x$$

to within 1% accuracy. For large x,

$$-Ei(-x) = -\frac{e^{-x}}{x}\left(1 - \frac{1}{x} + \frac{2!}{x^2} - \frac{3!}{x^3} + \cdots\right)$$

The National Bureau of Standards published in 1940 *Tables of Exponential Integrals,* Vol. 1, in which values of $-Ei(-x)$ can be found. IEC has also published nomograms from which $-Ei(-x)$ can be obtained (IEC 853-2, 1989).

Under steady-state conditions, $t \rightarrow \infty$ and x approaches zero. In this case, Equation (1.62) becomes

$$\theta_e(\infty) = W_t \frac{\rho_s}{2\pi} \ln \frac{4L^*}{D_e^*} \tag{1.63}$$

From this equation, we can define the external thermal resistance in the steady-state calculations as

$$T_4 = \frac{\rho_s}{2\pi} \ln \frac{4L^*}{D_e^*} \tag{1.64}$$

For cables in air it, is unnecessary to calculate a separate response for the cable environment. The complete transient $\theta(t)$ is obtained from Equation (1.62) but the external thermal resistance T_4, computed as described in Section 1.6.6.5, is included in the cable network.

1.5.1.4 Groups of Equally or Unequally Loaded Cables.
In a typical installation, several power cables are laid in a trench. The mutual heating effect reduces the current-carrying capacity of the cables, and this effect must be taken into account in rating computations. For groups of cables, the temperature for each cable is obtained at each point in time by adding to its own temperature the temperature rise caused by other nearby cables. To achieve better accuracy in calculations with multiple time steps, the effect of other cables should be added at each time step so that their effect can be included with that caused by the temperature rise of the cable itself. Thus, the temperature rise in the cable of interest "p" due to one other adjacent cable "k" can be computed from:

$$\theta_{pk}(t) = W_{lk} \frac{\rho_s}{4\pi}\left[-Ei\left(-\frac{d_{pk}^{*2}}{4\delta t}\right) + Ei\left(-\frac{d_{pk}^{*'2}}{4\delta t}\right)\right] \tag{1.65}$$

in which W_{lk} is the total joule losses in cable k, and d_{pk}^* and $d_{pk}^{*'}$ (m) denote the distance from the center of cable p to the center of cable k and its image, respectively, as shown in Figure 1-10.

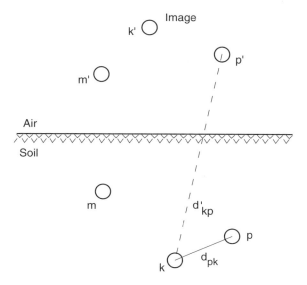

Figure 1-10 Example of cable configuration and image cables.

1.5.1.5 Total Temperature Rise.

The total transient temperature rise of a cable at any time is the sum of the rise due to its own losses, given by its own network, and the rises due to mutual heating given by the networks of other cables and image sources, as appropriate. Thus, the final temperature rise at any layer of the cable of interest (that is, at any node of the equivalent network) at time t after the beginning of the load step is obtained from

$$\theta_{ptot}(t) = \theta_i(t) + \alpha(t)\theta_e(t) + \theta_d(t) + \alpha(t)\sum_{k=1}^{N-1}[\theta_{pk}(t) + \theta_{pdk}(t)] \qquad (1.66)$$

where $\alpha(t)$ is the attainment factor for the transient temperature rise between the conductor and outside surface of the cable and N is the number of cables. $\theta_{pdk}(t)$ is the temperature rise in cable p caused by the dielectric losses in cable k and it is multiplied by $\alpha(t)$ only if cable k is energized at time $t = 0$. $\theta_d(t)$ is the internal temperature rise caused by the dielectric losses in cable p. In Equation (1.66), θ_{ptot} is defined for any layer of the cable, and in the above formulation, only $\theta_i(t)$ is different for each layer.

The attainment factor varies in time, and a reasonable approach for obtaining $\alpha(t)$ is to use (Morello, 1958)

$$\alpha(t) = \frac{\text{Temperature across cable at time } t}{\text{Steady-state temperature rise across the cable}} \qquad (1.67)$$

The conductor attainment factor is computed from this definition using Equations (1.54) and (1.42):

$$\alpha(t) = \frac{\theta_c(t)}{\theta_c(\infty)} = \frac{\dot{W}_c \sum_{j=1}^{n} T_{cj}(1 - e^{P_j t})}{W_c T} = \frac{\sum_{j=1}^{n} T_{cj}(1 - e^{P_j t})}{T} \qquad (1.68)$$

The attainment factor associated with dielectric losses is obtained from a similar equation with the network parameters reflecting the presence of dielectric loss only.

1.5.2 Transient Temperature Rise under Variable Loading

In order to perform computations for variable loading, a daily load curve is divided into a series of steps of constant magnitude. For different successive steps, the computations are done repeatedly, and the final result is obtained using the principle of superposition. The temperature rise above ambient at time τ can be represented as

$$\theta(\tau) - \theta(\tau - 1) \qquad (1.69)$$

1.5.3 Conductor Resistance Variations during Transients

Since the conductor electrical resistance, as well as the resistance of other metallic parts of the cable, changes with temperature, the effect of these changes should be taken into account when computing conductor and sheath losses. Goldenberg (1967, 1971) developed a technique for obtaining arbitrarily close upper and lower bounds for the temperature rise of the conductor, taking into account the changes of the resistance of metallic parts with temperature. His upper-bound formula has been adopted by CIGRE (1972, 1976) and IEC (1989) and is given by the following equation:

$$\theta_a(t) = \frac{\theta(t)}{1 + a[\theta(\infty) - \theta(t)]} \qquad (1.70)$$

where:
$\theta(t)$ = conductor transient temperature rise above ambient without correction for variation in conductor loss, and is based on the conductor resistance at the end of the transient

$\theta(\infty)$ = conductor steady-state temperature rise above ambient

a = temperature coefficient of electrical resistivity of the conductor material at the start of the transient. $\alpha = 1/[\beta + \theta(0)]$, with β being a reciprocal of temperature coefficient at 0°C and $\theta(0)$ is the conductor temperature at the start of the transient

1.5.4 Cyclic Rating Factor

The complexity of cyclic rating computations varies depending on the shape of the load curve and the amount of detail known for the load cycle. If only the load-loss factor or a daily load factor is known, a method proposed by Neher and McGrath (1957) can be used. This method involves modification of the cable external thermal resistance as discussed in Anders (1997) and in Section 1.6.6.5. This modified value is then used in Equation (1.51). Neher and McGrath's approach continues to be the basis for the majority of cyclic loading computations performed in North America.

If a more detailed analysis is required, the algorithm introduced by Goldenberg (1957, 1958) and later adopted by the IEC (1985, 1989) can be used. This approach, applicable to a single cable or a cable system composed of identical, equally loaded cables located in a uniform medium, requires computation of a cyclic rating factor M by which the permissible steady-state rated current (100% load factor) may be multiplied to obtain the permissible peak value of current during a daily (24 h) cycle such that the conductor temperature attains, but does not exceed, the standard permissible maximum temperature during the cycle. A factor derived in this way uses the steady-state temperature, which is usually the permitted maximum temperature, as its reference. The cyclic rating factor depends only on the shape of the daily cycle and is independent of the actual magnitudes of the current.

Several sections in this book address the issue of cyclic rating computations under various circumstances. Therefore, the procedure for the calculation of this factor is reviewed below and Chapter 6 presents a complete discussion of the relevance of the load-loss factor used in time-dependent rating calculations.

First, we will consider a single cable and then extend the analysis to a group of identical cables. The details of the development of the equations presented here can be found in Anders (1997). We will ignore the variation of the conductor resistance with temperature. This is consistent with the IEC approach for cyclic loading calculations. Following the IEC standard approach, only the six hourly steps before the time when the temperature reaches its highest value are involved; the remaining hourly steps being represented by a representative load, as illustrated in Figure 3-13.

The cyclic rating factor for uniform soil conditions is defined as (Anders, 1997; IEC, 1989)

$$M = \cfrac{1}{\sqrt{\mu\left[1 - \dfrac{\theta_R(6)}{\theta_R(\infty)}\right] + \sum_{i=0}^{5} Y_i\left[\dfrac{\theta_R(i+1)}{\theta_R(\infty)} - \dfrac{\theta_R(i)}{\theta_R(\infty)}\right]}} \tag{1.71}$$

with the following notation, where the subscript R corresponds to the steady state rated current:

$$\frac{\theta_R(t)}{\theta_R(\infty)} = [1 - k + k\beta(t)]\alpha(t) \qquad \text{for } i \geq 1$$
$$\theta_R(0) = 0 \tag{1.72}$$

$$k = \frac{\theta_e(\infty)}{\theta_R(\infty)} = \frac{W_I T_4}{W_c T + W_I T_4} \tag{1.73}$$

$$\alpha(t) = \frac{\theta_c(t)}{\theta_c(\infty)} \tag{1.74}$$

$$\beta(t) = \frac{-Ei\left(-\dfrac{D_e^{*2}}{16t\delta}\right) + Ei\left(-\dfrac{L^{*2}}{t \cdot \delta}\right)}{2 \ln\left(\dfrac{4L^*}{D_e^*}\right)} \tag{1.75}$$

where
D_e^* = external diameter of the cable or duct, m
L^* = depth of laying, m
W_I = the total joule loss in the cable, W/m
δ = soil thermal diffusivity, m²/s
T = internal thermal resistance of the cable K · m/W
μ = load loss factor

If the hourly load values are denoted by I_i, $i = 1, \ldots, 24$, then the load-loss factor is defined by:

$$\mu = \frac{1}{24}\frac{\sum\limits_{i=0}^{23} I_i^2}{I_{\max}^2} = \frac{1}{24}\sum_{i=0}^{23} Y_i \tag{1.76}$$

(See more discussion on the definition of the load-loss factor in Chapter 6.)

Calculations are simplified considerably when the conductor attainment factor can be assumed to be equal to one. In this case, Equation (1.71) takes the form

$$M = \frac{1}{\sqrt{(1-k)Y_0 + k\{B + \mu[1 - \beta(6)]\}}} \tag{1.77}$$

where:

$$B = \sum_{i=0}^{5} Y_i \Phi_i; \qquad \Phi_m = \beta(m+1) - \beta(m) \tag{1.78}$$

IEC (1989) identifies the following cases when the internal cable capacitances can be neglected. If the period from the initiation of the thermal transient is longer than

1. 12 h for all cables,
2. The product $\Sigma T \cdot \Sigma Q$; when dealing with fluid-pressure, pipe-type cables and all types of self-contained cables where the product $\Sigma T \cdot \Sigma Q \le 2$ h
3. The product $2\Sigma T \cdot \Sigma Q$; when dealing with gas-pressure, pipe-type cables and all types of self-contained cables where the product $\Sigma T \cdot \Sigma Q > 2$ h

where ΣT and ΣQ are the total internal thermal resistance (simple sum of all resistances) and capacitance (simple sum of all capacitances), respectively, of the cable.

Table 1-1, based on design values commonly used at present for the determination of cable dimensions, shows when cases 2 and 3 apply (IEC, 1989).

We will now consider groups of N cables with equal losses; the cables or ducts do not touch. In this case, $\theta_R(i)$ is the conductor temperature rise of the hottest cable

Table 1-1 Cases when the attainment factor can be assumed to be equal to 1

Type of cable	Case b)	Case c)
Fluid-filled cables	1) All voltages < 220 kV 2) 220 kV, sections ≤ 150 mm²	1) 220 kV, sections > 150 mm² 2) All voltages > 220 kV
Pipe-type, fluid-pressure cables	1) All voltages < 220 kV 2) 220 kV, sections ≤ 800 mm²	1) 220 kV, sections > 800 mm² 2) All voltages > 220 kV
Pipe-type, gas-pressure cables		1) ≤ 220 kV 2) Sections ≤ 1000 mm²
Cables with extruded insulation	1) All voltages < 60 kV 2) 60 kV, sections ≤ 150 mm²	1) 60 kV, sections > 150 mm² 2) All voltages > 60 kV

in the group. The external thermal resistance of the hottest cable in Equations (1.73) and (1.75) will now include the effect of the other $(N-1)$ cables and will be denoted by $T_4 + \Delta T_4$. We now obtain the following new form of Equation (1.75):

$$\beta_1(t) = \frac{\rho_s}{4\pi} \frac{-Ei\left(\frac{D_e^{*2}}{16t\delta}\right) + Ei\left(\frac{L^{*2}}{t\delta}\right) + \sum_{\substack{k=1 \\ k \neq p}}^{N}\left[-Ei\left(\frac{d_{pk}^2}{4t\delta}\right) + Ei\left(\frac{d_{pk}'^2}{4t\delta}\right)\right]}{(T_4 + \Delta T_4)} \tag{1.79}$$

The value of ΔT_4 is equal to (Anders, 1997)

$$\Delta T_4 = \frac{\rho_s \ln F}{2\pi} \tag{1.80}$$

where

$$F = \frac{d_{p1}' \cdot d_{p2}' \cdot \dots \cdot d_{pk}' \cdot \dots \cdot d_{pN}'}{d_{p1} \cdot d_{p2} \cdot \dots \cdot d_{pk} \cdot d_{pN}} \tag{1.81}$$

with factor d_{pp}'/d_{pp} excluded, leaving $(N-1)$ factors in Equation (1.81). The distances d_{pv} and d_{pv}' represent the distance between cables p and v and between the cable v and the image of cable p, respectively.

Introducing the notation

$$d_f = \frac{4L^*}{F^{1/(N-1)}} \tag{1.82}$$

Equation (1.79) can be approximated by

$$\beta_1(t) = \frac{-Ei\left(\frac{D_e^{*2}}{16t\delta}\right) + Ei\left(\frac{L^{*2}}{t\delta}\right) + (N-1)\left[-Ei\left(\frac{d_f^2}{16t\delta}\right) + Ei\left(\frac{L^{*2}}{t\delta}\right)\right]}{2 \ln\frac{4L^*F}{D_e^*}} \tag{1.83}$$

Also, Equation (1.73) becomes

$$k_1 = \frac{W_f(T_4 + \Delta T_4)}{W_c T + W_f(T_4 + \Delta T_4)} \tag{1.84}$$

The cyclic rating factor is given by Equation (1.71) with

$$\frac{\theta_R(t)}{\theta_R(\infty)} = [1 - k_1 + k_1 \beta_1(t)]\alpha(t) \tag{1.85}$$

1.6 EVALUATION OF PARAMETERS

Rating equations discussed in the previous sections contain many different parameters whose values need to be estimated first before ampacity calculations can proceed. For some of the parameters, analytical expressions can be derived; for others, empirical equations or curves have been proposed. In this section, we will summarize the calculation of the important parameters. For a detailed discussion on their derivation, the reader is referred again to Anders (1997).

We will start with a list of the symbols used in various expressions presented below. To facilitate the presentation, we will divide all the symbols into logical groups related to cable construction and laying conditions. This will be followed by a series of equations, tables, and charts that are needed for ampacity calculations. The order of the presentation follows the usual steps in cable rating calculations.

1.6.1 List of Symbols

1.6.1.1 General Data
f (Hz) = system frequency
U (V) = cable operating voltage (phase-to-phase)
LF = daily load factor
θ = conductor temperature[10]
θ_{amb} = ambient temperature

1.6.1.2 Cable Parameters

Conductor
S (mm^2) = cross-sectional area of conductor
D_e (mm) = external diameter of cable, or equivalent diameter of a group of cores in pipe-type cable
D_e^* (m) = external diameter of cable, or equivalent diameter of cable (cables in air)
d_c (mm) = external diameter of conductor

---

[10]subscripts t, s, and a will be used to denote tape, sheath, and armor, respectively.

d'_c (mm) = conductor diameter of equivalent solid conductor having the same central oil duct

d_i (mm) = conductor inside diameter

n = number of conductors in a cable

Three-conductor cables

d_x (mm) = Diameter of an equivalent circular conductor having the same cross-sectional area and degree of compactness as the shaped one

c (mm) = Distance between the axes of conductors and the axis of the cable for three-core cables (= $0.55\, r_1 + 0.29\, t$ for sector-shaped conductors)

r_1 (mm) = Circumscribing radius of three-sector shaped conductors in three-conductor cable

Insulation

D_i (mm) = diameter over insulation

t_1 (mm) = insulation thickness between conductors and sheath

ρ (K · m/W) = thermal resistivity of the material[11]

Three-conductor cables

t (mm) = insulation thickness between conductors

t_i (mm) = thickness of core insulation, including screening tapes plus half the thickness of any nonmetallic tapes over the laid-up cores

Sheath

D_s (mm) = sheath diameter

d (mm) = sheath mean diameter

t_s (mm) = sheath thickness

$\left.\begin{array}{l}p_2\\q_2\end{array}\right\}$ = ratios of minor section lengths, where minor section lengths are: a, p_2a, and q_2a, and a is the shortest section

$\varsigma(\Omega m)$ = electrical resistivity of sheath material at operating temperature

Armor or reinforcement

A (mm²) = cross-sectional area of the armor

D_a (mm) = external diameter of armor

d_a (mm) = mean diameter of armor

d_f (mm) = diameter of armor wires

d_2 (mm) = mean diameter of reinforcement

n_a = number of armor wires

n_t = number of tapes

ℓ_a (mm) = length of lay of a steel wire along a cable

ℓ_T (mm) = length of lay of a tape

t_t (mm) = thickness of tape

w_t (mm) = width of tape

---

[11]The same symbol is used for thermal resistivity of various materials. The appropriate numerical value will correspond to the material considered.

Jacket/Serving
t_j (mm) = thickness of the jacket
t_3 (mm) = thickness of the serving

Pipe-type cables
D_d (mm) = external diameter of the pipe
D_o (mm) = external diameter of pipe coating
D_{sm} (mm) = moisture barrier mean diameter
D_{sw} (mm) = diameter of skid wire
D_t (mm) = diameter of moisture barrier assembly
d_d (mm) = internal diameter of pipe
n_{sw} = number of skid wires
n_t = number of moisture barrier metallic tapes
ℓ_{sw} (mm) = lay of length of skid wires
ℓ_T (mm) = lay of moisture barrier metallic tapes
t_3 (mm) = thickness of pipe coating
t_t = thickness of moisture barrier metallic tape in pipe-type cable
w_t (mm) = width of moisture barrier metallic tape

1.6.1.3   Installation Conditions

Cables in Air
H (W/m^2) = solar radiation

Buried Cables
L (mm) = depth of burial of cables
D_x (mm) = fictitious diameter at which effect of loss factor commences
ρ_s (K · m/W) = thermal resistivity of soil
s (mm) = spacing between conductors of the same circuit
s_2 (mm) = axial separation of cables; for cables in flat formation, s_2 is the geometric
  mean of the three spacings

Duct Bank/Thermal Backfill
L_G (mm) = distance from the soil surface to the center of a duct bank
x, y (mm) = sides of duct bank/backfill ($y > x$)
N = number of loaded cables in a duct bank/backfill
ρ_c (K · m/W) = thermal resistivity of concrete used for a duct bank or backfill
ρ_e (K · m/W) = thermal resistivity of earth surrounding a duct bank/backfill

Cables in Ducts
D_d (mm) = internal diameter of the duct
D_o (mm) = external diameter of the duct
θ_m (°C) = mean temperature of duct filling medium

1.6.2   Conductor ac Resistance

Conductor resistance is calculated in two stages. First, the dc value R' (ohm/m) is
obtained from the following expression:

$$R' = \frac{1.02 \cdot 10^6 \, \rho_{20}}{S} [1 + \alpha_{20}(\theta - 20)]$$

In the second stage, the dc value is modified to take into account the skin and proximity effects. The resistance of a conductor when carrying an alternating current is higher than that of the conductor when carrying a direct current. The principal reasons for the increase are: skin effect, proximity effect, hysteresis and eddy current losses in nearby ferromagnetic materials, and induced losses in short-circuited non-ferromagnetic materials nearby. The degree of complexity of the calculations that can economically be justified varies considerably. Except in very high voltage cables consisting of large segmental conductors, it is common to consider only skin effect, proximity effect, and in some cases, an approximation of the effect of metallic sheath and/or conduit. The relevant expressions are:

$$R = R'(1 + y_s + y_p)$$

For cables in magnetic pipes and conduits:

$$R = R'[1 + 1.5(y_s + y_p)]$$

Material properties and the expressions for the skin and proximity factors are:

Material	Resistivity $(\rho_{20}) \cdot 10^{-8}$ $\Omega \cdot$ m at 20°C	Temperature coefficient $(\alpha_{20}) \cdot 10^{-3}$ per K at 20°C
Copper	1.7241	3.93
Aluminum	2.8264	4.03

Skin and proximity factors are computed from the following expressions:

$$y_s = \frac{x_s^4}{192 + 0.8x_s^4}$$

where

$$x_s^2 = F_k \cdot k_s \qquad F_k = \frac{8\omega f \cdot 10^{-7}}{R'}$$

The proximity factor is obtained from

$$y_p = F_p \left(\frac{d_c}{s}\right)^2 \left[0.312 \left(\frac{d_c}{s}\right)^2 + \frac{1.18}{F_p + 0.27}\right] \qquad x_p^2 = F_k \cdot k_p \qquad F_p = \frac{x_p^4}{192 + 0.8x_p^4}$$

For sector-shaped conductors:

$$s = d_x + t$$
$$d_c = d_x$$
$$y_p = 2y_p/3$$

For oval conductors:

$$d_c = \sqrt{d_{cminor} \cdot d_{cmajor}}$$

The above expressions apply when $x_p \leq 2.8$ (a majority of the cases). Otherwise, Equations (7.26) or (7.27) in Anders (1997) should be used.

Constants k_s and k_p are given in Table 1-2 (IEC 60287, 1994).

1.6.3 Dielectric Losses

When paper and solid dielectric insulations are subjected to alternating voltage, they act as large capacitors and charging currents flow in them. The work required to effect the realignment of electrons each time the voltage direction changes (i.e., 50 or 60 times a second) produces heat and results in a loss of real power that is called dielectric loss, which should be distinguished from reactive loss. For a unit length of a cable, the magnitude of the required charging current is a function of the dielectric constant of the insulation, the dimensions of the cable, and the operating voltage. For some cable constructions, notably for high-voltage, paper-insulated cables, this loss can have a significant effect on the cable rating. The dielectric losses are computed from the following expression:

Table 1-2 Values of skin and proximity factors

Type of conductor	Whether dried and impregnated or not	k_s	k_p
Copper			
Round, stranded	Yes	1	0.8
Round, stranded	No	1	1
Round, compact	Yes	1	0.8
Round, compact	No	1	1
Round, segmental		0.435	0.37
Hollow, helical stranded	Yes	Eq. (7.12) in Anders (1997)	0.8
Sector-shaped	Yes	1	0.8
Sector-shaped	No	1	1
Aluminum			*
Round, stranded	either	1	
Round, 4 segment	Either	0.28	
Round, 5 segment	Either	0.19	
Round, 6 segment	Either	0.12	
Segmental with peripheral strands	Either	Eq. (7.17) in Anders (1997)	

*Since there are no accepted experimental results dealing specifically with aluminum stranded conductors, IEC 60287 recommends that the values of k_p given in this table for copper conductors also be applied to aluminum stranded conductor of similar design as the copper ones.

$$W_d = 2\pi f \cdot C \cdot U_o^2 \cdot \tan \delta$$

where the electrical capacitance and the phase-to-to-ground voltage are obtained from

$$C = \frac{\varepsilon}{18 \ln\left(\dfrac{D_i}{d_c}\right)} \cdot 10^{-9} \qquad U_o = \frac{U}{\sqrt{3}}$$

The dielectric constant ε and the loss factor $\tan \delta$ are taken from Table 1-3.

1.6.4 Sheath Loss Factor

Sheath losses are current dependent, and can be divided into two categories according to the type of bonding. These are losses due to circulating currents that flow in the sheaths of single-core cables if the sheaths are bonded together at two points, and losses due to eddy currents, which circulate radially (skin effect) and azimuthally (proximity effect). Eddy current losses occur in both three-core and single-core cables, irrespective of the method of bonding. Eddy current losses in the sheaths of single-core cables, which are solidly bonded are considerably smaller

Table 1-3   Values of the dielectric constant and the loss factor

Type of cable	ε	$\tan \delta$
Cables insulated with impregnated paper		
Solid type, fully impregnated, preimpregnated, or mass-impregnated nondraining	4	0.01
Oil-filled, low-pressure		
up to $U_o = 36$ kV	3.6	0.0035
up to $U_o = 87$ kV	3.6	0.0033
up to $U_o = 160$ kV	3.5	0.0030
up to $U_o = 220$ kV	3.5	0.0028
Oil-pressure, pipe-type	3.7	0.0045
Internal gas-pressure	3.4	0.0045
External gas-pressure	3.6	0.0040
Cables with other kinds of insulation		
Butyl rubber	4	0.050
EPR, up to 18/30 kV	3	0.020
EPR, above 18/30 kV	3	0.005
PVC	8	0.1
PE (HD and LD)	2.3	0.001
XLPE up to and including 18/30 (36) kV, unfilled	2.5	0.004
XLPE above 18/30 (36) kV, unfilled	2.5	0.001
XLPE above 18/30 (36) kV, filled	3	0.005
Paper–polypropylene–paper (PPL)	2.8	0.001

than circulating current losses, and are ignored except for cables with large segmental conductors.

Losses in protective armoring also fall into several categories depending on the cable type, the material of the armor, and installation methods. Armored single-core cables without a metallic sheath generally have a nonmagnetic armor because the losses in steel-wire or tape armor would be unacceptably high. For cables with nonmagnetic armor, the armor loss is calculated as if it were a cable sheath, and the calculation method depends on whether the armor is single-point bonded or solidly bonded. For cables having a metallic sheath and nonmagnetic armor, the losses are calculated as for sheath losses, but using the combined resistance of the sheath and armor in parallel and a mean diameter equal to the rms value or the armor and sheath diameters. The same procedure applies to two- and three-core cables having a metallic sheath and nonmagnetic armor. For two- and three-core cables having metallic sheath and magnetic wire armor, eddy current losses in the armor must be considered. For two- and three-core cables having steel-tape armor, both eddy current losses and hysteresis losses in the tape must be considered together with the effect of armor on sheath losses.

Submarine cables require special consideration. Single-core ac cables for submarine power connections differ in many respects from underground cables, buried directly or in ducts. In fact, submarine cables are generally armored, can be manufactured in very long lengths, and are laid with a very large distance between them. For these reasons, calculation methods described in the IEC 60287 (1994) must be supplemented and modified in some points.

1.6.4.1 Sheath Bonding Arrangements.   Sheath losses in single-core cables depend on a number of factors, one of which is the sheath bonding arrangement. In fact, the bonding arrangement is the second most important parameter in cable ampacity computations after the external thermal resistance of the cable. For safety reasons, cable sheaths must be earthed, and hence bonded, at least at one point in a run. There are three basic options for bonding sheaths of single-core cables. These are: single-point bonding, solid bonding, and cross bonding (ANSI/IEEE, 1988).

In a single-point-bonded system, the considerable heating effect of circulating currents is avoided, but voltages will be induced along the length of the cable. These voltages are proportional to the conductor current and length of run, and increase as the cable spacing increases. Particular care must be taken to insulate and provide surge protection at the free end of the sheath to avoid danger from the induced voltages.

One way of eliminating the induced voltages is to bond the sheath at both ends of the run (solid bonding). The disadvantage of this is that the circulating currents that then flow in the sheaths reduce the current-carrying capacity of the cable.

Cross bonding of single-core cable sheaths is a method of avoiding circulating currents and excessive sheath voltages while permitting increased cable spacing and long run lengths. The increase in cable spacing increases the thermal independence of each cable and, hence, increases its current-carrying capacity. The cross bonding divides the cable run into three sections, and cross connects the sheaths in such a manner that the induced voltages cancel. One disadvantage of this system is that it is very expensive and, therefore, is applied mostly in high-voltage installations. Figure 1-11 gives a diagrammatic representation of the cross connections.

The cable route is divided into three equal lengths, and the sheath continuity is broken at each joint. The induced sheath voltages in each section of each phase are equal in magnitude and 120° out of phase. When the sheaths are cross connected, as shown in Figure 1-11, each sheath circuit contains one section from each phase such that the total voltage in each sheath circuit sums to zero. If the sheaths are then bonded and earthed at the end of the run, the net voltage in the loop and the circulating currents will be zero and the only sheath losses will be those caused by eddy currents.

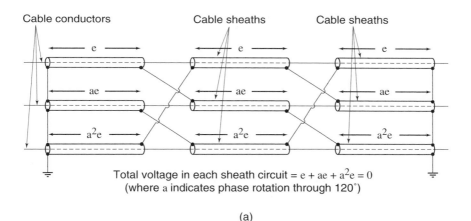

Figure 1-11 Diagrammatic representation of a cross bonded cable system. (a) Cables are not transposed. (b) Cables are transposed.

This method of bonding allows the cables to be spaced to take advantage of improved heat dissipation without incurring the penalty of increased circulating current losses. In practice, the lengths and cable spacings in each section may not be identical, and, therefore, some circulating currents will be present. The length of each section and cable spacings are limited by the voltages that exist between the sheaths and between the sheaths and earth at each cross-bonding position. For long runs, the route is divided into a number of lengths, each of which is divided into three sections. Cross bonding as described above can be applied to each length independently.

The cross-bonding scheme described above assumes that the cables are arranged symmetrically; that is, in a trefoil pattern. It is usual that single-core cables are laid in a flat configuration. In this case, it is a common practice in long-cable circuits or heavily loaded cable lines to transpose the cables as shown in Figure 1-11b so that each cable occupies each position for a third of the run.

All of the equations for sheath loss factors given in this section assume that the phase currents are balanced. The equations also require knowledge of the temperature of the sheath, which cannot be calculated until the cable rating is known, and, therefore, an iterative process is required. For the first calculation, the sheath temperature must be estimated; this estimate can be checked later after the current rating has been calculated. If necessary, the sheath losses, and, hence, the current rating, must be recalculated with the revised sheath temperature.

As discussed above, the power loss in the sheath or screen (λ_1) consists of losses caused by circulating currents (λ_1') and eddy currents (λ_1''). Thus,

$$\lambda_1 = \lambda_1' + \lambda_1'' \tag{1.86}$$

The loss factor in armor is also composed of two components: that due to circulating currents (λ_2') and, for magnetic armor, that caused by hysteresis (λ_2''). Thus,

$$\lambda_2 = \lambda_2' + \lambda_2'' \tag{1.87}$$

As mentioned above, for single-core cables with sheaths bonded at both ends of an electrical section, only losses caused by circulating currents are considered. An electrical section is defined as a portion of the route between points at which the sheaths or screens of all cables are solidly bonded. Circulating current losses are much greater than eddy current losses, and they completely dominate the calculations. Of course, there are no circulating currents when the sheaths are isolated or bonded at one point only.

The first step in computing the sheath loss factor is to obtain the sheath resistance and reactance. If the sheath is reinforced with a nonmagnetic tape or tapes, a parallel combination of both resistances is required. Once the sheath resistance is obtained, the sheath reactance needs to be computed. The appropriate formulae for the calculation of the sheath reactance are as follows.

For single-conductor and pipe-type cables:

$$X = 4\pi f \cdot 10^{-7} \cdot \ln \frac{2s}{d}$$

For single-conductor cables in flat formation, regularly transposed, sheaths bonded at both ends:

$$X_1 = 4\pi f \cdot 10^{-7} \cdot \ln\left[2 \cdot \sqrt[3]{2}\left(\frac{s}{d}\right)\right]$$

For single-conductor cables in flat configuration with sheaths solidly bonded at both ends, the sheath loss factor depends on the spacing. If it is not possible to maintain the same spacing in the electrical section (i.e., between points at which the sheaths of all cables are bonded), the following allowances should be made.

1). If the spacings are known, the value of X is computed from

$$X = \frac{l_a X_a + l_b X_b + \ldots l_n X_n}{l_a + l_b + \ldots l_n}$$

where
$l_a, l_b, \ldots, l_n$ are lengths with different spacing along an electrical section
$X_a, X_b, \ldots, X_n$ are the reactances per unit length of cable, given by equations for X
and X_1, where appropriate values of spacing $s_a, s_b, \ldots, s_n$ are used
2. If the spacings are not known, the value of λ'_1 calculated below should be increased by 25%:

$$X_m = 8.71 \cdot 10^{-7} \cdot f$$

The following equations provide sheath loss factors for various bonding arrangements and cable configurations.

1.6.4.2 Loss Factors for Single-Conductor Cables

1. Sheath bonded both ends, triangular configuration:

$$\lambda'_1 = \frac{R_s}{R} \cdot \frac{1}{1 + \left(\frac{R_s}{X_1}\right)^2}; \qquad \lambda''_1 = 0.$$

2. Sheath bonded both ends, flat configuration, regular transposition:

$$\lambda'_1 = \frac{R_s}{R} \cdot \frac{1}{1 + \left(\frac{R_s}{X}\right)^2}; \qquad \lambda''_1 = 0$$

3. Sheath bonded both ends, flat configuration, no transposition. Center cable equidistant from other cables:

$$\lambda'_{11} = \frac{R_s}{R} \left[\frac{\frac{1}{4}Q^2}{R_s^2 + Q^2} + \frac{\frac{3}{4}P^2}{R_s^2 + P^2} - \frac{2R_s PQX_m}{\sqrt{3}(R_s^2 + Q^2)(R_s^2 + P^2)} \right] \quad \text{in the leading phase}$$

$$\lambda'_{1m} = \frac{R_s}{R} \frac{Q^2}{R_s^2 + Q^2} \quad \text{in the middle cable}$$

$$\lambda'_{12} = \frac{R_s}{R} \left[\frac{\frac{1}{4}Q^2}{R_s^2 + Q^2} + \frac{\frac{3}{4}P^2}{R_s^2 + P^2} + \frac{2R_s PQX_m}{\sqrt{3}(R_s^2 + Q^2)(R_s^2 + P^2)} \right] \quad \text{in the lagging phase}$$

$$\lambda''_1 = 0$$

where P and Q are defined by

$$P = X_m + X \qquad Q = X - \frac{X_m}{3}$$

Ratings for cables in air should be calculated using λ'_{11}.

Large Segmental Conductors. When the conductor proximity effect is reduced, for example, by large conductors having insulated segments, λ''_1 cannot be ignored and is calculated by multiplying the value of the eddy current sheath loss factor calculated below by F:

$$F = \frac{4M^2N^2 + (M + N)^2}{4(M^2 + 1)(N^2 + 1)}$$

where

$$M = N = \frac{R_s}{X}$$

for cables in trefoil formation, and

$$M = \frac{R_s}{X + X_m} \qquad N = \frac{R_s}{X - \frac{X_m}{3}}$$

for cables in flat formation with equidistant spacing.

Sheaths Single-Point Bonded or Cross Bonded. The sheath loss factor is obtained in this case from the following formula:

$$\lambda''_1 = \frac{R_s}{R} \left[g_s \lambda_0 (1 + \Delta_1 + \Delta_2) + \frac{(\beta_1 t_s)^4}{12} \cdot 10^{-12} \right]$$

and the parameters in this expression depend on the cable arrangement, as follows. For lead-sheathed cables.

$$\beta_1 = 0 \qquad g_s = 1$$

For corrugated sheaths, the mean outside diameter should be used:

$$\beta_1 = \sqrt{\frac{4\pi\omega}{10^7 \varsigma}}$$

$$\omega = 2\pi f$$

$$g_s = 1 + \left(\frac{t_s}{D_s}\right)^{1.74}(\beta_1 D_s \times 10^{-3} - 1.6)$$

$$m = \frac{2\pi f}{R_s} \cdot 10^{-7}$$

If $m \le 0.1$, $\Delta_1 = 0$, $\Delta_2 = 0$

Three Single-Conductor Cables in Triangular Configuration

$$\lambda_0 = 3\left(\frac{d}{2s}\right)^2 \frac{m^2}{1+m^2} \qquad \Delta_1 = (1.44m^{2.45} + 0.33)\left(\frac{d}{2s}\right)^{0.92m+1.66} \qquad \Delta_2 = 0$$

Three Single-Conductor Cables in Flat Configuration

1. Center cable:

$$\lambda_0 = 6\left(\frac{d}{2s}\right)^2 \frac{m^2}{1+m^2} \qquad \Delta_1 = 0.86m^{3.08}\left(\frac{d}{2s}\right)^{1.4m+0.7} \qquad \Delta_2 = 0$$

2. Outer cable leading phase:

$$\lambda_0 = 1.5\left(\frac{d}{2s}\right)^2 \frac{m^2}{1+m^2} \qquad \Delta_1 = 4.7m^{0.7}\left(\frac{d}{2s}\right)^{0.16m+2} \qquad \Delta_2 = 21m^{3.3}\left(\frac{d}{2s}\right)^{1.47m+5.06}$$

3. Outer cable lagging phase:

$$\lambda_0 = 1.5\left(\frac{d}{2s}\right)^2 \frac{m^2}{1+m^2} \qquad \Delta_1 = \frac{0.74(m+2)m^{0.5}}{2+(m-0.3)^2}\left(\frac{d}{2s}\right)^{m+1} \qquad \Delta_2 = 0.92m^{3.7}\left(\frac{d}{2s}\right)^{m+2}$$

Sheaths Cross Bonded. The ideal cross-bonded system will have equal lengths and spacing in each of the three sections. If the section lengths are different, the induced voltages will not sum to zero and circulating currents will be present. These circulating currents are taken account of by calculating the circulating current loss factor λ_1', assuming the cables were not cross bonded, and multiplying this value by a factor to take account of the length variations. This factor F_c is given by[12]

$$F_c = \frac{p_2^2 + q_2^2 + 1 - p_2 - p_2 q_2 - q_2}{(p_2 + q_2 + 1)^2}$$

where:
$p_2 a$ = length of the longest section
$q_2 a$ = length of the second longest section
a = length of the shortest section

This formula deals only with differences in the length of minor sections. Any deviations in spacing must also be taken into account.

Where lengths of the minor sections are not known, IEC 287-2-1 (1994) recommends that the value for λ_1' based on experience with carefully installed circuits be

$\lambda_1' = 0.03$ for cables laid directly in the ground
$\lambda_1' = 0.05$ for cables installed in ducts

1.6.4.3 Three-Conductor Cables. The sheath loss factor depends in this case on the value of the sheath resistance as follows.

Round or Oval Conductors in Common Sheath, No Armor
$R_s \leq 100 \ \mu\Omega/m$

$$\lambda_1'' = \frac{3R_s}{R}\left[\left(\frac{2c}{d}\right)^2 \frac{1}{1 + \left(\frac{R_s \ 10^7}{\omega}\right)^2} + \left(\frac{2c}{d}\right)^4 \frac{1}{1 + 4\left(\frac{R_s \ 10^7}{\omega}\right)^2}\right]$$

$R_s > 100 \ \mu\Omega/m$

$$\lambda_1'' = \frac{3.2\omega^2}{RR_s}\left(\frac{2c}{d}\right)^2 10^{-14}$$

Sector-Shaped Conductors

$$\lambda_1'' = 0.94\frac{R_s}{R}\left(\frac{2r_1 + t}{d}\right)^2 \frac{1}{1 + \left(\frac{R_s \ 10^7}{\omega}\right)^2}$$

Three-Conductor Cables with Steel-Tape Armor. The value for λ_1' calculated above should be multiplied by the factor F_t:

---

[12]This formula is somewhat different from the one appearing in the latest IEC Standard 60287 from 1994. It was developed recently in EDF and will be introduced into the standard during the next maintenance cycle.

$$F_t = \left[1 + \left(\frac{d}{d_a} \right)^2 \frac{1}{1 + \frac{d_a}{\mu \delta_0}} \right] \qquad \delta_0 = \frac{A}{\pi d_a}$$

μ is usually taken as 300.

Cables with Separate Lead Sheath (SL Type) with Armor

$$\lambda_1' = \frac{R_s}{R} \frac{1.5}{1 + \left(\frac{R_s}{X} \right)^2}$$

1.6.4.4 Pipe-Type Cables. Nonmagnetic core screens, copper or lead, of pipe-type cables will carry circulating currents induced in the same manner as in the sheaths of solidly bonded single-core cables. As with SL type cables, there is an increase in screen losses due to the presence of the steel pipe. The loss factor is given by:

$$\lambda_1' = \frac{R_s}{R} \frac{1.5}{1 + \left(\frac{R_s}{X} \right)^2}$$

If additional reinforcement is applied over the core screens, the above formula is applied to the combination of sheath and reinforcement. In this case, R_s is replaced by the resistance of the parallel combination of sheath and reinforcement, and the diameter is taken as the rms diameter d', where

$$d' = \sqrt{\frac{d^2 + d_2^2}{2}}$$

with
d = mean diameter of the screen or sheath, mm
d_2 = mean diameter of the reinforcement, mm

1.6.5 Armor Loss Factor

Armored single-core cables for general use in ac systems usually have nonmagnetic armor. This is because of the very high losses that would occur in closely spaced single-core cables with magnetic armor. On the other hand, when magnetic armor is used, losses due to eddy currents and hysteresis in the steel must be considered.

The armoring or reinforcement on two-core or three-core cables can be either magnetic or nonmagnetic. These cases are treated separately in what follows. Steel wires or tapes are generally used for magnetic armor.

When nonmagnetic armor is used, the losses are calculated as a combination of sheath and armor losses. The equations set out above for sheath losses are applied, but the resistance used is that of the parallel combination of sheath and armor, and the sheath diameter is replaced by the rms value of the mean armor and sheath diameters.

For nonmagnetic tape reinforcement where the tapes do not overlap, the resistance of the reinforcement is a function of the lay length of the tape. The advice given in IEC 60287 to deal with this is as follows.

1. If the tapes have a very long lay length, that is, are almost longitudinal tapes, the resistance taken is that of the equivalent tube, that is, a tube having the same mass per unit length and the same internal diameter as the tapes.
2. If the tapes are wound at about 54° to the axis of the cable, the resistance is taken to be twice the equivalent tube resistance.
3. If the tapes are wound with a very short lay, the resistance is assumed to be infinite; hence, the reinforcement has no effect on the losses.
4. If there are two or more layers of tape in contact with each other and having a very short lay, the resistance is taken to be twice the equivalent tube resistance. This is intended to take account of the effect of the contact resistance between the tapes.

The loss factor is then given by the following expressions.

1.6.5.1 Single-Conductor Cables

With Nonmagnetic Wire Armor

$$\lambda_1 = \frac{R_A}{R} \cdot \chi \qquad \lambda_2 = \frac{R_s}{R} \cdot \chi$$

where

$$\chi = \frac{R_s R_A}{(R_s + R_A)^2}$$

With Magnetic Wire Armor. The armor losses are lowest when the armor and sheath are bonded together at both ends of a run; thus, this condition is selected for the calculations below. The method gives a combined sheath and armor loss for cables that are very widely spaced (greater than 10 m). It has been applied for submarine cables where the cable spacing may be very wide and there is a need for the mechanical protection provided by the steel wire armor. The following method does not take into account the possible influence of the surrounding media, which may be appreciable, in particular for cables laid under water. It gives values of sheath and armor losses that are usually higher than the actual ones, so that ratings are on the safe side.

The ac resistance of the armor wires will vary between about 1.2 and 1.4 times its dc resistance, depending on the wire diameter, but this variation is not critical because the sheath resistance is generally considerably lower than that of the armor wires. For magnetic wire, the armor loss factor is obtained from

$$\lambda_1' = \lambda_2 = \frac{R_e}{R}\left[\frac{B_2^2 + B_1^2 + R_e B_2}{(R_e + B_2)^2 + B_1^2}\right]$$

where

$$B_1 = \omega(H_s + H_1 + H_3) \qquad B_2 = \omega H_2$$

and

$$H_s = 2 \times 10^{-7} \ln\left(\frac{2s_2}{d}\right) \qquad H_1 = \pi\mu_e\left(\frac{n_a d_f^2}{\ell_a d_a}\right)10^{-7}\sin\beta\cos\gamma$$

$$H_2 = \pi\mu_e\left(\frac{n_a d_f^2}{\ell_a d_a}\right)10^{-7}\sin\beta\sin\gamma \qquad H_3 = 0.4(\mu_t\cos^2\beta - 1)\left(\frac{d_f}{d_a}\right)10^{-6}$$

Average values for magnetic properties of armor wires with diameters in the range of 4–6 mm and tensile strengths on the order of 40 Mpa:

$\mu_e = 400$
$\mu_t = 10$ for armor wires in contact
$\mu_t = 1$ for armor wires that are spaced
$\gamma = \pi/4$

1.6.5.2 Three-Conductor Cables, Steel-Wire Armor. The loss factor depends on the armor construction as follows.

Round Conductor Cable

$$\lambda_2 = 1.23\frac{R_A}{R}\left(\frac{2c}{d_a}\right)^2\frac{1}{\left(\dfrac{2.77\,R_A\,10^6}{\omega}\right)^2 + 1}$$

For the SL type cables, λ_2 calculated above should be multiplied by $(1 - \lambda_1')$, where λ_1' is calculated in the section on the sheath loss factor for SL type cables.

Sector-Shaped Conductors

$$\lambda_2 = 0.358\frac{R_A}{R}\left(\frac{2r_1}{d_a}\right)^2\frac{1}{\left(\dfrac{2.77\,R_A\,10^6}{\omega}\right)^2 + 1}$$

1.6.5.3 Three-Conductor Cables, Steel-Tape Armor or Reinforcement. The two components of the armor loss factor are computed as follows:

$$\lambda_2' = \frac{s^2 k^2 10^{-7}}{R d_a \delta_0} \qquad \lambda_2'' = \frac{2.23 s^2 k^2 \delta_0 10^{-8}}{R d_a}$$

with

$$\delta_0 = \frac{A}{\pi d_a} \qquad k = \frac{1}{1 + \dfrac{d_a}{\mu \delta_0}} \times \left(\frac{f}{50}\right)$$

μ is usually taken as 300.

Finally,

$$\lambda_2 = \lambda_2' + \lambda_2''$$

1.6.5.4 Pipe-Type Cables in Steel Pipe, Pipe Loss Factor. For a pipe type cable, the loss factor in the metallic pipe depends on the cable arrangement in the pipe. The loss factor is given by

$$\lambda_2 = \left(\frac{f}{60}\right)^{1.5} \left(\frac{A \cdot s + B \cdot d_d}{R}\right) 10^{-5}$$

where constants A and B are taken from Table 1-4.

1.6.6 Thermal Resistances

The internal thermal resistances and capacitances are characteristics of a given cable construction and were defined in Section 1.3. Without loss of accuracy, we will assume that these quantities are constant and independent of the component temperature. Where screening layers are present, we will also assume that for thermal calculations, metallic tapes are part of the conductor or sheath, whereas semiconducting layers (including metallized carbon-paper tapes) are part of insulation.

The units of the thermal resistance are K/W for a specified length. Since the length considered here is 1 m, the thermal resistance of a cable component is expressed in K/W per meter, which is most often written as K · m/W. This should be

Table 1-4   Constants for loss factor calculations in pipe-type cables

Configuration	A	B
Cradled	0.00438	0.00226
Triangular bottom of pipe	0.0115	−0.001485
Mean between trefoil and cradled	0.00794	0.00039
Three-core cable	0.0199	−0.001485

distinguished from the unit of thermal resistivity, which is also expressed as K · m/W. In transient computations discussed in Section 1.5, thermal capacitances associated with the same parts of the cable were identified. The unit of thermal capacitance is J/K · m and the unit of the specific heat of the material is J/K · m³. The thermal resistivities and specific heats of materials used for insulation and for protective coverings are given in Table 1-5.

Table 1-5 Values of thermal resistivity and capacity for cable materials (IEC 60287, 1994)

Material	Thermal resistivity (ρ) (K · m/W)	Thermal capacity ($c \cdot 10^{-6}$) [J/(m³ · K)]
Insulating materials*		
Paper insulation in solid-type cables	6.0	2.0
Paper insulation in oil-filed cables	5.0	2.0
Paper insulation in cables with external gas pressure	5.5	2.0
Paper insulation in cables with internal gas pressure		
a) preimpregnated	6.5	2.0
b) mass impregnated	6.0	2.0
PE	3.5	2.0
XLPE	3.5	2.0
Polyvinyl chloride		
up to and including 3 kV cables	5.0	1.7
greater than 3 kV cables	6.0	1.7
EPR		
up to and including 3 kV cables	3.5	2.0
greater than 3 kV cables	5.0	2.0
Butyl rubber	5.0	2.4
Rubber	5.0	2.4
Paper–polypropylene–paper (PPL)	6.5	2.0
Protective coverings		
Compounded jute and fibrous materials	6.0	2.0
Rubber sandwich protection	6.0	2.0
Polychlroprene	5.5	2.0
PVC		
up to and including 35 kV cables	5.0	1.7
greater than 35 kV cables	6.0	1.7
PVC/bitumen on corrugated aluminum sheaths	6.0	1.7
PE	3.5	2.4
Materials for duct installations		
Concrete	1.0	2.3
Fiber	4.8	2.0
Asbestos	2.0	2.0
Earthenware	1.2	1.8
PVC	6.0	1.7
PE	3.5	2.4

*For the purpose of current rating computations, the semiconducting screening materials are assumed to have the same thermal properties as the adjacent dielectric materials.

The thermal resistances of the insulation and the external environment of a cable have the greatest influence on cable rating. In fact, for the majority of buried cables, the external thermal resistance accounts for more than 70% of the temperature rise of the conductor. For cables in air, the external thermal resistance has a smaller effect on cable rating than in the case of buried cables.

The calculation of thermal resistances of the internal components of cables for single-core cables, whether based on rigorous mathematical computations or empirical investigations, is straightforward. For three-core cables the calculations are somewhat more involved. Also, the calculation of the external thermal resistance requires particular attention. The relevant formulae are given below.

1.6.6.1 Thermal Resistance of the Insulation

Single-Conductor Cables. The thermal resistance between one conductor and the sheath is computed from

$$T_1 = \frac{\rho}{2\pi} \ln\left(1 + \frac{2t_1}{d_c}\right)$$

For oval-shaped conductors, the diameter over the insulation is the geometric mean of the minor and major diameters over the insulation.

For corrugated sheaths, t_1 is based on the mean internal diameter of the sheath which is equal to $[(D_{it} + D_{oc})/2] - t_s$.

The dimensions of the cable occur in $\ln[1 + (2t_1/d_c)]$ and, therefore, this expression plays the role of a geometric factor or shape modulus, and has been termed the geometric factor.

Three-Conductor Belted Cables. The computation of the internal thermal resistance of three-core cables is more complicated than for the single-core case. Rigorous mathematical formulas cannot be determined, although mathematical expressions to fit the conditions have been derived either experimentally or numerically. The general method of computation employs geometric factor (G) in place of the logarithmic term in the equation given above; that is,

$$T_1 = \frac{\rho}{2\pi} G$$

The value of G is obtained from Figure 1-12.
For cables with sector-shaped conductors,

$$G = F_1 \ln\left(\frac{d_a}{2r_1}\right) \quad \text{and} \quad F_1 = 3 + \frac{9t}{2\pi(d_x + t) - t}$$

Also, d_a = external diameter of the belt insulation, mm.
All three-core cables require fillers to fill the space between insulated cores and the belt insulation or a sheath. In the past, when impregnated paper was used to in-

sulate the conductors, the resistivity of the filler material very closely matched that of the paper (around $6 \text{ K} \cdot \text{m/W}$). Because polyethylene insulation has much lower thermal resistivity ($3.5 \text{ K} \cdot \text{m/W}$), the higher thermal resistivity of the filler may have a significant influence on the overall value of T_1 and, hence, on the cable rating. For these cables, the following approximating formula is applied:

$$T_1^{filler} = \frac{\rho_i}{2\pi} G + 0.031(\rho_f - \rho_i)e^{0.67t/d_c}$$

where ρ_f and ρ_i are the thermal resistivities of filler and insulation, respectively, and G is the geometric factor obtained from Figure 1-12, assuming $\rho_f = \rho_i$.

Three-Conductor Shielded Cables. Screening reduces the thermal resistance of a cable by providing additional heat paths along the screening material of high thermal

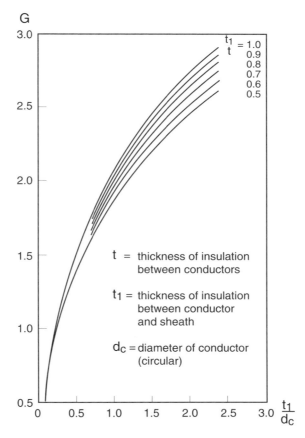

Figure 1-12 Geometric factor for three-core belted cables with circular conductors (IEC 60287, 1994).

conductivity, in parallel with the path through the dielectric. The thermal resistance of the insulation is thus obtained in two steps. First, the cables of this type are considered as belted cables for which $t_1/t = 0.5$. Then, in order to take account of the thermal conductivity of the metallic screens, the results are multiplied by a factor K, called the screening factor, values of which are obtained from Figure 1-13. Thus, we have

$$T_1 = K\frac{\rho}{2\pi}G$$

As discussed above, filler resistivity may have a significant influence on the value of T_1 for plastic-insulated cables. However, for screened cables, IEC advises that no additional modifications of the above formula be taken.

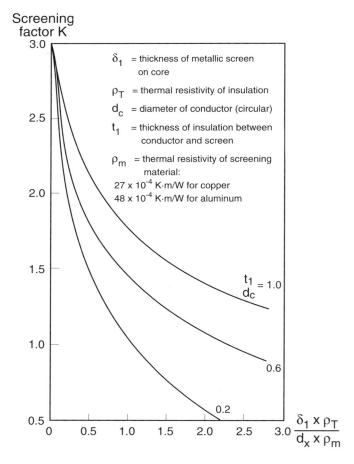

Figure 1-13   Thermal resistance of three-core screened cables with circular conductors compared to that of a corresponding unscreened cable (IEC 287, 1994).

For cables with sector-shaped conductors,

$$G = F_1 \ln\left(\frac{d_a}{2r_1}\right) \quad \text{and} \quad F_1 = 3 + \frac{9t}{2\pi(d_x + t) - t}$$

The screening factor is obtained from Figure 1-14.

Oil-Filled Cables with Round Conductors and Round Oil Ducts Between Cores. In cables of this construction, oil ducts are provided by laying up an open spiral duct of metal strip in each filler space. The expression for the thermal resistance between one core and the sheath was obtained experimentally:

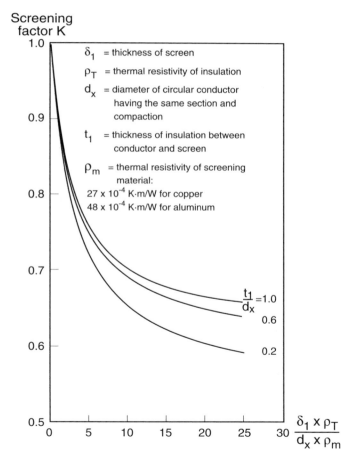

Figure 1-14 Thermal resistance of three-core screened cables with sector-shaped conductors compared to that of a corresponding unscreened cable (IEC 287-2-1, 1994).

$$T_1 = 0.358\rho\left(\frac{2t_i}{d_c + 2t_i}\right)$$

where t_i (mm) is the thickness of core insulation, including carbon black and metallized paper tapes plus half of any nonmetallic tapes over the three laid-up cores. This equation assumes that the space occupied by the metal ducts and the oil inside them has a very high thermal conductance compared with that of the insulation; the equation, therefore, applies irrespective of the metal used to form the duct or its thickness.

Three-Core Cables with Circular Conductors and Metal-Tape Core Screens and Circular Oil Ducts between the Cores. The thermal resistance between one conductor and the sheath is

$$T_1 = 0.35\rho\left(0.923 - \frac{2t_i}{d_c + 2t_i}\right)$$

where t_i (mm) is the thickness of core insulation, including the metal screening tapes plus half of any nonmetallic tapes over the three laid-up cores.

SL Type Cables. In SL type cables the lead sheath around each core may be assumed to be isothermal. The thermal resistance T_1 is calculated the same way as for single-core cables.

1.6.6.2 Thermal Resistance between Sheath and Armor, T_2

Single-Core, Two-Core, and Three-Core Cables Having a Common Metallic Sheath. The thermal resistance between sheath and armor is obtained from Equation (1.14), representing thermal resistance of any concentric layer. With the notation applicable to this part of the cable, we have

$$T_2 = \frac{\rho}{2\pi} \ln\left(1 + \frac{2t_2}{D_s}\right)$$

SL-Type Cables. In these cables, the thermal resistance of the fillers between sheaths and armoring is given by

$$T_2 = \frac{\rho}{6\pi}\overline{G}$$

where $\overline{G}$ is the geometric factor given in Figure 1-15.

1.6.6.3 Thermal Resistance of Outer Covering (Serving), T_3. The external servings are generally in the form of concentric layers, and the thermal resistance T_3 is given by

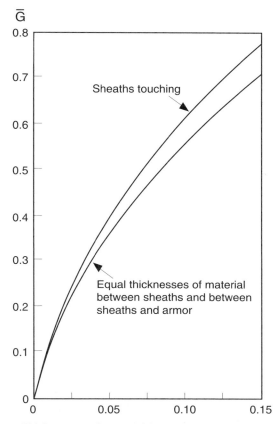

Figure 1-15 Geometric factor for obtaining the thermal resistances of the filling material between the sheaths and armor of SL-type cables (IEC 287-1-1, 1994).

$$T_3 = \frac{\rho}{2\pi} \ln\left(1 + \frac{2t_3}{D_a'}\right)$$

For corrugated sheaths,

$$T_3 = \frac{\rho}{2\pi} \ln\left[\frac{D_{oc} + 2t_3}{\left(\dfrac{D_{oc} + D_{it}}{2}\right) + t_s} \right]$$

1.6.6.4 Pipe-Type Cables. For these three-core cables, the following computational rules apply:

1. The thermal resistance T_1 of the insulation of each core between the conductor and the screen is calculated from the equation for single-core cables.

2. The thermal resistance T_2 is made up of two parts:

   (a) The thermal resistance of any serving over the screen or sheath of each core. The value to be substituted for part of T_2 in the rating equation is the value per cable; that is, the value for a three-core cable is one-third of the value of a single core. The value per core is calculated by the method given in Section 1.6.6.2 for the bedding of single-core cables. For oval cores, the geometric mean of the major and minor diameters $\sqrt{d_M d_m}$ is used in place of the diameter for a circular core assembly.

   (b) The thermal resistance of the gas or liquid between the surface of the cores and the pipe. This resistance is calculated in the same way as part T_4, which is between a cable and the internal surface of a duct, as given in Section 1.6.6.5. The value calculated will be per cable and should be added to the quantity calculated in (a) above, before substituting for T_2 in the rating equation .

3. The thermal resistance T_3 of any external covering on the pipe is dealt with as in Section 1.6.6.3. The thermal resistance of the metallic pipe itself is negligible.

1.6.6.5 *External Thermal Resistance.*

The current-carrying capability of cables depends to a large extent on the thermal resistance of the medium surrounding the cable. For a cable laid underground, this resistance accounts for more than 70% of the temperature rise of the conductor. For underground installations, the external thermal resistance depends on the thermal characteristics of the soil, the diameter of the cable, the depth of laying, mode of installation (e.g., directly buried, in thermal backfill, in pipe or duct, etc.), and on the thermal field generated by neighboring cables. For cables in air, the external thermal resistance has a smaller effect on the cable rating. For aerial cables, the effect of installation conditions (e.g., indoors or outdoors, proximity of walls and other cables, etc.) is an important factor in the computation of the external thermal resistance. In the following sections, we will describe how the external thermal resistance of buried and aerial cables is computed.

External Thermal Resistance of Buried Cables—Directly Buried Cables. For buried cables, two values of the external thermal resistance are calculated: T_4, corresponding to dielectric losses (100% load factor); and $T_{4\mu}$, the thermal resistance corresponding to the joule losses, where allowance is made for the daily load factor (LF) and the corresponding loss factor μ (a more detailed discussion about the load loss factor is offered in Chapter 5):

$$\mu = 0.3 \cdot (LF) + 0.7 \cdot (LF)^2$$

The effect of the loss factor is considered to start outside a diameter D_x, defined as $D_x = 61200\sqrt{\delta(\text{length of cycle in hours})}$ where δ is soil diffusivity (m^2/h). For a daily load cycle and typical value of soil diffusivity of 0.5×10^{-6} m^2/s, D_x is equal to 211 mm (or 8.3 in). The value of D_x is valid even when the diameter of the cable or pipe is greater than D_x.

The following expressions are used to calculate the external thermal resistance of a single cable or a group of equally loaded identical cables:

$$T_2 = \frac{\rho_s}{2\pi} \ln \frac{4L \cdot F}{D_e} \qquad T_{4\mu} = \frac{\rho_s}{2\pi} \left\langle \ln \frac{D_x}{D_e} + \mu \cdot \ln \frac{4L \cdot F}{D_x} \right\rangle$$

For cables in pipes, one should use D_o in place of D_e in the above formulas.

A factor F accounts for the mutual heating effect of the other cables or cable pipes in a system of equally loaded, identical cables or cable pipes. For several loaded cables placed underground, we must deal with superimposed heat fields. The principle of superposition is applicable if we assume that each cable acts as a line source and does not distort the heat field of the other cables. Therefore, in the following subsections, we will assume that the cables are spaced sufficiently apart so that this assumption is approximately valid. The axial separation of the cables should be at least two cable diameters. The case in which the superposition principle is not applicable is discussed in Section 9.6.3 of Anders (1997).

The distances needed to compute factor F are defined in Figure 1-16. These are center-to-center distances.

For cable p:

$$F = \left(\frac{d'_{p1}}{d_{p1}} \right) \left(\frac{d'_{p2}}{d_{p2}} \right) \cdots \left(\frac{d'_{pk}}{d_{pk}} \right) \cdots \left(\frac{d'_{pq}}{d_{pq}} \right)$$

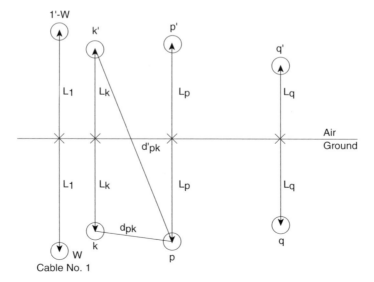

Figure 1-16 Illustration of the development of an equation for the external thermal resistance of a single cable buried under an isothermal plane.

There are $(q - 1)$ terms, with the term d'_{pp}/d_{pp} excluded. The rating of the cable system is determined by the rating of the hottest cable or cable pipe, usually the cable with the largest ratio, L/D_o. For a single isolated cable or cable pipe, $F = 1$.

When the losses in the sheaths of single-core cables laid in a horizontal plane are appreciable, and the sheaths are laid without transposition and/or are bonded at all joints, the inequality of losses affects the external thermal resistance of the cables. In such cases, the value of the F factor used to calculate T_4 and $T_{4\mu}$ is modified by first computing the sheath factor (SHF):

$$(SHF) = \frac{1 + 0.5(\lambda'_{11} + \lambda'_{12})}{1 + \lambda'_m}$$

and then calculating

$$F' = F^{(SHT)}$$

External Thermal Resistance of Buried Cables—Cables or Cable Pipes Buried in Thermal Backfill. In many North American cities, medium- and low-voltage cables are often located in duct banks in order to allow a large number of circuits to be laid in the same trench. The ducts are first installed in layers with the aid of distance pieces, and then a bedding of filler material is compacted after each layer is positioned. Concrete is the material most often used as a filler. High- and extra-high-voltage cables are, on the other hand, often placed in an envelope of well-conducting backfill to improve heat dissipation. What both methods of installation have in common is the presence of a material that has a different thermal resistivity from that of the native soil. The first attempt to model the presence of a duct bank or a backfill was presented by Neher and McGrath (1957) and later adopted in IEC Standard 60287 (1982). Later, the basic method of Neher and McGrath was extended to take into account backfills and duct banks of elongated rectangular shapes, and to remove the assumption that the external perimeter of the rectangle is isothermal.

The application of a thermal backfill is discussed in some detail in Chapter 4 of this book. The formulae presented below and used in Chapter 4 apply to the case in which the longer to shorter length ratio is smaller than 3.

When the cables are installed in a backfill or duct bank, the following modification of the expressions for the external thermal resistance are applied:

$$T_4 = \frac{\rho_c}{2\pi} \ln \frac{4L \cdot F}{D_e} + \frac{N}{2\pi}(\rho_e - \rho_c)G_b$$

$$T_{4\mu} = \frac{\rho_c}{2\pi}\left(\ln \frac{D_x}{D_e} + \mu \ln \frac{4L \cdot F}{D_x}\right) + \mu\frac{N}{2\pi}(\rho_e - \rho_c)G_b$$

where

$$G_b = \ln[u + \sqrt{(u^2 - 1)}] \qquad u = \frac{L_G}{r_b} \qquad \text{and} \qquad \ln r_b = \frac{1}{2}\frac{x}{y}\left(\frac{4}{\pi} - \frac{x}{y}\right)\ln\left(1 + \frac{y^2}{x^2}\right) + \ln\frac{x}{2}$$

External Thermal Resistance of Buried Cables—Cables in Ducts and Pipes.
This section deals with the external thermal resistance of cables in ducts or pipes
filled with air or a liquid. Cables in ducts that have been completely filled with a
pumpable material having a thermal resistivity not exceeding that of the surrounding soil, either in the dry state or when sealed to preserve the moisture content of the
filling material, may be treated as directly buried cables.

The external thermal resistance of a cable in duct or pipe consists of three parts:

1. The thermal resistance of the air or liquid between the cable surface and the
   duct internal surface, T_4'.
2. The thermal resistance of the duct itself, T_4''. The thermal resistance of a metal pipe is negligible.
3. The external thermal resistance of the duct, T_4'''

The value of T_4 to be substituted in the current rating equation will be the sum of
the individual parts; that is:

$$T_4 = T_4' + T_4'' + T_4''' \qquad T_{4\mu} = T_4' + T_4'' + T_{4\mu}'''$$

The constituent thermal resistances are computed from the following formulae:

$$T_4' = \frac{U}{1 + 0.1(V + Y\theta_m)D_e}$$

Constants U, V, and Y are read from Table 1-6.

Table 1-6   Values of contants U, V, and Y (IEC 60287, 1994)

Installation condition	U	V	Y
In metallic conduit	5.2	1.4	0.011
In fiber duct in air	5.2	0.83	0.006
In fiber duct in concrete	5.2	0.91	0.010
In asbestos cement			
duct in air	5.2	1.2	0.006
duct in concrete	5.2	1.1	0.011
Earthenware ducts	1.87	0.28	0.0036
Gas pressure cable in pipe	0.95	0.46	0.0021
Oil-pressure, pipe-type cable	0.26	0.0	0.0026

$$T''_4 = \frac{\rho}{2\pi} \ln \frac{D_o}{D_d}$$

where ρ is the thermal resistivity of duct material. For metal ducts, $T''_4 = 0$ and

$$T'_4 = \frac{\rho_c}{2\pi} \left(\ln \frac{D_x}{D_d} + \mu \ln \frac{4L \cdot F}{D_x} \right) + \mu \frac{N}{2\pi}(\rho_e - \rho_c)G_b$$

External Thermal Resistance of Buried Cables—Unequally Loaded or Dissimilar Cables. The method suggested for the calculation of ratings of a group of cables set apart is to calculate the temperature rise at the surface of the cable under consideration caused by the other cables of the group, and to subtract this rise from the value of $\Delta\theta$ used in Equation (1.40) for the rated current. An estimate of the power dissipated per unit length of each cable must be made beforehand, and this can be subsequently amended as a result of the calculation where it becomes necessary.

For cable j, the losses are

$$W_j = n[I_j^2 R_j(1 + \lambda_1 + \lambda_2)\mu_j + W_{dj}]$$

The thermal resistance between cable j and cable i, the cable being studied, is

$$T_{ij} = \frac{\rho_s}{2\pi} \ln \frac{d'_{ij}}{d_{ij}} \qquad \text{for directly buried cables}$$

$$T_{ij} = \frac{\rho_c}{2\pi} \ln \frac{d'_{ij}}{d_{ij}} + \frac{N}{2\pi}(\rho_e - \rho_c)G_b \qquad \text{for cables in backfill or duct bank}$$

The temperature rise at the surface of cable i due to the losses in cable j is given by

$$\Delta\theta_{ij} = W_j \cdot T_{ij}$$

and the temperature rise at the surface of cable i due to all other cables in the group is computed from

$$\Delta\theta_{\text{int}} = \sum_{\substack{j=1 \\ j \neq i}}^{N} \Delta\theta_{ij}$$

External Thermal Resistance of Cables in Air—Simple Configurations. Heat transfer phenomena are more complex for cables installed in free air than for those located underground. The external thermal resistance of cables in air can be written as

$$T_4 = \frac{1}{\pi D_e^* h_t}$$

where h_t is the total heat transfer coefficient defined in Equation (1.21). This coefficient is a nonlinear function of the cable surface temperature and one approximation, proposed in IEC 60287 (1994), is

$$T_4 = \frac{1}{\pi D_e^* h (\Delta \theta_s)^{1/4}}$$

where h (W/m^{2}K$^{5/4}$) is the heat transfer coefficient embodying convection, radiation, conduction, and mutual heating and is given by the following analytical expression:

$$h = \frac{Z}{(D_e^*)^g} + E$$

Constants Z, E, and g are given in Table 1-7. Served cables and cables having a nonmetallic surface are considered to have a black surface. Unserved cables, either plain lead or armored, should be given a value of h equal to 88% of the value for the black surface.

The cable surface temperature rise is obtained iteratively from the following equation:

$$(\Delta \theta_s)_{n+1}^{1/4} = \left[\frac{\Delta \theta + \Delta \theta_d}{1 + K_A (\Delta \theta_s)_n^{1/4}} \right]^{1/4}$$

The iterations are started by setting the initial value of $(\Delta \theta_s)^{1/4} = 2$ and reiterating until $(\Delta \theta_s)_{n+1}^{1/4} - (\Delta \theta_s)_n^{1/4} \le 0.001$.

The variable K_A is defined as

$$K_A = \frac{\pi D_e^* h}{1 + \lambda_1 + \lambda_2} \left[\frac{T_1}{n} + (1 + \lambda_1) T_2 + (1 + \lambda_1 + \lambda_2) T_3 \right]$$

$\Delta \theta_d$ is the temperature rise caused by dielectric losses and is obtained from Equations (1.41) and (1.44) as

$$\Delta \theta_d = W_d \left[\left(\frac{1}{1 + \lambda_1 + \lambda_2} - \frac{1}{2} \right) T_1 - \frac{n \lambda_2 T_2}{1 + \lambda_1 + \lambda_2} \right]$$

If the dielectric losses are neglected, $\Delta \theta_d = 0$.

When the cable is exposed to solar radiation, the following expression applies:

$$(\Delta \theta_s)_{n+1}^{1/4} = \left[\frac{\Delta \theta + \Delta \theta_d + \sigma H K_A / \pi h}{1 + K_A (\Delta \theta_s)_n^{1/4}} \right]^{1/4}$$

Table 1-7 Values for constants Z, E, and g for black surfaces of cables

No.	Installation	Z	E	G	Mode
1	Single cable[a]	0.21	3.94	0.60	$\geq 0.3\ D_e^*$
2	Two cables touching, horizontal	0.21	2.35	0.50	$\geq 0.5\ D_e^*$
3	Three cables trefoil pattern	0.96	1.25	0.20	$\geq 0.5\ D_e^*$
4	Three cables touching, horizontal	0.62	1.95	0.25	$\geq 0.5\ D_e^*$
5	Two cables touching, vertical	1.42	0.86	0.25	$\geq 0.5\ D_e^*$
6	Two cables spaced D_e^*, vertical	0.75	2.80	0.30	$\geq 0.5\ D_e^*$ D_e^*
7	Three cables touching, vertical	1.61	0.43	0.20	$\geq 1.0\ D_e^*$
8	Three cables spaced D_e^*, vertical	1.31	2.00	0.20	$\geq 1.0\ D_e^*$ D_e^* D_e^*
9	Single cable	1.69	0.63	0.25	
10	Three cables in trefoil	0.94	0.79	0.20	

[a]Values for a "single cable" also apply to each cable of a group when they are spaced horizontally with a clearance between cables of at least 0.75 times the cable overall diemeter.

The solar absorption coefficients for various covering materials are given in Table 1-8.

Plain lead or unarmored cables should be assigned a value of h equal to 80% of the value for a cable with a black surface.

Ambient temperature should be increased by $\Delta\theta_{sr}$, where

$$\Delta\theta_{sr} = \sigma D_e^* H T_4$$

External Thermal Resistance of Cables in Air—Derating Factors for Groups of Cables in IEC 601042 (1991). An approach to dealing with groups of identical cables in air, shown in Figure 1-17, is presented in IEC60287-2-2 (1995) and discussed in Anders (1997). The following limitations apply in this case:

1. A maximum of nine cables in a square formation (the last arrangement in Figure 1-17a)
2. A maximum of six circuits, each comprising three cables mounted in a trefoil pattern, with up to three circuits placed side by side or two circuits placed one above the other (the last arrangement in Figure 1-17b)
3. Cables for which dielectric losses can be neglected (usually, only lower voltage polymeric cables are installed in groups).

When cables are installed in groups as shown in Figure 1-17, the rating of the hottest cable will be lower than in the case when the same cable is installed in isolation. This reduction is caused by mutual heating. A simple method to account for this mutual heating effect is to calculate the rating of a single cable or circuit using the method described in the previous section and apply a reduction factor. This is defined as follows:

$$I_g = F_g \cdot I_1$$

where
I_g = rating of the hottest cable in the group, A
I_1 = rating of the same cable or circuit isolated, A
F_g = group reduction factor

Table 1-8 Absorption coefficient of solar radiation values for cable surfaces

Material	σ
Bitumen/jute serving	0.8
Polychloroprene	0.8
PVC	0.6
PE	0.4
Lead	0.6

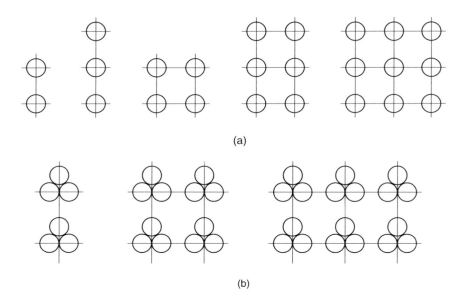

(a)

(b)

Figure 1-17   Typical groups of (a) multicore cables, and (b) trefoil circuits.

The reduction in current is a result of an increase in the external thermal resistance of a cable or circuit in a group as compared to the case in which the cable or circuit is installed in isolation. A basic criterion is that the temperature rise above ambient of the conductor be the same for the grouped and isolated cases.

The reduction factor is computed from

$$F_g = \sqrt{\frac{1}{1 - k_1 + k_1(T_{4g}/T_{4I})}}$$

where

$$k_1 = \frac{W_t \cdot T_{4I}}{\theta_c - \theta_{amb}}$$

and

T_{4g} = external thermal resistance of the hottest cable in the group, K · m/W

T_{4I} = external thermal resistance of one cable, assumed to be isolated, used to compute the rated current I, K · m/W

W_t = power loss from one multicore cable or one single-core cable mounted in a trefoil pattern, assumed to be isolated, when carrying the current I_1, W/m

The term T_{4g}/T_{4I} can be calculated from the ratio (h_I/h_g) by using the iterative relationship

$$\left(\frac{T_{4g}}{T_{4I}}\right)_{n+1} = \frac{h_I}{h_g}\left[\frac{1-k_1}{(T_{4g}/T_{4I})_n} + k_1\right]^{0.25}$$

and starting with $(T_{4g}/T_{4I})_1 = (h_I/h_g)$. The above equation converges quickly, and one iteration with $(T_{4g}/T_{4I})_1 = (h_I/h_g)$ is usually sufficient. Alternatively, when (h_I/h_g) is less than 1.4, it is sufficient to substitute (h_I/h_g) for (T_{4g}/T_{4I}) in the equation for F_g.

Values for the ratio (h_I/h_g) have been obtained empirically and are given in of Table 1-9 (IEC 1042, 1991).

If a clearance exceeding the appropriate value given in column 2 of Table 1-9 cannot be maintained with confidence throughout the length of the cable, the reduction coefficient is determined as follows.

1. For horizontal clearances, we assume that the cables are touching each other or the vertical surface. Appropriate values are given in column 4 of Table 1-9.

2. For vertical clearances, the reduction coefficient due to grouping is derived according to the value of the expected clearance:

   a) Where the clearance is less than the appropriate value given in column 2 of Table 1-9, but can be maintained at a value equal or exceeding the minimum given in column 3, the appropriate value of (h_I/h_g) is obtained from the formula in column 4.

   b) Where the clearance is less than the minimum given in column 3, we assume that the cables are touching each other. Suitable values of (h_I/h_g) are provided in column 4 of Table 1-9.

The values in column 4 are the average values for cables having diameters from 13 to 76 mm. More precise values for multicore cables may be evaluated for a specific cable diameter, both inside and outside this range, by consulting Table 1-7.

1.6.7 Thermal Capacitances

In Section 1.3.1.2, thermal capacitance was defined as the product of the volume of the material and its specific heat. In the following section, formulas for special and concentric layers are presented. In all formulas, Q is the thermal capacitance (J/K · m) and c is the specific heat of the material (J/K · m^3).

1.6.7.1 Oil in the Conductor. The thermal capacitance of oil inside the conductor is obtained from

$$Q_o = \left(\frac{\pi d_i^{*2}}{4} + S \cdot F\right)c$$

where F is a factor representing the unfilled cross section (usually 0.36).

Table 1-9 Data for calculating reduction factors for grouped cables

Arrangement of cable	Thermal proximity effect is negligible if e/D_e is greater then or equal to:	Thermal proximity effect is not negligible	
		If e/D_e is less then	Average value of $h_l/h_g{}^{a,b}$
Side by Side			
2 multicore	0.5	0.5	1.41
3 multicore	0.75	0.75	1.65
2 trefoils	1.0	1.0	1.2
3 trefoils	1.5	1.5	1.25
One above the other			
2 multicore	2	2 or 0.5	$1.085\,(e/D_e)^{-0.128}$ or 1.35
3 multicore	4	4 or 0.5	$1.19\,(e/D_e)^{-0.136}$ or 1.57
2 trefoils	4	4 or 0.5	$1.106\,(e/D_e)^{-0.078}$ or 1.39
Near to a vertical surface or to a horizontal surface below the cable	0.5	0.5	1.23

[a]The formulae for (h_l/h_g) given in column 4 of this table shall not be used for values of e/D_e less than 0.5 or greater than the appropriate values given in column 1.

[b]Average values for cables having diameter from 13 mm to 76 mm. More precise values of (h_l/h_g) for multicore cables may be evaluated for a specific cable diameter, both inside and outside this range, by consulting Table V of IEC 287.

1.6.7.2 Conductor. The thermal capacitance of the conductor is given by

$$Q_c = S \cdot c$$

1.6.7.3 Insulation. Representation of the capacitance of the insulation depends on the duration of the transient. For transients lasting longer than about 1 h, the capacitance of the insulation is divided between the conductor and the sheath positions according to a method given by Van Wormer (1955) (see Section 1.3.1.2). This assumes logarithmic temperature distribution across the dielectric during the transient, as in the steady state. The total thermal capacitance is obtained as

$$Q_i = \frac{\pi}{4}(D_i^{*2} - d_c^{*2})c$$

The portion pQ_i is placed at the conductor and the portion $(1-p)Q_i$ at the sheath, where p is the Van Wormer coefficient, defined in Equation (1.24) as

$$p = \frac{1}{2 \ln\left(\dfrac{D_i}{d_c}\right)} - \frac{1}{\left(\dfrac{D_i}{d_c}\right)^2 - 1}$$

From the thermal point of view, the thickness of the dielectric includes any non-metallic semiconducting layer either on the conductor or on the insulation.

For shorter durations (less than 1 h), it has been found necessary to divide the insulation into two portions having equal thermal resistances. The thermal capacitance of each portion of the insulation is then assumed to be located at its boundaries, using the Van Wormer coefficient to split the capacitances as shown in Figure 1-2. The thermal capacitance of the first part of the insulation is given by

$$Q_n = \frac{\pi}{4}(D_i^* \cdot d_c^* - d_c^{*2})c$$

This capacitance is split into two parts using the Van Wormer coefficient as follows:

$$Q_{i1} = p^* Q_n, \qquad Q_{i2} = (1-p^*)Q_n \qquad p^* = \frac{1}{\ln \dfrac{D_i}{d_c}} - \frac{1}{\dfrac{D_i}{d_c} - 1}$$

The total thermal capacitance of the second part is given by

$$Q_{i2} = \frac{\pi}{4}(D_i^{*2} - D_i^* \cdot d_c^*)c$$

which leads to the third and fourth part of the thermal capacitance of the insulation, defined as

$$Q_{i3} = p^* Q_{l2}, \qquad Q_{i4} = (1 - p^*) Q_{l2}$$

In the case of dielectric losses, the cable thermal circuit is the same as shown in Figure 1-2, with the Van Wormer coefficient apportioning to the conductor a fraction of the dielectric thermal capacitance given by Equation (1.24). For cables at voltages higher than 275 kV, the Van Wormer coefficient is given by Equation (1.29).

1.6.7.4 Metallic Sheath or Any Other Concentric Layer.
The thermal capacitance of all other concentric layers of cable components such as sheath, armor, jacket, armor bedding, or serving is computed by using Equation (1.23). However, one should remember that the thermal capacitances of nonmetallic layers have to be divided into two parts using the Van Wormer factor given by Equation (1.25). The appropriate dimensions for the inner and outer diameters must be used in order to attain sufficient accuracy for short-duration transients.

1.6.7.5 Reinforcing Tapes.
The thermal capacitance of the reinforcing tapes over the sheath is

$$Q_T = n_t \left(w_t^* t_t^* \sqrt{\ell_t^{*2} + (\pi d_2^*)^2} \right) c$$

1.6.7.6 Armor.
The thermal capacitance of the armor is obtained from

$$Q_a = n_1 \left(\frac{\pi d_f^{*2}}{4} \sqrt{\ell_a d_a^*} \right) c$$

1.6.7.7 Pipe-Type Cables.
The thermal capacitances of cable components are computed as described above, and the thermal capacitance of the oil in the pipe is obtained from

$$Q_{op} = \frac{\pi}{4} (D_d^{*2} - 3D_d^{*2}) c$$

The capacitance of the oil is divided into two equal parts.

The thermal capacitance of the skid wires is generally neglected, but may be computed using the equation for the capacitance of the armor, if needed. The thermal capacitance of the metallic pipe and of the external covering are computed using Equation (1.23).

1.7 CONCLUDING REMARKS

The rating calculations described in this chapter will serve as a basis for the new developments presented in the following chapters. In many applications discussed further, the external thermal resistance of the cable will be modified. In Chapter 2, we will distinguish the thermal properties of two different regions through which the cable route may pass, whereas in Chapter 3 we will look at the mutual thermal resistance between the rated cable and an external heat source. The probabilistic nature of the external thermal resistivity of buried cables will be explored in Chapter 4. In the same chapter, we will investigate the optimal dimensions of a thermal backfill. The calculation of the load loss factor appearing in the equation for the external thermal resistance with a nonunity load factor is addressed in Chapter 5. For cables in air, discussed in Chapter 6, several alternative approaches to the computation of the convective and radiative heat transfer coefficients will be introduced.

REFERENCES

AEIC CS1-90, (1990) *"Specifications for Impregnated Paper-Insulated, Metallic-Sheathed Cable, Solid Type,"* 11th edition.

AEIC CS2-90, (1990) *"Specifications for Impregnated Paper and Laminated Paper Polypropylene Insulated Cable, High-Pressure Pipe Type,"* 5th edition.

AEIC CS3-90, (1990) *"Specifications for Impregnated Paper-Insulated, Metallic-Sheathed Cable, Low-Pressure Gas-Filled Type,"* 3rd edition.

AEIC CS4-93, (1993) *"Specifications for Impregnated-Paper-Insulated Low and Medium Pressure Self-Contained Liquid Filled Cable,"* 8th edition .

AEIC CS5-94, (1994) *"Specifications for Cross-linked Polyethylene Insulated Shielded Power Cables Rated 5 Through 46 kV,"* 10th edition.

AEIC CS6-87, (1987) *"Specifications for Ethylene Propylene Rubber Insulated Shielded Power Cables Rated 5 Through 69 kV,"* 5th edition.

AEIC CS7-93, (1993) *"Specifications for Cross-linked Polyethylene Insulated Shielded Power Cables Rated 69 Through 138 kV,"* 3rd edition.

AEIC CS8-00 (2000) "Specifications for Extruded Dielectric Shielded Power Cables Rated 5 through 46 kV," 1st edition.

AEIC CS31-84, (1984) *"Specifications for Electrically Insulated Low Viscosity Pipe Filling Liquids for High-Pressure Pipe-Type Cable,"* 2nd edition.

AEIC G1-68, (1968) *"Guide for Application of AEIC Maximum Insulation Temperatures at the Conductor for Impregnated-Paper-Insulated Cables,"* 2nd edition.

AEIC G2-72, (1972) *"Guide for Electrical Tests of Cable Joints 138 kV and Above,"* 1st edition.

AEIC G4-90 (1990) *"Guide for Installation of Extruded Dielectric Insulated Power Cable Systems Rated 69 kV Through 138 kV,"* 1st edition.

AEIC G5-90, (1990) *"Underground Extruded Power Cable Pulling Guide,"* 1st edition.

AEIC G6-95, (1995) *"Guide for Establishing the Maximum Operating Temperatures of Extruded Dielectric Insulated Shielded Power Cables,"* 1st edition.

AEIC G7-90, (1990) *"Guide for Replacement and Life Extension of Extruded Dielectric 5-35 kV Underground Distribution Cables,"* 1st edition.

AEIC G8-95, (1995) *"Guide for an Electric Utility Quality Assurance Program for Extruded Dielectric Power Cables,"* 1st edition.

Anders, G.J. (1997), *Rating of Electric Power Cables—Ampacity Calculations for Transmission, Distribution and Industrial Applications,* IEEE Press, New York, McGraw-Hill, (1998).

ANSI/IEEE Standard 575 (1988), *"Application of Sheath-Bonding Methods for Single Conductor Cables and the Calculation of Induced Voltages and Currents in Cable Sheaths."*

Baudoux, A., Verbisselet, J., Fremineur, A. (1962) "Etude des Regimes Thermiques des Machines Electriques par des Modeles Analogiques, Application aux Cables et aux Transformateurs," Rapport CIGRE, No. 126.

Buller, F.H. (1951), "Thermal Transients on Buried Cables," *Trans. Amer. Inst. Elect. Engrs.,* Vol. 70, pp. 45–52.

CIGRE (1972), "Current Ratings of Cables for Cyclic and Emergency Loads. Part 1. Cyclic Ratings (Load Factor less than 100%) and Response to a Step Function," *Electra,* No. 24, October 1972, pp. 63–96.

CIGRE (1976), "Current Ratings of Cables for Cyclic and Emergency Loads. Part 2. Emergency Ratings and Short Duration Response to a Step Function," *Electra,* No. 44, Jan. 1976, pp. 3–16.

CIGRE (1986), "Current Ratings of Cables Buried in Partially Dried Out Soil. Part 1: Simplified Method that Can be Used with Minimal Soil Information (100% Load Factor)," *Electra,* No. 104, pp. 11–22.

Goldenberg, H., (1957), "The Calculation of Cyclic Rating Factors for Cables Laid Direct or in Ducts," *Proc. IEE,* Vol. 104, Pt. C, pp. 154–166.

Goldenberg, H., (1958), "The Calculation of Cyclic Rating Factors and Emergency Loading for One or More Cables Laid Direct or in Ducts," *Proc. IEE,* Vol. 105, Pt. C, pp. 46–54.

Goldenberg, H., (1967), "Thermal Transients in Linear Systems with Heat Generation Linearly Temperature-Dependent: Application to Buried Cables," *Proc. IEE,* Vol. 114, pp. 375–377.

Goldenberg, H. (1971), "Emergency Loading of Buried Cable with Temperature-Dependent Conductor Resistance," *Proc. IEE,* Vol. 118, pp. 1807–1810.

IEC Standard 60287 (1969, 1982, 1994), *"Calculation of the Continuous Current Rating of Cables (100% load factor),"* 1st edition 1969, 2nd edition 1982, 3rd edition 1994–1995.

IEC Standard 60853-1 (1985)," *Calculation of the Cyclic and Emergency Current Ratings of Cables. Part 1: Cyclic Rating Factor for Cables up to and Including 18/30 (36) kV,"* Publication 853-1.

IEC Standard 60853-2 (1989), *"Calculation of the Cyclic and Emergency Current Ratings of Cables. Part 2: Cyclic Rating Factor of Cables Greater than 18/30 (36) kV and Emergency Ratings for Cables of All Voltages,"* Publication 853-2.

IEC Standard 601042 (1991), *"A method for Calculating Reduction Factors for Groups of Cables in Free Air, Protected from Solar Radiation."*

IEEE (1994), Standard 835, *"IEEE Standard—Power Cable Ampacity Tables,"* IEEE Press, New York.

Morello, A. (1958) "Variazioni Transitorie die Temperatura Nei Cavi per Energia," *Elettrotecnica,* Vol. 45, pp. 213–222.

Neher, J.H., and McGrath, M.H. (1957), "Calculation of the Temperature Rise and Load Capability of Cable Systems," *AIEE Trans.,* Vol. 76, Part 3, pp. 755–772.

Neher, J.H. (1964), "The Transient Temperature Rise of Buried Power Cable Systems," *IEEE Trans.,* Vol. PAS-83, pp. 102–111.

Van Valkenburg, M.E. (1964), *Network Analysis,* Prentice-Hall Inc., Englewood Cliffs, NJ.

Van Wormer, F.C. (1955), "An Improved Approximate Technique for Calculating Cable Temperature Transients," *Trans. Amer. Inst. Elect. Engrs,* Vol. 74, Pt. 3, pp. 277–280.

Wollaston, F.O. (1949), "Transient Temperature Phenomena of 3 conductor cables," *AIEE Trans.,* Vol. 68, Pt. 2, pp. 1248–1297.

CHAPTER 2

Ampacity Reduction Factors for Cables Crossing Thermally Unfavorable Regions

When power cables cross regions with unfavorable thermal conditions, temperatures higher than the design value can occur. If the region is wide enough, the rating of the cable will usually be based on the assumption that the entire route is characterized by the same conditions. In a majority of cases, the unfavorable thermal environment will be very short, usually a few meters (e.g., a street crossing). In these cases, the effect of the crossing is usually ignored. However, the conductor temperature in such cases may be much higher than in the remainder of the route, and cable derating is required. Often, the cables crossing streets are laid in metallic or nonmetallic pipes. In such cases, a smaller ampacity reduction can be achieved if ventilated pipes are installed. In this chapter, an analytical solution for the computation of the derating factors is presented and the effect of pipe ventilation is examined. The major reference sources on this subject are Brakelmann, (1985, 1999a,b), Brakelmann and Anders (2004a,b), and Weedy (1988).

2.1 CABLES CROSSING THERMALLY UNFAVORABLE REGIONS

2.1.1 Introduction

Power cables are usually laid in congested urban areas and very seldom are the thermal conditions along their route constant. Cables are usually laid along the street length. Occasionally, they are located in a greenbelt and cross streets at right angles. Even if no street crossing occurs, the presence of the trees will considerably change the local thermal environment. When a rated cable crosses an area of unfavorable thermal characteristics, its temperature will be higher than if it was in a uniform environment for which the rating is normally calculated. A common occurrence of cables crossing a paved parking lot may raise the conductor temperature by as much as 10°C if the soil composition is the same as outside the crossing (Williams, 1997). As will be shown below, the additional conductor temperature rise may exceed 20°C under some particularly unfavorable conditions. Therefore, the ampacity of the rated cable should be reduced accordingly. Up to 40% reduction may be re-

Rating of Electric Power Cables in Unfavorable Thermal Environment. By George J. Anders
ISBN 0-471-67909-7 © 2004 the Institute of Electrical and Electronics Engineers.

quired, as illustrated in the numerical example in Section 2.1.4. This is much larger than the up to 5% reduction considered by some utilities. On the other hand, many utilities do not derate cables crossing unfavorable thermal environments. This practice may lead to premature cable failures as the circuits become more heavily loaded.

The magnitude of the required cable derating will depend on the following factors:

- The length of the unfavorable portion of the route
- The difference in the thermal resistivities of the native soil and the crossed area
- The difference in the ambient temperature at the two sections
- The laying condition of the cable in the unfavorable environment

The derating model developed in the next section will take into account all of these factors. The numerical examples in Section 2.1.4 will show how important each of these factors is in the cable rating process.

2.1.2   Ampacity Reduction Modeling

In the development of a mathematical model, we will use the cable laying example shown in Figure 2-1. In this figure, the cable is rated for the conditions characterizing Region 2, and we assume that it crosses a short section of unfavorable thermal conditions, represented by Region 1. The length of this section is equal to b_0 m. In addition, the ambient temperature may be different in both sections, as illustrated in Figure 2-1 with superscripts 1 and 2, respectively. When crossing streets, cables are often placed in short sections of pipes. This is also shown in Figure 2-1.

To find the temperature at any point of the rated cable, taking into account its longitudinal heat flux and the changing environment, we will discretize the length of the rated circuit along the z-axis as shown in Figure 2-2. The notation for the loss factors and the thermal resistances is the same as in Section 1.2 with the exception of the longitudinal heat flow q_L and the conductor thermal resistance T_L, which are discussed next.

2.1.3   Temperature Distribution Along the Rated Cable

In Figure 2-2, the thermal resistances with subscripts 1, 2, and 3 are the internal quantities for the cable as defined in the IEC Standard 287 (1994). ΔT_L [K/(W-m)] is the longitudinal thermal resistance of the cable computed from

$$T_L = \frac{\rho_c}{A} \tag{2.1}$$

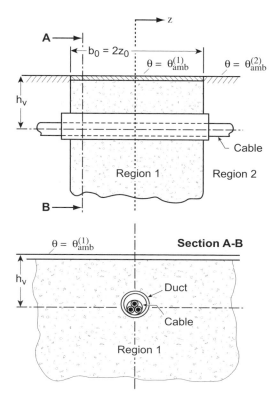

Figure 2-1 Cable crossing thermally unfavorable region.

where A (m^2) is the cross-sectional area of the conductor. ρ_c is the thermal resistivity of the material (K · m/W).

For cables located in ducts or pipes, the longitudinal thermal resistance of the air or oil in the pipe can be neglected. The thermal resistance of the air in the duct or oil in the pipe due to a radial heat transfer is included in the value of T_4, as described below.

The only difference between the two regions of the diagram is in the value of the external thermal resistance and the ambient temperature. The radial external thermal resistance in region 1, $T_4^{(1)}$, is composed of three components:

1. The thermal resistance of the air or liquid between the cable surface and the duct internal surface, $T_4'^{(1)}$

2. The thermal resistance of the duct itself, $T_4''^{(1)}$. The thermal resistance of a metal pipe is negligible.

3. The external thermal resistance of the duct, $T_4'''^{(1)}$

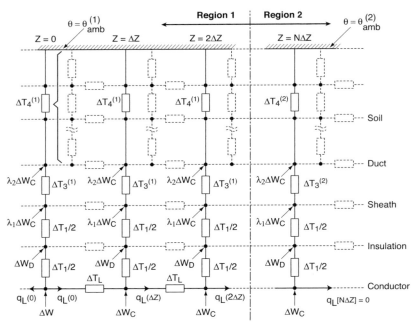

Figure 2-2 Discretized length $N \cdot \Delta z$ of the rated cable with the thermal resistances ΔT corresponding to the length Δz of the discretization interval (from Brakelmann and Anders 2004, with permission from IEEE Press).

Thus,

$$T_4^{(1)} = T_4'^{(1)} + T_4''^{(1)} + T_4'''^{(1)} \tag{2.2}$$

The details of the computations of thermal resistances and other cable parameters are discussed in Anders (1997); a summary of the relevant formulas is offered in Chapter 1.

The network representation in Figure 2-2 allows us to consider the heat flow in two directions—radial and longitudinal—in what essentially is a three-dimensional problem.

Our first goal is to obtain a differential equation describing the heat transfer in both regions [Equation (2.16)] followed by the development of the solution to this equation resulting in the conductor temperature profile along the cable route [Equations (2.30) and (2.31)]. In the development of the influence of the changing environment on the cable rating, we will consider a small element of the cable with the length dz, as shown in Figure 2-3 where $W_c(z)$ is the heat generated by the short section of the cable conductor.

In the steady state, from the energy conservation principle, we have

$$W_c(z) \cdot \Delta z + W_L \cdot (z - \Delta z) = W_r(z) \cdot \Delta z + W_L(z) \tag{2.3}$$

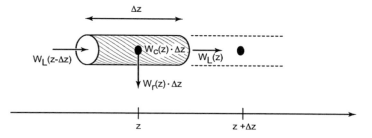

Figure 2-3   An element of the conductor of length Δz.

From Equation (2.3), we obtain

$$(W_r - W_c) + \frac{\partial W_L}{\partial z} = 0 \tag{2.4}$$

The derivative in Equation (2.4) can be computed from the Fourier's Law [Equation (1.1)] as

$$W_L = -\frac{1}{T_L}\frac{d\theta}{dz} \tag{2.5}$$

Considering the heat flow in the radial direction, we have

$$\theta = \theta_{amb} + \Delta\theta_d + W_c \cdot T_t + (W_r - W_c) \cdot T_r \tag{2.6}$$

where

$$T_r = T_1 + n \cdot (T_2 + T_3 + T_4) \tag{2.7}$$

θ_{amb} = ambient temperature (°C)
W_c = the heat losses generated by the conductor current (W/m)

Defining the equivalent radial thermal resistance of the cable in section v ($v = 1$, 2) as

$$T_t^{(v)} = T_1 + n \cdot (1 + \lambda_1) \cdot T_2 + n \cdot (1 + \lambda_1 + \lambda_2) \cdot (T_3 + T_4^{(v)}) \tag{2.8}$$

the conductor temperature in each region is obtained from Equation (2.6) as:

$$\theta^{(v)} = \theta_{amb}^{(v)} + \Delta\theta_d^{(v)} + W_c \cdot T_t^{(v)} + (W_r - W_c) \cdot T_r^{(v)} \tag{2.9}$$

where
$\Delta\theta_d$ = temperature rise due to dielectric losses (°C)
θ_{amb} = ambient temperature (°C)

n = number of conductors in the cable
λ_1, λ_2 = sheath and armor loss factors

$$T_r^{(v)} = T_1 + n \cdot (T_2 + T_3 + T_4^{(v)}) \tag{2.10}$$

The conductor joule losses, W_c, are temperature dependent. This dependence can be written as

$$W_c = W_{c0}[1 + \alpha_T(\theta - \theta_0)] = W_{ct} + \Delta W \cdot \theta \tag{2.11}$$

with

$$W_{ct} = W_{c0}(1 - \alpha_T\theta_0) \quad \text{and} \quad \Delta W = \alpha_T W_{c0} \tag{2.12}$$

where
α_T = conductor temperature coefficient of resistance (1/K)
W_{c0} = heat rate in the rated cable at the reference temperature θ_0 (usually 20°C) (W/m)

Substituting Equations (2.4), (2.5), and (2.11) into (2.9), we obtain the following differential equation describing the conductor temperature for soil v (v = 1, 2):

$$\theta^{(v)} - \frac{1}{\gamma_v^2} \cdot \frac{\partial^2 \theta^{(v)}}{\partial z^2} = \frac{W_{ct} \cdot T_t^{(v)} + \Delta\theta_d^{(v)} + \theta_{amb}^{(v)}}{(1 - \Delta W \cdot T_t^{(v)})} \tag{2.13}$$

where

$$\gamma_v^2 = T_L \cdot (1 - \Delta W \cdot T_t^{(v)})/T_r^{(v)} \tag{2.14}$$

The term on the right-hand side of Equation (2.13) is the conductor temperature in the steady state. Introducing the notation

$$\theta_\infty^{(v)} = \frac{W_{ct} \cdot T_t^{(v)} + \Delta\theta_d^{(v)} + \theta_{amb}^{(v)}}{1 - \Delta W \cdot T_t^{(v)}}$$

the differential Equation (2.13) becomes

$$\theta^{(v)} - \frac{1}{\gamma_v^2} \frac{d^2\theta^{(v)}}{dz^2} = \theta_\infty^{(v)} \tag{2.16}$$

The solution of this equation is sought next, leading to Equations (2.30) and (2.31) representing the temperature distributions in both regions.

A general solution of Equation (2.16) can be written in a form of the following hyperbolic function (Brakelmann, 1985):

$$\theta^{(v)} = \theta_\infty^{(v)} + A_v \sinh(\gamma_v \cdot z) + B_v \cosh(\gamma_v \cdot z) \tag{2.17}$$

The constants A_v and B_v are obtained by taking into account the following boundary conditions:

- The longitudinal heat flux disappears at the center of section 1 ($z = 0$) and for $z = \infty$.
- The temperature is a continuous function when it crosses the boundary.
- The heat flux is a continuous function when it crosses the boundary.

We will review these boundary conditions below.

The longitudinal heat flux and, hence, the heat transfer rate (the heat transfer rate is obtained by multiplying heat flux by the area) at the center of zone 1 is equal to zero. Hence,

$$W_L^{(1)}(z) = -A_1 \frac{\gamma_1}{T_L} = 0 \Rightarrow A_1 = 0 \tag{2.18}$$

The longitudinal heat flux tends to zero when $z \rightarrow \infty$. Since the longitudinal heat rate in zone 2 is given by

$$W_L^{(2)}(z) = -\frac{\gamma_2}{T_L} \cdot [A_2 \cosh(\gamma_2 \cdot z) + B_2 \sinh(\gamma_2 \cdot z)] \tag{2.19}$$

we have

$$W_L^{(2)}(z \rightarrow \infty) = 0 \Rightarrow A_2 = -B_2 \tag{2.20}$$

Substituting these values in Equation (2.17), we obtain

$$\theta^{(1)}(z) = \theta_\infty^{(1)} + B_1 \cosh(\gamma_2 \cdot z) \tag{2.21}$$

and

$$\theta^{(2)}(z) = \theta_\infty^{(2)} - A_2 \cdot e^{-\gamma_2 \cdot z} \tag{2.22}$$

The longitudinal heat rates in both regions now become

$$W_L^{(1)}(z) = -\frac{\gamma_1}{T_L} B_1 \cdot \sinh(\gamma_1 \cdot z) \tag{2.23}$$

$$W_L^{(2)}(z) = -\frac{\gamma_2}{T_L} \cdot A_2 \cdot e^{-\gamma_2 \cdot z} \tag{2.24}$$

Since the temperature is a continues function when it crosses the boundary, that is, $\theta^{(1)}(z_0) = \theta^{(2)}(z_0)$, we have

$$\theta_\infty^{(1)} + B_1 \cosh(\gamma_1 \cdot z_0) = \theta_\infty^{(2)} - A_2 e^{-\gamma_2 \cdot z_0} \tag{2.25}$$

Also, since the heat flux is a continuous function when it crosses the boundary, that is $W_L^{(1)}(z_0) = W_L^{(2)}(z_0)$, from Equations (2.23) and (2.24) we obtain

$$\gamma_1 \cdot B_1 \cdot \sinh(\gamma_1 \cdot z_0) = \gamma_2 \cdot A_2 \cdot e^{-\gamma_2 \cdot z_0} \tag{2.26}$$

From Equations (2.25) and (2.26) we can compute

$$B_1 = -\frac{\theta_\infty^{(1)} - \theta_\infty^{(2)}}{\cosh(\gamma_1 \cdot z_0) + \dfrac{\gamma_1}{\gamma_2} \sinh(\gamma_1 \cdot z_0)} = -\tau \cdot \Delta\theta_\infty \tag{2.27}$$

with

$$\tau = \frac{1}{\cosh(\gamma_1 \cdot z_0) + \dfrac{\gamma_1}{\gamma_2} \sinh(\gamma_1 \cdot z_0)} \geq 1 \qquad \text{for } z_0 \geq 0 \tag{2.28}$$

$$\Delta\theta_\infty = \theta_\infty^{(1)} - \theta_\infty^{(2)}$$

and

$$A_2 = -e^{\gamma_2 \cdot z_0} \cdot \frac{\gamma_1}{\gamma_2} \cdot \sinh(\gamma_1 \cdot z_0) \cdot \tau \cdot \Delta\theta_\infty = -\frac{\Delta\theta_\infty \cdot e^{\gamma_2 \cdot z_0}}{1 + \dfrac{\gamma_1}{\gamma_2} \coth(\gamma_1 \cdot z_0)} \tag{2.29}$$

Thus, the temperature distributions in both intervals are obtained by substituting Equations (2.27) and (2.29) into (2.21) and (2.22), respectively:

$$\theta^{(1)}(z) = \theta_\infty^{(1)} - (\theta_\infty^{(1)} - \theta_\infty^{(2)}) \cdot \tau \cdot \cosh(\gamma_1 \cdot z) \tag{2.30}$$

$$\theta^{(2)}(z) = \theta_\infty^{(2)} + (\theta_\infty^{(1)} - \theta_\infty^{(2)}) \cdot \tau \cdot \frac{\gamma_1}{\gamma_2} \cdot \sinh(\gamma_1 \cdot z_0) \cdot e^{-\gamma_2(z-z_0)} \tag{2.31}$$

with $b_0 = 2 \cdot z_0$, the width of Region 1.

2.1.4 Derating Factor

The rating of the cable in Figure 2-1 is obtained on the basis of the temperature in the center of the unfavorable zone (Region 1), that is, from Equation (2.30),

$$\theta^{(1)}(0) = \theta_\infty^{(1)} - (\theta_\infty^{(1)} - \theta_\infty^{(2)}) \cdot \tau \tag{2.32}$$

Let θ_{max} be the maximum permissible conductor temperature. Then, with $\theta^{(1)}(0)$ = θ_{max}, the rating of the cable is obtained by substituting Equation (2.15) into Equation (2.32) and solving the resulting quadratic equation for W_{c0}, which, in turn, is equal to $W_{c0} = R_0 I^2$. Thus,

$$I^2 = \frac{(b/a) - \sqrt{(b/a)^2 - (c/a)}}{R_0} \tag{2.33}$$

where

$$a = \alpha_T \cdot f_a \cdot T_t^{(1)} \cdot T_t^{(2)} \tag{2.34}$$

$$2b = \alpha_T \cdot (\theta_{max} - \theta_A^{(1)}) \cdot T_t^{(2)} + f_a \cdot T_t^{(1)} - \tau \cdot [(g + \alpha_T \theta_A^{(2)}) \cdot T_t^{(1)} - (g + \alpha_T \theta_A^{(1)}) \cdot T_t^{(2)}] \tag{2.35}$$

$$c = \theta_{max} - \theta_A^{(1)} + \tau \cdot (\theta_A^{(1)} - \theta_A^{(2)}) \tag{2.36}$$

$$f_\alpha = 1 + \alpha_T(\theta_{max} - \theta_0) \tag{2.37}$$

$$g = 1 - \alpha_T \theta_0 \tag{2.38}$$

$$\theta_A^{(v)} = \Delta \theta_d^{(v)} + \theta_{amb}^{(v)} \tag{2.39}$$

and R_0 is the conductor resistance at $\theta = \theta_0$.

If the cable were laid in the soil of Region 2 throughout the entire route, its rating would be based on the external thermal resistance $T_4^{(2)}$ and the ambient temperature $\theta_{amb}^{(2)}$. Thus, the derating factor for this cable is equal to

$$DF^2 = \frac{b - \sqrt{b^2 - a \cdot c}}{\alpha_T \cdot T_t^{(1)} \cdot (\theta_{max} - \theta_A^{(2)})} \tag{2.40}$$

The coefficients γ_1 and γ_2 in Equations (2.30) and (2.31) are a function of conductor current, hence an iterative procedure is required to obtain the derating factor.

In the following example, we will investigate the dependence of the derating factor on the four parameters that differentiate the two zones in which the cable is laid.

Example 2.1
We will consider a street crossing of a 138 kV high-pressure, oil-filled, pipe-type cable, model 3. The ampacity of this cable for the uniform laying conditions is equal to 902 A. The cable crosses a street 10 m wide with the soil thermal resistivity under the asphalt equal to 2.5 K · m/W. The ambient temperature in this section is equal to 25°C.

For pipe-type cables, it is common to denote the thermal resistance of the oil by T_2 and the covering of the pipe by T_3. The value of T_2 will change with the value of the current and it will be different in the two regions; hence, it will be adjusted during the iterations. The following equation is used to calculate T_2 (Anders, 1997):

$$T_2 = T'_4 = \frac{0.26}{1 + 0.00056 \cdot \theta_m \cdot D_e}$$

where θ_m is the mean oil temperature (°C) and D_e (mm) is an outside diameter of one core.

The cable parameters in Table 2-1, taken from Table A1, are required in the analysis.

As noted above, the thermal resistance of the oil is temperature dependent. The values for sections 1 and 2 are given for the current of 902 A. The external thermal resistances of the two regions are also different and are specified above.

We will start the iterative computations by assuming conductor current equal to that of the steady-state normal conditions. We compute the joule losses at 20°C:

$$W_{c0} = \frac{I^2 R}{1 + \alpha_T \cdot (\theta_{max} - 20)} = \frac{19.93}{1 + 0.00393 \cdot 50} = 16.66 \text{ W/m}$$

Thus, from Equation (2.12),

$$W_{ct} = W_{c0}(1 - \alpha_T \theta_0) = 16.66(1 - 0.00393 \cdot 20) = 15.35 \text{ W/m}$$

$$\Delta W = \alpha_T W_{c0} = 0.00393 \cdot 16.66 = 0.0655 \text{ W/m}$$

For the first iteration, the total thermal resistances are obtained from Equations (2.10) and (2.8) as

Table 2-1 Values for Example 2-1

	Computed parameters	
Cable ampacity	I (A)	902
Conductor resistance at θ_{max}	R (ohm/km)	0.0781
Concentric/skid wires loss factor	λ_1	0.010
Pipe loss factor	λ_2	0.311
Thermal resistance of insulation	T_1 (K · m/W)	0.422
Thermal resistance of oil, Region 1	T^1_2 (K · m/W)	0.046
Thermal resistance of oil, Region 2	T^2_2 (K · m/W)	0.082
Thermal resistance of jacket/pipe covering	T_3 (K · m/W)	0.017
External thermal resistance 100% LF, Region 1	T^1_4 (K · m/W)	1.073
External thermal resistance 100% LF, Region 2	T^2_4 (K · m/W)	0.343
External thermal resistance nonunity LF, Region 1	T^1_4 (K · m/W)	0.903
External thermal resistance nonunity LF, Region 2	T^2_4 (K · m/W)	0.289
Losses in the conductor	W_c (W/m)	19.93
Dielectric losses	W_d (W/m)	4.83

$$T_r^{(1)} = T_1 + n \cdot (T_2 + T_3 + T_4^{(1)}) = 0.422 + 3 \cdot (0.046 + 0.017 + 0.903) = 3.32 \text{ K} \cdot \text{m/W}$$

$$T_r^{(2)} = 0.422 + 3 \cdot (0.082 + 0.017 + 0.289) = 1.59 \text{ K} \cdot \text{m/W}$$

$$T_t^{(1)} = T_1 + n \cdot [(1 + \lambda_1)T_2 + (1 + \lambda_1 + \lambda_2)(T_3 + T_4^{(1)})]$$

$$= 0.422 + 3 \cdot [1.01 \cdot 0.046 + 1.321(0.017 + 0.903)] = 4.21 \text{ K} \cdot \text{m/W}$$

$$T_t^{(2)} = 0.422 + 3 \cdot [1.01 \cdot 0.082 + 1.321(0.017 + 0.289)] = 1.88 \text{ K} \cdot \text{m/W}$$

The longitudinal thermal resistance of the conductor is (other, parallel paths for the longitudinal heat transfer are neglected here)

$$T_{L2} = \frac{0.0026}{0.00101} = 2.57 \text{ K/(m-W)}$$

The characteristic constant γ is computed from Equation (2.14):

$$\gamma_1 = \sqrt{T_L \cdot (1 - \Delta W \cdot T_t^{(v)})/T_r^{(v)}} = \sqrt{\frac{2.57 \cdot (1 - 0.0655 \cdot 4.21)}{3.32}} = 0.75 \text{ 1/m}$$

$$\gamma_2 = \sqrt{\frac{2.57 \cdot (1 - 0.0655 \cdot 1.88)}{1.59}} = 1.19 \text{ 1/m}$$

The temperature rise caused by the dielectric losses is equal to

$$\Delta\theta_d^{(1)} = W_d(T_1/2n + T_2^{(1)} + T_3 + T_4^{(1)})$$

$$= 3 \cdot 4.83(0.422/6 + 0.046 + 0.017 + 1.073) = 17.48 \text{ K}$$

$$\Delta\theta_d^{(2)} = W_d(T_1/2n + T_2^{(2)} + T_3 + T_4^{(2)})$$

$$= 3 \cdot 4.83(0.422/6 + 0.082 + 0.017 + 0.343) = 7.42 \text{ K}$$

The normalized steady-state temperatures at rated current are obtained from Equation (2.15):

$$\theta_\infty^{(1)} = \frac{W_{ct}T_t^{(1)} + \Delta\theta_d^{(1)} + \theta_{amb}^{(1)}}{1 - \Delta W \cdot T_t^{(1)}} = \frac{15.35 \cdot 4.21 + 17.48 + 25}{1 - 0.0655 \cdot 4.21} = 147.7°C$$

$$\theta_\infty^{(2)} = \frac{15.35 \cdot 1.88 + 7.42 + 25}{1 - 0.0655 \cdot 1.88} = 70.0°C$$

We can now compute the conductor temperature distribution using Equations (2.30) and (2.31). The results are shown in Figure 2-4 for the rated current of 902 A

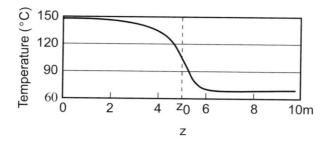

Figure 2-4 Temperature distribution from the center of street crossing with rated current of 902 A.

with the highest value at the center of the street crossing. At this point, the temperature reaches 147°C, and only three meters from the edge of the street (at $z \approx 3$ m) the conductor temperature attains its design value of 70°C.

To calculate the derating factor, auxiliary variables given by Equation (2.33) are computed first from Equations (2.34) to (2.39):

$$f_\alpha = 1 + \alpha_T(\theta_{max} - \theta_0) = 1 + 0.00393 \cdot (70 - 20) = 1.20$$

$$g = 1 - \alpha_T\theta_0 = 1 - 0.00393 \cdot 20 = 0.92$$

$$\tau = \frac{1}{\cosh(\gamma_1 \cdot z_0) + \dfrac{\gamma_1}{\gamma_2} \cdot \sinh(\gamma_1 \cdot z_0)} = \frac{1}{\cosh(0.75 \cdot 5) + \dfrac{0.75}{1.19} \cdot \sinh(0.75 \cdot 5)} = 0.029$$

$$\theta_A^{(1)} = \Delta\theta_d^{(1)} + \theta_{amb}^{(1)} = 17.48 + 25 = 42.48°C$$

$$\theta_A^{(2)} = 7.42 + 25 = 32.42°C$$

$$a = \alpha_T \cdot f_\alpha \cdot T_t^{(1)} \cdot T_t^{(2)} = 0.00393 \cdot 1.20 \cdot 4.21 \cdot 1.88 = 0.037$$

$$2b = \alpha_T \cdot (\theta_{max} - \theta_A^{(1)}) \cdot T_t^{(2)} + f_\alpha \cdot T_t^{(1)} - \tau \cdot [(g + \alpha_T\theta_A^{(2)}) \cdot T_t^{(1)} - (g + \alpha_T\theta_A^{(1)}) \cdot T_t^{(2)}]$$

$$= 0.00393 \cdot (70 - 42.48) \cdot 1.88 + 1.20 \cdot 4.21 - 0.029$$

$$\cdot \, [(0.92 + 0.00393 \cdot 32.42) \cdot 4.21 - (0.92 + 0.00393 \cdot 42.48) \cdot 1.88] = 5.19$$

$$b = 2.58$$

$$c = \theta_{max} - \theta_A^{(1)} + \tau \cdot (\theta_A^{(1)} - \theta_A^{(2)}) = 70 - 42.48 + 0.029 \cdot (42.48 - 32.42) = 27.81$$

The derating factor is now obtained from Equation (2.40):

$$DF^2 = \frac{b - \sqrt{b^2 - ac}}{\alpha_T \cdot T_t^{(1)}(\theta_{max} - \theta_A^{(2)})} = \frac{2.58 - \sqrt{2.58^2 - 0.037 \cdot 27.81}}{0.00393 \cdot 4.21 \cdot (70 - 32.42)} = 0.336$$

$$DF = 0.58$$

The derated current is thus equal to 523 A. This current is used in the second iteration to compute the losses and the thermal resistance of the oil. The process converges after nine iterations with the following results:

$$DF = 0.579 \qquad I = 522 \text{ A}$$

The reduction of ampacity is very significant. In the reminder of this example, we will explore the variation of the derating factor with changes of soil thermal resistivity and the ambient temperature in section 1. Let us define two parameters:

$$x = \frac{T_4^{(1)}}{T_4^{(2)}} \qquad \text{and} \qquad y = \frac{\theta_{amb}^{(1)}}{\theta_{amb}^{(2)}}$$

In practical cable installations, the value of x can reach 3 and the ambient temperature in the street crossings can easily be 5°C higher than the normal ambient. Figure 2-5 shows the dependence of the derating factor on parameters x and y and as a function of the width of the crossing with $z_0 = b_0/2$.

We can observe from Figure 2-5 that the derating factor becomes almost constant when the width of the road equal 3 m. The required reduction in cable rating becomes very significant and reaches 50% when the soil thermal resistivity is three times greater than in the rest of the route and the ambient temperature is increased by 5°C.

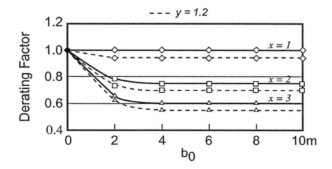

Figure 2-5 Derating factor as a function of the width of the crossing, thermal resistance, and ambient temperature of the crossing area (from Brakelmann and Anders 2004, with permission from IEEE Press).

For comparison purposes, the ratings of this cable located entirely in the soil with unfavourable conditions specified above are summarized in Table 2-2.

Thus, if the pipe-type cable crosses a street 10 m wide, and the thermal resistivity of the soil is two or three times higher than the rest of the route, the rating can be based on the unfavorable route conditions alone. However, if the unfavorable region is only 3 m wide (e.g., cable passes in the vicinity of a tree), then the cable rating should be reduced by about 30% assuming that the thermal resistivity of the soil is three times higher in this region than outside. ∎

2.1.5 Cyclic Rating for Cable Crossing an Unfavorable Region

As indicated in Section 1.5.4, in cyclic loading calculations, the detail of the load cycle is only needed over a period of 6 hours before the time of maximum temperature, and earlier values can be represented with sufficient accuracy by using an average. The loss-load factor μ provides this average. Estimation of the time of maximum temperature is done by inspection, bearing in mind that although it usually occurs at the end of the period of maximum current, this may not always be the case.

Since in Equation (1.77) only parameter k depends on the external thermal conditions, we can modify Equation (1.73) for each of the two thermal environments as

$$k^{(v)} = \frac{W_I T_4^{(v)}}{W_c T + W_I T_4^{(v)}} = \frac{W_I T_4^{(v)}}{W_c T_t^{(v)}} \tag{2.41}$$

When one value of the coefficient k has been computed, the other can be easily obtained as follows. Assume that a cyclic rating factor M has been derived for the cable system located in the soil with thermal resistivity $\rho^{(2)}$ assumed to be that of the native soil. We will determine a new cyclic factor for the same cable system with the soil thermal resistivity equal to $\rho^{(1)}$. The factor k in Equation (2.41) is dependent on the depth of laying and soil resistivity. We will first rewrite Equation (2.41) as follows:

$$k = \frac{W_I T_4}{W_c T + W_I T_4} = \frac{W_I T_4}{W_I \dfrac{T}{1 + \lambda_1 + \lambda_2} + W_I T_4} = \frac{W_I T_4}{W_I T_c + W_I T_4} = \frac{T_4}{T_c + T_4} \tag{2.42}$$

Table 2-2 Rating and ampacity reduction factor of the pipe-type cable for specified soil paramters

Ambient temperature (°C)	Soil thermal resistivity (K · m/W)		
	0.8	1.6	2.4
25	902	665	522
	$DF = 1.00$	$DF = 0.74$	$DF = 0.58$
30	841	612	472
	0.93	$DF = 0.68$	$DF = 0.52$

where T_c is the cable internal thermal resistance as if all the joule losses were generated at the conductor.

If the laying conditions in region $v = 2$, for which a value of k is known, change to conditions in region 1, then

$$k^{(1)} = \frac{T_4^{(1)}}{T_c + T_4^{(1)}} \tag{2.43}$$

and

$$k^{(2)} = \frac{T_4^{(2)}}{T_c + T_4^{(2)}} \tag{2.44}$$

Computing T_c from Equation (2.44) and substituting this in Equation (2.43), we obtain

$$k^{(1)} = \frac{1}{1 + \dfrac{T_4^{(2)}}{T_4^{(1)}}\left[\dfrac{1 - k^{(2)}}{k^{(2)}}\right]} \tag{2.45}$$

and the cyclic rating factors in Equations (1.71) and (1.77) become, respectively,

$$M^{(v)} = \frac{1}{\sqrt{\mu\left[1 - \dfrac{\theta_R^{(v)}(6)}{\theta_R^{(v)}(\infty)}\right] + \displaystyle\sum_{i=0}^{5} Y_i\left[\dfrac{\theta_R^{(v)}(i + 1)}{\theta_R^{(v)}(\infty)} - \dfrac{\theta_R^{(v)}(i)}{\theta_R^{(v)}(\infty)}\right]}} \tag{2.46}$$

$$M^{(v)} = \frac{1}{\sqrt{(1 - k^{(v)})Y_0 + k^{(v)}\{B + \mu[1 - \beta(6)]\}}} \tag{2.47}$$

Because of the very involved nature of the calculation of the temperature ratios in Equation (2.46), it might be more convenient to use Equation (2.47) in cyclic rating calculations when a cable crosses an unfavorable region.

Example 2.2
In this example, we present calculations for a circuit of three cables (cable Model No.1) in flat formation, not touching. The parameters of this cable are given in Appendix A as $\lambda_1 = 0.09$, $T_1 = 0.214$ K $\cdot$ m/W, $T_3 = 0.104$ K $\cdot$ m/W, and $T_4 = 1.933$ K $\cdot$ m/W. Laying conditions are as follows. Cables are located 1 m below the ground in a flat configuration. Spacing between cables is equal to one cable diameter (spacing between centers equal to two cable diameters). Ambient soil temperature is 15°C. The cables are solidly bonded and not transposed. The circuit crosses a boulevard in a vicinity of a large tree, as illustrated in Figure 2-6.

The thermal resistivity of the native soil is equal to 1.0 K $\cdot$ m/W and in the vicinity of the tree it is 2.5 K $\cdot$ m/W. The length of the unfavorable region is 5 m.

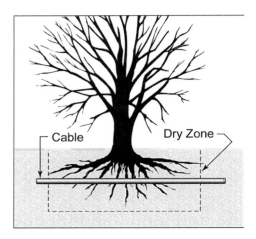

Figure 2-6   10 kV circuit laid in a vicinity of a large tree.

We will begin the calculations with the uniform soil conditions as in Region 2. We will compute the transient temperature response of the center cable for the first 24 h after the application of a step function of rated current and then calculate the cyclic rating factors. The temperature rise due to the dielectric loss is negligible for this cable. This part of the example is extracted from Examples 5.3, 5.5, and 5.6 in Anders (1997) in order to present a complete picture of the required initial developments before the cyclic rating factor in the unfavorable region is computed.

Since we will perform computations without the help of a computer, we will use a two-section equivalent network, neglecting for the moment the superscript (v), and remembering that the calculations pertain to Region 2.

An equivalent network, which represents the cable with sufficient accuracy, is derived with two sections $T_A Q_A$ and $T_B Q_B$, as shown in Figure 2-7. The calculation of the equivalent thermal resistances T_A and T_B and capacitances Q_A and Q_B is discussed in Section 1.3.3.2.

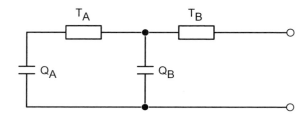

Figure 2-7   Two-loop equivalent network.

1. The Thermal Circuit of the Cable

The thermal circuit for this cable has been derived in Example 3.6 in Anders (1997). The following values were obtained there:

$$T_A = 0.214 \text{ K} \cdot \text{m/W} \qquad Q_A = 1435.1 \text{ J/K} \cdot \text{m}$$

$$T_B = 0.113 \text{ K} \cdot \text{m/W} \qquad Q_B = 692.3 \text{ J/K} \cdot \text{m}$$

2. Calculation of the Response of the Cable Circuit

The response of the cable circuit to a step function of rated current is computed from the following equation:

$$\theta(t) = W_c[T_a(1 - e^{-at}) + T_b(1 - e^{-bt})] \qquad (2.48)$$

where the parameters are computed as shown below (see Example 1.4):

$$T_a = \frac{1}{a-b}\left[\frac{1}{Q_A} - b(T_A + T_B)\right] \qquad \text{and} \qquad T_b = T_A + T_B - T_a$$

$$a = \frac{M_0 + \sqrt{M_0^2 - N_0}}{N_0} \qquad b = \frac{M_0 - \sqrt{M_0^2 - N_0}}{N_0}$$

$$M_0 = 0.5(T_A Q_A + T_B Q_B + T_B Q_A) \qquad N_0 = T_A Q_A T_B Q_B$$

Substituting numerical values, we obtain:

$$M_0 = \tfrac{1}{2}(0.214 \cdot 1435.1 + 0.113 \cdot 692.3 + 0.113 \cdot 1435.1) = 273.8 \text{ s}$$

$$N_0 = 0.214 \cdot 1435.1 \cdot 0.113 \cdot 692.3 = 24025.3 \text{ s}^2$$

$$a = \frac{273.8 + \sqrt{273.8^2 - 24025.3}}{24025.3} = 0.0208 \text{ s}^{-1}$$

$$b = \frac{273.8 - \sqrt{273.8^2 - 24025.3}}{24025.3} = 0.0020 \text{ s}^{-1}$$

$$T_a = \frac{1}{0.0208 - 0.0020}\left[\frac{1}{1435.1} - 0.0020(0.214 + 0.113)\right] = 0.0023 \text{ K} \cdot \text{m/W}$$

$$T_b = 0.214 + 0.113 - 0.0023 = 0.3247 \text{ K} \cdot \text{m/W}$$

The transient temperature rise for each step is determined from Equation (2.48); it is calculated for 1 h (3600 s) below; the results for other times are tabulated in Table 2-4.

Table 2-3 Components for cable environment partial transient

x	Item	Time (h)							
		1	2	3	4	5	6	12	24
$\dfrac{D_e^{*2}}{16t\delta}$	x	0.045	0.022	0.015	0.011	0.009	0.007	0.004	0.002
	$-E(-x)$	2.579	3.25	3.648	3.932	4.153	4.334	5.024	5.715
$\dfrac{d_{pk}^{*2}}{4t\delta}$	x	0.720	0.36	0.24	0.18	0.144	0.12	0.06	0.03
	$-E(-x)$	0.365	0.775	1.076	1.31	1.50	1.66	2.295	2.959

$$\theta_c(1) = W_c[T_a(1 - e^{-at}) + T_b(1 - e^{-bt})] = 30.82[0.0023(1 - e^{-0.0208 \cdot 1 \cdot 3600})$$
$$+ 0.3247(1 - e^{-0.002 \cdot 1 \cdot 3600})] = 10.1 \text{ K}$$

The conductor-to-cable surface attainment factor is calculated for 1 h below; results for the other times are shown in Table 2-3.

$$\alpha(1) = \frac{\theta_c(t)}{W_c(T_A + T_B)} = \frac{10.1}{30.82(0.214 + 0.113)} = 0.999$$

We can observe that because of the small value of the cable time constant, the conductor temperature rise above the cable surface temperature reaches its steady-state value within 1 h.

3. Calculation of the Response of Cable Environment
The response of the cable environment is given by Equation (1.65) as

$$\theta_{pk}(t) = W_{lk}\frac{\rho_s}{4\pi}\left[-Ei\left(-\frac{d_{pk}^{*2}}{4\delta t}\right) + Ei\left(-\frac{d_{pk}^{*'2}}{4\delta t}\right)\right] \tag{2.49}$$

A sample calculation for 1 h is presented below. The other values are given in Tables 2-3 and 2-4.

Table 2-4 Summary of temperature transient components

Time (h)	$\theta_c(t)$ (K)	$\alpha(t)$	$\theta_e(t)$ (K)	$\theta(t)$ (K)
1	10.1	0.999	8.8	18.9
2	10.1	1	12.8	22.9
3	10.1	1	15.4	25.5
4	10.1	1	17.4	27.5
5	10.1	1	19.0	29.1
6	10.1	1	20.3	30.4
12	10.1	1	25.5	35.6
24	10.1	1	30.9	41.0

$$d^*_{pk} = 0.072 \text{ m}$$

$$d^{*\prime}_{pk} = \sqrt{0.072^2 + 2^2} = 2.001 \text{ m}$$

$$\delta = 0.5 \cdot 10^{-6} \text{ m}^2/s$$

On a term-by-term basis for 1 h (3600 s), we have

$$x = \frac{D^{*2}_e}{16t\delta} = \frac{0.0358^2}{16 \cdot 3600 \cdot 0.5 \cdot 10^{-6}} = 0.0445, \qquad -Ei(-x) = 2.579$$

$$x = \frac{L^{*2}}{t\delta} = \frac{1^2}{3600 \cdot 0.5 \cdot 10^{-6}} = 555.6, \qquad -Ei(-x) = 0$$

$$x = \frac{d^{*2}_{pk}}{4t\delta} = \frac{0.072^2}{4 \cdot 3600 \cdot 0.5 \cdot 10^{-6}} = 0.72, \qquad -Ei(-x) = 0.365$$

$$x = \frac{(d^{*\prime}_{pk})^2}{4t\delta} = \frac{2.001^2}{4 \cdot 3600 \cdot 0.5 \cdot 10^{-6}} = 556.1, \qquad -Ei(-x) = 0$$

$$\theta_e(1) = \frac{1 \cdot 33.38}{4\pi}[(2.579 - 0) + 2(0.365 - 0)] = 8.8 \text{ K}$$

The second term in Equation (2.49) represents the effect of the image sources, and is usually negligible at normal depths of laying and for durations of less than 24 h.

4. Complete Temperature Rise

The complete transient is obtained by summing up to contributions from the internal and the external temperature rises. A sample calculation for 1 h is shown below. The values for this and other times are summarized in column 5 of Table 2-4.

$$\theta(1) = 10.1 + 0.999 \cdot 8.8 = 18.9 \text{ K}$$

Next, we will determine the cyclic rating factor of the cable system analyzed above. The load curve applied to this circuit is shown in Figure 2-8 and summarized in Table 2-5.

The load-loss factor is then equal to[1]

$$\mu = \frac{1}{24} \sum_{i=0}^{23} Y_i = \frac{0.091 + 0.061 + 0.052 + \ldots + 0.360}{24} = 0.504$$

First, we have to identify the time during the 24 h period at which we expect that the conductor temperature rise will reach its maximum value. From analysis of the

[1]Calculation of the load-loss factor is discussed in more detail in Chapter 5.

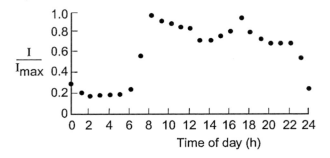

Figure 2-8 Cyclic load divided by highest load.

Table 2-5 Details of the load cycle in Figure 2-8 (from IEC 60853, with permission from IEC)

Time (h)	Load (p.u.)	Value of Y_i	Order of Y_i
0	0.302	0.091	Y_{17}
1	0.247	0.061	Y_{16}
2	0.227	0.052	Y_{15}
3	0.232	0.054	Y_{14}
4	0.235	0.056	Y_{13}
5	0.246	0.061	Y_{12}
6	0.290	0.084	Y_{11}
7	0.600	0.360	Y_{10}
8	1.000	1.000	Y_9
9	0.950	0.902	Y_8
10	0.940	0.884	Y_7
11	0.910	0.828	Y_6
12	0.892	<u>0.796</u>	$\underline{Y_5}$
13	0.770	<u>0.593</u>	$\underline{Y_4}$
14	0.772	<u>0.596</u>	$\underline{Y_3}$
15	0.800	<u>0.640</u>	$\underline{Y_2}$
16	0.853	<u>0.728</u>	$\underline{Y_1}$
17	0.996	<u>0.992</u>	$\underline{Y_0}$
18	0.853	0.728	Y_{23}
19	0.790	0.624	Y_{22}
20	0.740	0.548	Y_{21}
21	0.740	0.548	Y_{20}
22	0.722	0.521	Y_{19}
23	0.600	0.360	Y_{18}

values in Table 2-5, we select $i = 18$ as the hour at which this maximum occurs. Note that this is not the time of maximum load, and our selection is based on engineering judgment.[2] The six preceding hours are underlined in Table 2-5 and the values of Y reordered accordingly. The new order is shown in the last column of Table 2-5.

Much of the work for the determination of the cyclic rating factor for the cable system under consideration has already been performed. In particular, the conductor attainment factors $\alpha(t)$ and the exponential integral values have been computed for all hours. The next step is to evaluate the cable-surface attainment factors $\beta_1(t)$. A sample computation is performed below for $t = 1$ h, with the remaining values summarized in column 3 of Table 2-6.

$$D_e^* = 0.038 \text{ m} \qquad L^* = 1.0 \text{ m} \qquad d_{pk} = 0.072 \text{ m}$$

$$d'_{pk} = \sqrt{0.072^2 + 2.0^2} = 2.001 \text{ m} \qquad \delta = 0.5 \cdot 10^{-6} \text{ m}^2/s \qquad N = 3$$

The auxiliary variables defined in Equations (1.81) and (1.82) are equal to

$$F = \frac{d'_{p1} \cdot d'_{p2}}{d_{p1} \cdot d_{p2}} = \frac{2.001 \cdot 2.001}{0.072 \cdot 0.072} = 772.4 \qquad d_f = \frac{4L^*}{F^{1/(N-1)}} = \frac{4 \cdot 1.0}{\sqrt{772.4}} = 0.114$$

From Equation (1.83), Table 2-3 and, Table 2-4, we obtain

$$\beta_1(1) = \frac{-Ei\left(\dfrac{D_e^{*2}}{16t\delta}\right) + Ei\left(\dfrac{L^{*2}}{t\delta}\right) + (N-1)\left[-Ei\left(\dfrac{d_f^2}{16t\delta}\right) + Ei\left(\dfrac{L^{*2}}{t\delta}\right)\right]}{2 \ln \dfrac{4L^*F}{D_e^*}}$$

$$= \frac{(2.579 - 0) + 2 \cdot (0.365 - 0)}{2 \ln \dfrac{4 \cdot 1.0 \cdot 772.4}{0.0358}} = 0.146$$

Next, from Equation (1.84), we compute factor k_1 with the value of equivalent thermal resistance T_A and T_B and $T_4 + \Delta T_4$ given above (the external thermal resistance given in Table A1 already includes the mutual heating effect). We have

$$k_1 = \frac{W_i(T_4 + \Delta T_4)}{W_c T + W_i(T_4 + \Delta T_4)} = \frac{33.38 \cdot 1.933}{30.82(0.214 + 0.113) + 33.38 \cdot 1.933} = 0.865$$

The ratios $\theta_R(i)/\theta_R(\infty)$ are computed from Equation (1.85). As an example, we will calculate this ratio for $i = 1$. The remaining values, corresponding to the underlined Y values in Table 2-5, are shown in column 5 in Table 2-6.

$$\frac{\theta_R(1)}{\theta_R(\infty)} = [1 - k_1 + k_1\beta_1(t)]\alpha(t) = (1 - 0.865 + 0.865 \cdot 0.146) \cdot 1.0 = 0.261$$

---

[2]If no other information is available, the time of the highest loading should be selected.

Table 2-6 Evaluation of cyclic rating factor

Time (h)	Y_i	$\alpha(t)$	$\beta(t)$	$\dfrac{\theta_R(i)}{\theta_R(\infty)}$
0	0.992	0.999		0
1	0.728	1	0.146	0.261
2	0.640	1	0.211	0.318
3	0.596	1	0.255	0.356
4	0.593	1	0.288	0.384
5	0.796	1	0.315	0.407
6	—	1	0.337	0.426
			0.356	

The cyclic rating factor is now computed from Equation (1.71) as follows:

$$M = \frac{1}{\sqrt{\mu\left[1 - \dfrac{\theta_R(\tau)}{\theta_R(\infty)}\right] + \sum_{i=0}^{\tau-1} Y_i\left[\dfrac{\theta_R(i+1)}{\theta_R(\infty)} - \dfrac{\theta_R(i)}{\theta_R(\infty)}\right]}}$$

$$= [0.504(1 - 0.426) + 0.992(0.261) + 0.728(0.318 - 0.261) + 0.640(0.356 - 0.318)$$

$$+ 0.596(0.384 - 0.356) + 0.593(0.407 - 0.384) + 0.796(0.426 - 0.407)]^{-1/2} = 1.23$$

The same factor could be computed from Equation (1.77) since the attainment factor is equal to 1; that is,

$$B = \sum_{i=0}^{5} Y_i \Phi_i = 0.992 \cdot 0.146 + 0.728(0.211 - 0.146) + 0.640(0.255 - 0.211)$$

$$+ 0.596(0.288 - 0.255) + 0.593(0.315 - 0.288) + 0.796(0.337 - 0.315) = 0.274$$

$$M = \frac{1}{\sqrt{(1 - k_1)Y_0 + k_1\{B + \mu[1 - \beta(6)]\}}}$$

$$= \frac{1}{\sqrt{(1 - 0.865) \cdot 0.992 + 0.865 \cdot [0.274 + 0.504(1 - 0.337)]}} = 1.23$$

Thus, the permissible peak value of the cyclic load current is

$$1.23 \cdot 629 = 774 \text{ A}$$

We can now compute the cyclic rating factor in Region 1. In our example, the depth is not changed. Therefore, from Equation (2.45),

$$k_1^{(1)} = \cfrac{1}{1 + \cfrac{\rho^{(2)}}{\rho^{(1)}}\left[\cfrac{1 - k^{(2)}}{k^{(2)}}\right]} = \cfrac{1}{1 + \cfrac{1.0}{2.5}\left[\cfrac{1 - 0.865}{0.865}\right]} = 0.941$$

The cyclic loading factor in Region 1 is thus equal to

$$M^{(1)} = \cfrac{1}{\sqrt{(1 - k_1^{(1)})Y_0 + k_1^{(1)}\{B + \mu[1 - \beta(6)]\}}}$$

$$= \cfrac{1}{\sqrt{(1 - 0.941) \cdot 0.992 + 0.941 \cdot [0.274 + 0.504(1 - 0.337)]}} = 1.26$$

The steady-state ampacity of this cable system is now smaller than before. The new steady-state rating is obtained in a similar manner to that of Example 2.1 and is equal to 429 A. Therefore, the new cyclic rating is equal to $1.26 \cdot 429 = 540$ A. ∎

2.1.6 Soil Dryout Caused by Moisture Migration

The laying conditions examined in this chapter are particularly conducive to the formation of a dry zone around the cable. Under unfavorable conditions, the heat flux from the cable entering the soil may cause significant migration of moisture away from the cable. A dried-out zone may develop around the cable, in which the thermal conductivity can be reduced by a factor of three or more over the conductivity of the bulk. The drying-out conditions may occur in both regions of the route, but particularly in the region of high thermal resistivity.

The modeling of the dry zone around the cable proceeds in a very similar way to that discussed in Section 1.4.1.2. The developments presented there will be modified, leading to an expression for the conductor temperature taking into account the drying-out conditions in the unfavorable region.

We will start by observing that the a new term $(v - 1)\Delta\theta_x$ appears in Equation (1.49) for the temperature rise at the cable outside surface. This leads to a slight modification of Equation (2.9), yielding

$$\theta^{(v)} = \theta_{amb}^{(v)} + \Delta\theta_d^{\prime(v)} - (v^{\prime(v)} - 1) \cdot \Delta\theta_x + W_c \cdot T_t^{\prime(v)} + (W_r - W_c) \cdot T_r^{\prime(v)} \quad (2.50)$$

where $T_t^{\prime(v)}$ and $T_r^{\prime(v)}$ are the modified values of $T_t^{(v)}$ and $T_r^{(v)}$ defined in (2.8) and (2.10), respectively, obtained by replacing $T_r^{(v)}$ by $v^{(v)} \cdot T_4^{(v)}$:

$$T_r^{\prime(v)} = T_1 + n \cdot (1 + \lambda_1) \cdot T_2 + n \cdot (1 + \lambda_1 + \lambda_2) \cdot (T_3 + v^{(v)} \cdot T_4^{(v)})$$
$$\text{and } T_r^{\prime(v)} = T_1 + n \cdot (T_2 + T_3 + v^{(v)} \cdot T_4^{(v)}) \quad (2.51)$$

The resistivity ratio $v^{(v)}$ is defined separately for each region.

The temperature rise $\Delta\theta_d^\prime$ caused by the dielectric losses is obtained in a similar way by replacing $T_4^{(v)}$ by $v^{(v)} \cdot T_4^{(v)}$.

The differential Equation (2.13) becomes

$$\theta^{(v)} - \frac{1}{\gamma_v^2} \cdot \frac{\partial^2 \theta^{(v)}}{\partial z^2} = \frac{\theta_{amb}^{(v)} + \Delta\theta_d^{\prime(v)} - (v^{\prime(v)} - 1) \cdot \Delta\theta_x + W_{ct} \cdot T_t^{\prime(v)}}{(1 - \Delta W \cdot T_t^{\prime(v)})} = \theta_\infty^{\prime(v)} \quad (2.52)$$

where

$$\gamma_v^{\prime 2} = (1 - \Delta W \cdot T_t^{\prime(v)}) \cdot \frac{T_L}{T_r^{\prime(v)}} \quad (2.53)$$

Example 2.3
Let us assume that in the cable system examined in Example 2.1 a dry zone has developed in Region 1, with dry soil thermal resistivity of 3.0 K · m/W and critical temperature rise $\Delta\theta_x = 35°C$.

The calculations proceed in a similar way to that in Example 2.1 with the following modifications:

$$T_r^{\prime(1)} = T_1 + n \cdot (T_2 + T_3 + v^{\prime(1)}T_4^{(1)}) = 0.422 + 3 \cdot (0.046 + 0.017 + 1.2 \cdot 0.903)$$
$$= 3.86 \text{ K} \cdot \text{m/W}$$

$$T_t^{\prime(1)} = T_1 + n \cdot [(1 + \lambda_1)T_2 + (1 + \lambda_1 + \lambda_2)(T_3 + v^{\prime(1)}T_4^{(1)})]$$
$$= 0.422 + 3 \cdot [1.01 \cdot 0.046 + 1.321(0.017 + 1.2 \cdot 0.903)] = 4.92 \text{ K} \cdot \text{m/W}$$

The characteristic constant is computed from Equation (2.53):

$$\gamma_1' = \sqrt{T_L \cdot (1 - \Delta W \cdot T_t^{\prime(v)})/T_r^{\prime(v)}} = \sqrt{\frac{2.57 \cdot (1 - 0.0655 \cdot 4.92)}{3.86}} = 0.67 \text{ 1/m}$$

The temperature rise caused by the dielectric losses is equal to

$$\Delta\theta_d^{\prime(1)} = W_d(T_1/2n + T_2^{(1)} + T_3 + v^{\prime(1)}T_4^{\prime(1)})$$
$$= 3 \cdot 4.83(0.422/6 + 0.046 + 0.017 + 1.2 \cdot 1.073) = 20.59 \text{ K}$$

The new auxiliary variables become

$$\tau' = \frac{1}{\cosh(\gamma_1' \cdot z_0) + \dfrac{\gamma_1'}{\gamma_2} \cdot \sinh(\gamma_1' \cdot z_0)} = \frac{1}{\cosh(0.67 \cdot 5) + \dfrac{0.67}{1.19} \cdot \sinh(0.67 \cdot 5)}$$
$$= 0.044$$

$$\theta_A^{\prime(1)} = \Delta\theta_d^{\prime(1)} + \theta_{amb}^{(1)} = 20.59 + 25 = 45.59°C$$

$$a' = \alpha_T \cdot f_a \cdot T_t^{\prime(1)} \cdot T_t^{(2)} = 0.00393 \cdot 1.20 \cdot 4.92 \cdot 1.88 = 0.044$$

$$2b' = \alpha_T \cdot (\theta_{\max} - \theta_A^{\prime(1)}) \cdot T_t^{(2)} + f_a \cdot T_t^{\prime(1)} - \tau' \cdot [(g + \alpha_T \theta_A^{(2)}) \cdot T_t^{\prime(1)} - (g + \alpha_T \theta_A^{\prime(1)}) \cdot T_t^{(2)}]$$
$$= 0.00393 \cdot (70 - 45.59) \cdot 1.88 + 1.20 \cdot 4.92 - 0.044 \cdot [(0.92 + 0.00393 \cdot 32.42)$$
$$\cdot 4.92 - (0.92 + 0.00393 \cdot 45.59) \cdot 1.88] = 5.93$$

$$b = 2.97$$

$$c' = \theta_{\max} - \theta_A^{\prime(1)} + \tau' \cdot (\theta_A^{\prime(1)} - \theta_A^{(2)}) = 70 - 45.59 + 0.044 \cdot (45.59 - 32.42) = 25.0$$

The derating factor is now obtained from Equation (2.40):

$$DF'^2 = \frac{b' - \sqrt{b'^2 - a'c'}}{\alpha_T \cdot T_t^{\prime(1)}(\theta_{\max} - \theta_A^{(2)})} = \frac{2.97 - \sqrt{2.97^2 - 0.044 \cdot 25.0}}{0.00393 \cdot 4.92 \cdot (70 - 32.42)} = 0.261$$

$$DF = 0.51$$

The derated current is thus equal to 461 A. This current is used in the second iteration to compute the losses and the thermal resistance of the oil. The process converges after nine iterations with the following results:

$$DF = 0.509 \qquad I = 459 \text{ A}$$

We can observe that the rating of the pipe type cable is almost half of the original value when the cable is laid in a uniform soil. For comparison purposes, we should mention that the rating of this cable laid in a uniform soil with the thermal resistivity of 3.0 K · m/W is 443 A. ∎

2.2 VENTILATED ROUTES

We have shown above that cables crossing streets or other thermally unfavorable regions will experience a substantial reduction of current-carrying capability. For a street crossing of a short length (5–10 m) the longitudinal heat flow along the metallic elements of the cable can limit the ampacity reduction. If the ampacity reduction is unacceptable, additional measures must be taken to counter the effect of unfavorable thermal conditions. Such measures may include the use of a thermal backfill discussed in Chapter 4 or the laying of heat pipes (Weedy, 1988; Iwata et al., 1992). In this section, we will show that a substantial increase in cable ampacity can be achieved if air-filled pipes are used to help with the air movement around the loaded cables. Those pipes can either contain the cables or can be laid parallel to the cable route and led to the soil surface at the ends to interact with the ambient air. Then, the temperature gradient along the cable route causes self-supporting air convection in the pipe which, in turn, produces substantial cooling effect.

An additional advantage of using metallic rather than plastic pipes is that they may lead to a substantial reduction in the magnetic field. We will examine this issue at the end of the chapter.

2.2.1 Cable Laid in a Pipe with Air Convection

For short, thermally adverse sections of the cable route (Brakelmann, 1999a,b), as well as for longer sections (Brakelmann, 1999c; Brakelmann and Anders, 2004b) a substantial increase in current ratings can be achieved if the cables are laid in a pipe that, at a certain distance, is directed up to the soil surface so that the air in the pipe can interact with the ambient air. For shorter cooling distances, the temperature gradient along the cable route causes self-supporting air convection in the pipe, which, in turn, produces a substantial cooling effect. For longer cooling distances (up to several kilometers), convection forced by air fans can be introduced.

This section is concerned with cables directly buried in the ground. By using air ventilating pipe, we not only reduce the effect of the magnetic field, but also achieve high ampacity values. In the developments presented below, we will compute the temperature change along the cable route with the effect of air movement caused by the temperature gradient.

2.2.1.1 Self-Supporting Air Convection.

Let us now consider an installation shown in Figure 2-9 with two types of pipe. In one case, we will consider a PE pipe and in the other a steel pipe. The pipes are laid at the depth of h_v meters, the cooling section has length l_0, and both ends of the pipe is lead up to the soil surface. The installation shown in Figure 2-9 is such that the left vertical pipe section ends at the soil surface, whereas the right vertical pipe section ends at a height h_z above the ground.

The loaded cables will heat up the surrounding soil as well as the air inside the pipe. Thermal asymmetries (e.g., wind forces at the inlet points) will induce the first movement of the air along the pipe, so that, for example, air of higher temperature is moved into the right vertical pipe section. Now, at both ends of the cooling section, two vertical columns of air are present with different temperatures and, therefore, with different densities. This causes different gravitational forces, which will enforce the initial air movement, which will finally result in a stationary circulation.

This circulation is defined by an air movement with constant velocity and with a temperature rise $\Delta\theta$ of the air along the cable route, so that the left vertical air column with inlet temperature θ_{in} is opposed by the right vertical air column with outlet temperature $\theta_{out} = \theta_{in} + \Delta\theta$. The pressure rise (Pa · K/m) caused by the difference in densities of the air columns Δp (Pa · K/m), as well as the velocity of the air circulation, are calculated from (Brakelmann, 1999a,b,c):

$$\Delta p = g \cdot h_p \cdot [\rho(\theta_{in}) - \rho(\theta_{out})] \qquad (2.54)$$

with
g = gravitation constant, m/s²
$h_p = h_v + h_z$ = the height of the right air column, m
θ_{in}^* = inlet-temperature of the air, K

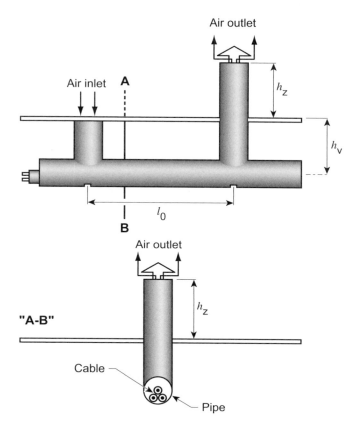

Figure 2-9 High-voltage cable system with cables in touching triangular formation inside a pipe.

θ^*_{out} = outlet-temperature, K

ρ = density of air, kg/m³, and its dependence on the temperature θ (K) is given by[3]

$$\rho(\theta) = 352.64/\theta^* \qquad (2.55)$$

Hence, Equation (2.54) will take the form

$$\Delta p = 3,459.4 \cdot h_p \cdot (1/\theta^*_{in} - 1/\theta^*_{out}) \qquad (2.56)$$

This pressure rise causes air circulation with the velocity

$$w = \sqrt{\frac{2 \cdot \Delta p \cdot d_i}{l_p \cdot \rho(\overline{\theta}) \cdot f}} \qquad (2.57)$$

[3]An asterisk denotes that the temperature is expressed in degrees Kelvin. Otherwise, it is given in degrees Celsius.

with the total pipe length

$$l_p = l_0 + 2 \cdot h_v + h_z \tag{2.58}$$

and

d_i = inner pipe diameter, m
$\rho(\overline{\theta})$ = density of air (median value of the total length), kg/m^3
f = friction coefficient of the air flow

The friction coefficient is also temperature dependent and can be obtained from (Ozisik, 1985)

$$f = \alpha \cdot (182 \log \mathrm{Re} - 1.64)^{-2} \tag{2.59}$$

where

Re = Reynolds number
α = 0.85 for metallic pipes, =1 for nonmetallic pipes, and = 2 for concrete/stone surfaces

The Reynolds number is obtained from the following equation:

$$\mathrm{Re} = \frac{w \cdot d_i}{v} \tag{2.60}$$

were v is the kinematic viscosity of air. This quantity as well as air thermal conductivity are temperature dependent and are computed from (Anders, 1997)

$$v = 1.32 \cdot 10^{-5} + 9.5 \cdot 10^{-8} \cdot \theta \tag{2.61}$$

$$k = 2.42 \cdot 10^{-2} + 7.2 \cdot 10^{-5} \cdot \theta \tag{2.62}$$

The Reynolds number is computed iteratively since the velocity of air is unknown initially. Also, the current rating is calculated by an iterative process, where for each value of the current, the temperature distribution as well as the losses absorbed by the flowing air and the losses induced by the cooling pipe into the surrounding soil along the cable route are considered. Similar to the procedure for cables with lateral cooling, (Brakelmann, 1980, 1985) the cable route is longitudinally split into a sufficiently large number of subsections (e.g., 50). Beginning with the inlet point of the air (inlet temperature θ_{in}), and proceeding subsection by subsection, in each cross-section of the cable trench, a field analysis is performed in which the air temperature and the temperature-dependent cable losses are the input variables. Results of each field analysis are the temperatures in the cable trench, and, hence, the conductor temperatures of all cables, as well as the losses per unit length absorbed by the cooling pipe.

This field analysis is based on the charge simulation method (CSM); it considers the nonlinear behavior of the soil due to a possible drying out by means of a thermal potential (Brakelmann, 1985; Heinhold, 1987, Anders, 1997).

Temperature-dependent characteristic values of the air flow, for example, the thermal resistance between cables and air and between air and pipe wall as well as the thermal capacitance and the friction coefficient of the air column, are calculated as described in (Ihle, 1997 and Haubrich, 1973), or a simplified approach can occasionally be used based on the IEC 60287 (1994) standard (see Example 2.4).

From the losses absorbed by the flowing air, the temperature rise $\Delta\theta(x)$ along the subsection length Δx can be calculated from

$$\Delta\theta_L(x) = \frac{W_L(x) \cdot \Delta x}{w \cdot \frac{\pi}{4}d_i^2 \cdot \rho \cdot c_L} \tag{2.63}$$

where
$W_L(x)$ = longitudinal losses generated in the cable and dissipated in the air in the duct, W/m
c_L = volumetric specific heat of air, J/m³ °C

From this, we obtain the air temperature in the next subsection:

$$\theta_L(x + \Delta x) = \theta_L(x) + \Delta\theta_L(x) \tag{2.64}$$

This temperature is used as an input variable for the field analysis in the next cross section of the cable trench.

Arriving at the cooling section end, the calculation delivers the outlet temperature of the air θ_{out} as well as the maximum conductor temperatures of the cable conductor. This calculation procedure is repeated for varying load currents until one of the cables reaches its permissible conductor temperature.

The above procedure can be summarized in the following algorithm.

1. Set the current (outer iteration loop).
2. Assume the air outlet temperature (inner iteration loop) and determine from this the flow velocity.
3. Start at the inlet point with given air temperature.
4. Use the above equations to determine temperatures and power flows in the cross section, and, additionally, the longitudinal power flow W_L.
5. Calculate the longitudinal temperature rise and, from this, the mean air temperature $\overline{\theta}_{air}(x)$ in the next cross section.
6. Repeat steps 4 and 5 until the end of the pipe is reached.
7. Compare the calculated air outlet temperature with the assumed one and iterate again. From this, a new flow velocity follows.
8. Repeat the steps 3 to 7 until the air outlet temperature and the flow velocity are found.
9. Look at the highest conductor temperature (at the outlet point) and compare it with the permissible one. Iterate the current.

The longitudinal heat flow W_L is computed by considering the thermal circuit in Figure 2-10. The notation in this figure is as follows:

$\bar{\theta}_{air}$ = mean air temperature in the pipe, °C

θ_c, θ_e = conductor and cable surface temperatures, respectively, °C

θ_i, θ_p = temperatures of the inner and outer walls of the pipe, respectively, °C

T = cable total internal thermal resistance, W · m/W

T'_4, T_4''' = thermal resistance of the pipe wall and the soil outside the pipe, respectively, K · m/W

W_L = longitudinal heat flow in the air in the pipe, W/m

W_t = total losses generated by the cable circuit, W/m

W_p = losses dissipated in the soil, W/m

Using information in Figure 2-10 we can compute the longitudinal heat flow as follows. From Figure 2-10, we have

$$\bar{\theta}_{air} = \theta_e - (T_{air}/2) \cdot W_t = \theta_i + (T_{air}/2) \cdot W_p \qquad (2.65)$$

$$\bar{\theta}_{air} = \theta_c - (T + T_{air}/2) \cdot W_t \qquad (2.66)$$

with

$$W_p = \frac{\theta_i - \theta_{amb}}{T'_4 + T_4'''} \qquad (2.67)$$

or

$$W_p = \frac{\bar{\theta}_{air} - \theta_{amb}}{T_{air}/2 + T'_4 + T_4'''} \qquad (2.68)$$

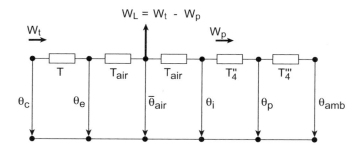

Figure 2-10 Thermal circuit for evaluation of the longitudinal heat flow in the pipe.

Finally,

$$W_L = W_t - W_p \tag{2.69}$$

The thermal resistance of air is temperature dependent. In many instances, T_{air} in Equations (2.65) and (266.) can be computed from a simplified formula given in the IEC Standard 60287 (1994). When more accurate results are needed, the air thermal resistance should be computed from

$$T_{air} = \frac{T_{rad} \cdot T_{conv}}{T_{rad} + T_{conv}} \tag{2.70}$$

The convective and radiative thermal resistances are evaluated as described in Anders (1997), Ihle (1997), or Haubrich (1973) and also discussed in Chapter 1.

The thermal resistance T_4'' is computed from

$$T_4'' = \frac{\rho_p}{2\pi} \ln \frac{D_o}{D_d} \tag{2.71}$$

where ρ_p is the thermal resistivity of the pipe material.

The external resistance of the pipe will depend whether soil drying out occurs or not. For the moist soil, we have

$$T_4''' = \frac{\rho_e}{2\pi} \ln \frac{4L}{D_o} \tag{2.72}$$

where ρ_e is the moist soil thermal resistivity and L is the depth of the center of the pipe.

When soil drying out occurs, the expression for the external thermal resistance takes the form

$$T_{4x}' = v \cdot T_4''' - \theta_x/W_p \tag{2.73}$$

with

$$v = (\rho_c/\rho_e) \qquad \text{for } \theta_o \geq \theta_{cr}$$

$$v = 1 \qquad \text{for } \theta_o \leq \theta_{cr} \tag{2.74}$$

$$\theta_x = \theta_{cr} \cdot (v - 1)$$

where
ρ_c = thermal resistivity of the dry soil, K $\cdot$ m/W
θ_{cr} = critical temperature at which drying out starts, °C

The calculation method described above is based on the conditions of a constant load and constant temperatures of the ambient air and the soil. In practice, these conditions will never be met. Therefore, a complete description of the thermal behavior must consider the time dependence and three-dimensional and nonlinear field conditions, taking into account the different situations of yearly and daily variation of the cable load as well as of the ambient soil and air temperatures, all of which are random parameters. However, for cable rating calculations a common practice is to consider the most pessimistic scenario. For cables laid in ventilated pipes, we shall adopt the same approach. In particular, we are combining the maximum load of the cable with the most adverse temperatures of ambient air and soil in the steady-state conditions.

The cooling pipes will draw in and circulate ambient air as long as the air temperature is lower than the soil temperature at the pipe surface. This means that the warmed-up cable trench is cooled by the system even during periods when the cable is lightly loaded or even not loaded at all.

2.2.1.2 Forced Air Convection.
To enable greater cooling section lengths, a fan in the cooling pipe or in the inlet pipe can force an air flow with a desired velocity. The fan can be powered by induction conductors (multicore, low-voltage cables) connected to a coil and laid parallel to the HV cable. This coil can take the necessary energy for the fan from the high-voltage cable system. In this case, the intensity of the air cooling is dependent on the momentary load of the cable system. Alternatively, any other supply can be used to produce the necessary flow velocity.

2.2.2 Numerical Example Illustrating the Cooling-Pipe Concepts.
In this section, we will illustrate through a numerical example the effect of various parameters influencing the rating calculations for cables cooled by ventilated pipes.

Example 2.4
We will consider the 110 kV cable system, model 6 with a unity load factor. The cable has a rating of 1098 A. We will refer to this condition as the base case.

Let us now consider a situation in which these cables are laid in a triangular configuration in a ventilated pipe of diameter $d_i = 300$ mm and the length l_0. The ambient soil temperature is 20°C and the inlet air temperature is 25°C.

Our aim is to compute the ampacity of this cable circuit when laid in a ventilated pipe, so that the maximum conductor temperature does not exceed 90°C. We will use the air velocity w, the height of the ventilated pipe h_z and the length l_0 as the parameters. We will show the calculations for the first section of the cable route with self-supporting convection, $l_0 = 30$ m and $h_z = 1$ m.

From the laying conditions with $h_v = 1.0$ m, the total pipe length is equal to 33 m. The parameters of this circuit for the flat formation are taken from Appendix A. The parameters for the triangular formation in a PE duct are also needed for this example. These are summarized in Table 2-7.

We will start by assuming that the cable is loaded to its full capability of 978 A. This will later be revised as the calculations proceed.

Table 2-7 Parameters for cable model 6

		Computed parameters (hottest cable	
		Triangular formation in a duct	Flat formation
Cable ampacity	I (A)	978	1098
Conductor resistance at θ_{max}	R (ohm/km)	0.0234	0.0231
Sheath loss factor	λ_1	0	0
Armor loss factor	λ_2	0	0
Thermal resistance of insulation	T_1 (K · m/W)	0.431	0.431
Thermal resistance of jacket	T_2 (K · m/W)	0.05	0.05
External thermal resistance, drying out considered (includes mutual heating effect)	T_4 (K · m/W)	1.18*	2.806
Losses in the conductor	W_c (W/m)	22.38	27.85
Dielectric losses	W_d (W/m)	0.27	0.27

*This value does not consider the mutual heating effect, whereas, for the flat formation, it is included.

In the analysis that follows, we will need the properties of air at the operating temperature. This temperature is unknown at this stage, and we will start by assuming an outlet air temperature of 40°C, which yields a mean air temperature in the pipe approximately equal to 33°C.

From Equations (2.55) and (2.61), we have

$$\rho = 352.64/\theta^* = 352.64/(33 + 273) = 1.15 \text{ kg/m}^3$$

$$v = 1.32 \cdot 10^{-5} + 9.5 \cdot 10^{-8} \cdot \theta = 1.32 \cdot 10^{-5} + 9.5 \cdot 10^{-8} \cdot 33 = 1.63 \cdot 10^{-5} \text{ m}^2/s$$

We first compute the pressure differential from Equation (2.56):

$$\Delta p = 3,459.4 \cdot h_p \cdot (1/\theta^*_{in} - 1/\theta^*_{out}) = 3,459.4 \cdot 2 \cdot \left(\frac{1}{25 + 273} - \frac{1}{40 + 273} \right)$$
$$= 1.11 \text{ N/m}^2$$

We now assume the air velocity of 1 m/s and compute the Reynolds number from the following equation:

$$\text{Re} = \frac{w \cdot d_i}{v} = \frac{1 \cdot 0.3}{1.6\text{E-}5} = 1.84 \cdot 10^4$$

The friction coefficient is obtained from Equation (2.59):

$$f = \alpha \cdot (1.82 \log \text{Re} - 1.64)^{-2} = 1 \cdot [1.82 \log(1.84\text{E}4) - 1.64]^{-2} = 0.0268$$

The air velocity is obtained from Equation (2.57) as

$$w = \sqrt{\frac{2 \cdot \Delta p \cdot d_i}{l_p \cdot \rho(\bar{\theta}) \cdot f}} = \sqrt{\frac{2 \cdot 1.11 \cdot 0.3}{33 \cdot 1.15 \cdot 0.0268}} = 0.81 \text{ m/s}$$

We could iterate at this stage as well since the computed air velocity is different than the assumed one. We will skip this iteration step here.

The thermal resistance of the air in the pipe is obtained using the model in IEC 60287 (1994).[4] Since the IEC standard does not have the constants for a PE duct, we will use quantities for an earthenware buried pipe. Also, since we have three cables in the duct, we will need an equivalent cable diameter, which, in this case is equal to $D_{cab} = 2.15 \cdot 90 = 193.5$ mm. In the first iteration with the air temperature of 25°C we have

$$T_{air} = \frac{U}{1 + 0.1(V + Y\bar{\theta}_{air})D_{cab}} = \frac{1.87}{1 + 0.01(0.28 + 0.0036 \cdot 25) \cdot 193.5} = 0.229 \; \frac{\text{K} \cdot \text{m}}{\text{W}}$$

From (2.6.6), we have

$$\theta_c = \bar{\theta}_{air} + (T + n \cdot T_{air}/2) \cdot W_t = 25 + (0.431 + 0.05 + 3 \cdot 0.229/2) = 43.7°\text{C}$$

Since the temperature of the conductor is lower than the one used to compute the total losses, the computer program performs another iteration, but we will skip this step here.

The thermal resistance of the pipe is obtained from Equation (2.71):

$$T_4'' = \frac{\rho}{2\pi} \ln\left(\frac{D_o}{D_d}\right) = \frac{7}{2\pi} \ln\left(\frac{0.31}{0.3}\right) = 0.037 \; \frac{\text{K} \cdot \text{m}}{\text{W}}$$

For the calculation of the soil thermal resistance, we need to determine whether drying out occurs under the given conditions. Assuming that the critical temperature rise is 35°C, the critical temperature at which the drying out occurs is equal to $35 + 20 = 55°\text{C}$. No drying out occurs in this case since the conductor temperature is already lower than the critical temperature, hence

$$T_4''' = \frac{\rho}{2\pi} \ln\left(\frac{4L}{D_0}\right) = \frac{1.0}{2\pi} \ln\left(\frac{4 \cdot 1.0}{0.31}\right) = 0.407 \; \frac{\text{K} \cdot \text{m}}{\text{W}}$$

From Equation (2.68), we also have

$$W_p = \frac{\bar{\theta}_{air} - \theta_{amb}}{T_{air}/2 + T_4'' + T_4'''} = \frac{25 - 20}{0.229/2 + 0.037 + 0.407} = 8.96 \text{ W/m}$$

---

[4]The Reynolds number indicates that the flow is turbulent. Therefore, the thermal resistance of the air in the pipe should actually be computed using heat balance equations with full heat correlations, as is done in the computer program. However, the method presented here is suitable for illustrative purposes since the algorithm is essentially the same and only the thermal resistance of the air is computed differently.

The longitudinal heat flow is thus equal to

$$W_L = W_t - W_p = 67.95 - 8.96 = 59.0 \text{ W/m}$$

We can now calculate the temperature in the second section of the pipe. From Equation (2.63) we compute the temperature rise in the first section with an assumed section length of 1 m:

$$\Delta \theta_L(x) = \frac{W_L(x) \cdot \Delta x}{w \cdot \frac{\pi}{4} d_i^2 \cdot \rho \cdot c_L} = \frac{59 \cdot 1.0}{0.81 \cdot \frac{3.14}{4} \cdot 0.3^2 \cdot 1.15 \cdot 1007} = 0.89°C$$

The air temperature in the second section is thus equal to 25.89°C.

The process now continues for the remaining sections of the pipe length. The computed air temperature at the pipe outlet is then used to compute a new mean air temperature, a new pressure difference, and a new air velocity, which are used in the second iteration of the inner loop. When the solution is reached, the outer loop calculations are restarted with an adjusted current, with the goal of reaching the maximum permissible conductor temperature.

The results of the calculations are shown in Figure 2-11. Dotted curves in the diagram show the continuous current ratings of the model 6 cable laying in a PE-pipe as a function of the cooling section length. Parameter is the height h_z of the outlet point. We can observe from Figure 2-11 that, depending on the outlet height h_z, cooling section lengths up to approximately 80 m can be used to permit steady-state rating of 1098 A.

The solid lines in Figure 2-11 show the continuous current ratings of the cable with the air velocity as the parameter. We can observe that in the case of forced circulation, the basic rating of 1098 A can be transmitted with cooling section lengths of, for example, 400 m (for w = 6 m/s), which is a normal distance of the cable joints. Another advantage of this construction is that the low inlet temperature of the air can be used for the cooling of the joints. Enlarging the inner pipe diameter (e.g., using concrete pipes), offers the possibility of a far greater cooling section lengths (Brakelmann, 1999c).

Additional benefits can be achieved if the cable system is laid in a steel pipe, as illustrated in Figure 2-12.

The dotted curves in Figure 2-12 represent the continuous current ratings of the 110 kV XLPE cable, laying now in the steel pipe, as a function of the cooling section length, again with the height h_z of the outlet point as a parameter. Because of the higher losses in this case, section lengths of between 5 and 40 m can be considered for the steady-state rating of 1098 A.

The solid lines in Figure 2-12 show the continuous current ratings of the 110 kV XLPE cable, laying in a ferromagnetic steel pipe. The rating is a function of the cooling section length with the air velocity w as the parameter. Because of the higher losses, the steady-state rating of 1098 A is achieved with cooling section lengths of between 200 and 310 m only when the wind velocity is greater than 6 m/s.

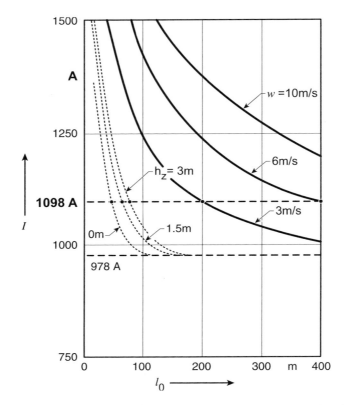

Figure 2-11   Continuous current rating as function of the cooling section length l_0. Dotted lines: self-supporting air convection parameter, outlet height h_z. Solid lines: forced air convection. Parameter: flow velocity w. Air inlet temperature $\theta_{in} = 25°C$. (From Brakelmann and Anders, 2004b, with permission from IEEE Press.)

Selection of the Pipe Size. We will now consider the effect of the size of the pipe on the cable rating. We will consider a cooling section length of 400 m and use air inlet temperature and air velocity as the parameters. The results for the cable system examined in Example 2.4 are shown in Figure 2-13.

Referring to this figure, for a maximum air velocity of $w = 10$ m/s and steady-state load current of 1098 A, the following inner pipe diameters can be considered:

- 370 mm for a steady-state inlet temperature of the air of 35°C
- 330 mm for a steady-state inlet temperature of the air of 25°C
- 300 mm for a steady-state inlet temperature of the air of 11°C, as a yearly mean value   ∎

2.2.3   *Reduction of a Magnetic Field.* As shown in the previous section, laying cables in ventilated pipes can lead to an increase thermal rating when the cables

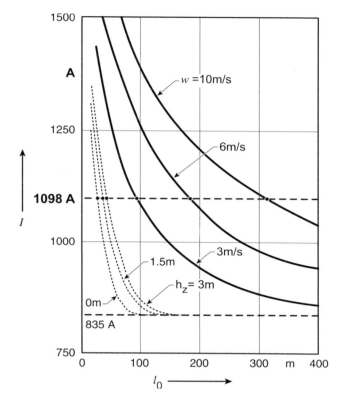

Figure 2-12 The same as Figure 2-11, but for steel pipe with inner diameter of 300 mm (from Brakelmann and Anders, 2004b, with permission from IEEE Press).

cross an unfavorable environment. Pipes made of steel are selected most often for pipe-type cables and can also be used in ventilated installations. Ferromagnetic pipes also have an additional benefit of reducing magnetic field in the vicinity of the loaded cable. We will explore this issue in this section.

Because of community concerns about possible health effects of low-frequency electromagnetic fields, many utilities have adopted a policy of "prudent avoidance" when planning new power transmission lines. Guideline for human exposure to magnetic fields (50 Hz/60 Hz) is a maximum value of 100 μT suggested by the World Health Organization and adopted by law in some countries.

High-voltage cables have no outer electric field, and their stationary magnetic induction above the ground surface is normally lower than 100 μT. Nevertheless, electronic systems, especially monitors with electron beams, may be disturbed by far weaker magnetic fields.

Measures to reduce the magnetic fields of power cables are described, for example, in Brakelmann (1995, 1996, 1998) and Bucea and Kent (1998). Effective measures include:

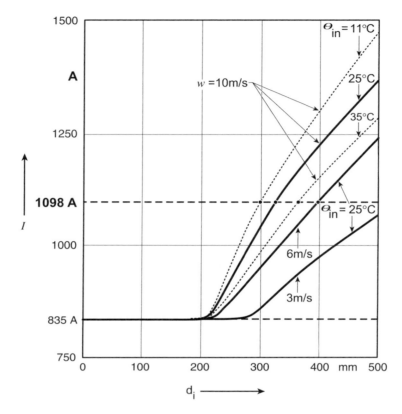

Figure 2-13   Continuous current rating as function of the inner diameter of the pipe (from Brakelmann and Anders, 2004b, with permission from IEEE Press).

- Suitable laying configurations (close, possibly bundled laying of cables cores; great laying depth)
- Parallel compensation conductors
- Encapsulating the cables in ferromagnetic material, for example, in steel pipes

Arranging steel plates in the cable trench seems to be problematic because of corrosion problems. Also, this method of shielding may not be very effective since the magnetic field near to the edges of the shielding plate can be significantly magnified (Pommerenke and Shen, 1993).

From the cable rating point of view, all shielding measures mentioned above reduce the transmission capacity of the cable by elevating their external thermal resistance and/or by creating additional losses. We will examine these effects below. We will also discuss alternative solutions that do not have this disadvantage and which can even lead to an increase of current ratings and to lower losses of the transmission system.

Example 2.5

We will continue the examination of the cable system discussed in Example 2.4. We recall that when the cables are laid at a depth of 1.0 m in flat formation with spacing between the center axes of 0.4 m, for uniform soil conditions and with 100% load factor, the cable has a rating of 1098 A. We will refer to this condition as the base case.

The results of the analysis are displayed in Table 2-8 (Brakelmann and Anders, 2004b). In this table, B_1 is the magnetic induction at 1 m height above the ground. W_{cab} are the power losses for different configurations. Row 4 corresponds to a PE pipe with 300 mm inner diameter, and row 5 corresponds to a steel pipe with 300 mm inner diameter. SPB and SB denote single-point and solid bonding of the sheaths, respectively.

For the base case conditions, the magnetic induction at a distance of 1 m above the ground (see Table 2-8, row 1) is $B_1 = 36.6$ µT (the magnetic induction at the soil surface is $B_1 = 131.2$ µT).

Arranging the cables in a touching triangular formation leads to a strongly reduced magnetic induction of $B_1 = 3.8$ µT if the sheaths are single-point bonded (Table 2-8, row 2), and of $B_1 = 3.1$ µT for solid bonding (Table 2-8, row 3), respectively. However, the current ratings are reduced by 15% down to 932 A or by 28% down to 791 A, respectively, in these cases.

An effective, but also expensive, method is to increase the laying depth of, example to 2.0 m. In this case, for the touching triangular configuration and single-point bonding, the magnetic induction becomes $B_1 = 1.5$ µT, whereas the current rating is reduced to 854 A.

Rows 6 to 11 in Table 2-8 summarize some results for the forced-air cooling scenario with two pipe-laying conditions. The nonmetallic pipe has an inner diameter of 300 mm (rows 6 to 9) and the steel pipe has an inner diameter of 330 mm (rows 10 and 11). The air inlet temperature θ_{in} and the air velocity w are used as parameters. All cases are based on a steady-state load current of 1098 A.

For the nonmetallic pipe, the magnetic induction becomes $B_1 = 4.5$ µT one meter above the ground, whereas it practically disappears for the ferromagnetic steel pipe with $B_1 \approx 0$ µT.

We can observe from Table 2-8 that lower air temperatures lead to lower losses, compared with the base-case conditions. The reductions reach about 10% of the value reported in row 1.

W_{cool} in Table 2-8 is that part of the cable losses that is absorbed by the air flow. If the heated-up air of the cooling system is used for heat recovery, for example, for the heating of buildings, this part of cable losses reduces the losses of the total system W_{tot}. For example, losses reported in row 9 of Table 2-8 constitute only about a quarter of the losses present in the base case.

Rows 10 and 11 in Table 2-8 show the corresponding results for the case of a steel pipe. For the yearly mean value of the inlet air temperature of 11°C, the cable losses are 111 W/m; that is, 32% greater than in the base-case conditions. If, however, the waste heat is recovered, the total losses can be reduced to approximately 26 W/m; that is, down to 31% of the losses in the base-case conditions (row 11).

Table 2-8 Continuous current rating and magnetic induction

Row	Configuration	μ_{in}, °C	w, m/s	I, A	B_1, μT	W_t, W/m	W_{cool}, W/m	W_{tot}, W/m
1	⊙ ⊙ ⊙ |0.4m|0.4m| SPB	—	—	1098	36.6	84.2	—	84.2
2	SPB	—	—	932	3.8	62.6	—	62.6
3	SB	—	—	791	3.1	64.5	—	64.5
4	$\mu_r = 1$ $\kappa = 0$ SPB	—	—	978	4.0	68.0	—	68.0
5	$\mu_r \gg 1$ $\kappa > 0$ SPB	—	—	835	0	69.0	—	69.0
With forced air circulation								
6	$d_i = 300$mm $l_0 = 400$m $\mu_r = 1; \kappa = 0$ SPB	25	6	1098	4.5	80.7	36.9	43.8
7	$d_i = 300$mm $l_0 = 400$m $\mu_r = 1; \kappa = 0$ SPB	25	10	1098	4.5	78.8	43.9	35.9
8	$d_i = 300$mm $l_0 = 400$m $\mu_r = 1; \kappa = 0$ SPB	11	6	1098	4.5	78.0	45.4	32.6
9	$d_i = 300$mm $l_0 = 400$m $\mu_r = 1; \kappa = 0$ SPB	11	10	1098	4.5	75.8	54.8	21.0
10	$d_i = 330$mm $l_0 = 400$m $\mu_r \gg 1; \kappa > 0$ SPB	25	10	1098	0	115.6	73.3	42.3
11	$d_i = 330$mm $l_0 = 400$m $\mu_r \gg 1; \kappa > 0$ SPB	11	10	1098	0	110.9	84.5	26.4

In summary, we can state that air cooling decreases the conductor temperatures and due to this (for the same current) the ohmic losses, accompanied by a reduced or even extinguished outer magnetic field. ∎

2.3 CONCLUDING REMARKS

From the information reported in this Chapter, it is quite clear that when a cable circuit crosses an area of unfavorable thermal conditions, derating of its current-carrying capacity is required. This derating can become very significant, even when the width of the unfavorable region is relatively small. If derating is not feasible, remedial actions should be taken, the simplest of which would be to make sure that the thermal properties of the soil in an unfavorable location match those outside the region.

Using Equation (2.40), derating curves similar to Figure 2-5 can be developed; the derating will be a function of the soil conditions and the size of the conductor. In order to derive such curves, some standard installation conditions would have to be assumed. A set of derating graphs could be developed depending on the width of the unfavorable region, its thermal properties, and also on the depth of burial. The amount of work involved would depend on the number of cases to be standardized.

In addition, diagrams similar to Figure 2-4 could be developed, showing the influence temperature as a function of the distance from the center of the crossing, to give guidance on the length of the area in which isolating boards, heat pipes, or special backfills should be applied to reduce the effect of unfavorable conditions.

Measures to reduce the magnetic field of underground cables normally lead to strong reductions of the current rating. We have shown in this chapter that this problem can be eliminated by laying the cables in a pipe that is cooled by free or forced convection of the ambient air. This system offers at the ends of its cooling sections a great part of the cable losses as utilizable heat. The heat recovery of these losses results in an essential improvement of the efficiency of power transmission, combined with a low or even disappearing outer magnetic field.

The expenses for such a cooling system seem to be acceptable, since

- No cooling stations are needed
- Current-dependent operation of fans by means of induction conductors is possible
- Techniques of microtunneling can be used

2.4 CHAPTER SUMMARY

The most important development presented in this chapter relates to the calculation of a derating factor for cables crossing unfavorable thermal environment. When a cable crosses from a normal design region, referred to with a superscript 2, to an unfavorable region 1 of length z_0 (the letter v is also used to denote region 1 or 2), the derating factor DF can be computed from

$$DF^2 = \frac{b - \sqrt{b^2 - a \cdot c}}{\alpha_T \cdot T_t^{(1)} \cdot (\theta_{max} - \theta_A^{(2)})}$$

where

$$T_t^{(v)} = T_1 + n \cdot (1 + \lambda_1) \cdot T_2 + n \cdot (1 + \lambda_1 + \lambda_2) \cdot (T_3 + T_4^{(v)})$$

$$T_r^{(v)} = T_1 + n \cdot (T_2 + T_3 + T_4^{(v)})$$

$$\gamma_v^2 = T_L \cdot (1 - \Delta W \cdot T_t^{(v)})/T_r^{(v)}$$

$$W_{ct} = W_{c0}(1 - \alpha_T\theta_0) \qquad \text{and} \qquad \Delta W = \alpha_T W_{c0}$$

$$\tau = \frac{1}{\cosh(\gamma_1 \cdot z_0) + \dfrac{\gamma_1}{\gamma_2} \sinh(\gamma_1 \cdot z_0)}$$

$$a = \alpha_T \cdot f_a \cdot T_t^{(1)} \cdot T_t^{(2)}$$

$$2b = \alpha_T \cdot (\theta_{max} - \theta_A^{(1)}) \cdot T_t^{(2)} + f_\alpha \cdot T_t^{(1)} - \tau \cdot [(g + \alpha_T\theta_A^{(2)}) \cdot T_t^{(1)} - (g + \alpha_T\theta_A^{(1)}) \cdot T_t^{(2)}]$$

$$c = \theta_{max} - \theta_A^{(1)} + \tau \cdot (\theta_A^{(1)} - \theta_A^{(2)})$$

$$f_\alpha = 1 + \alpha_T(\theta_{max} - \theta_0)$$

$$g = 1 - \alpha_T\theta_0$$

$$\theta_A^{(v)} = \Delta\theta_d^{(v)} + \theta_{amb}^{(v)}$$

with
α_T = conductor temperature coefficient of resistance (1/K)
W_{c0} = heat rate in the rated cable at the reference temperature θ_0 (usually 20°C) (W/m)
θ_{max} = maximum allowable conductor temperature (°C)
θ_{amb} = ambient temperature in respective regions (°C)
θ_d = temperature rise caused by dielectric losses in respective regions (K)

The remaining variables are the classical thermal resistances of cable components and the sheath and armor loss factors.

As γ depends upon the current in the rated cable, which is to be determined, an iterative solution is necessary, using as a first estimation of this current the rated current when the cable is located entirely in region 2.

REFERENCES

Anders, G.J. (1997), *"Rating of Electric Power Cables: Ampacity Computations for Transmission, Distribution and Industrial Applications,"* IEEE Press, New York, McGraw-Hill, New York (1998).

Anders, G.J., and Brakelmann, H. (1999a), "Cable Crossings—Derating Considerations. Part I—Derivation of the Derating Equations," *IEEE Trans. on Power Delivery,* Vol. PWRD-13, No.4, July 1999, pp. 709–714.

Anders, G.J., and Brakelmann, H. (1999b), "Cable Crossings—Derating Considerations. Part II—Example of Derivation of the Derating Curves," *IEEE Trans. on Power Delivery,* Vol. PWRD-13, No.4, July 1999, pp. 715–720.

Brakelman, H., and Anders, G.J. (2004b), "Increasing Ampacity for Cables by an Application of Ventilated Pipes," submitted for publication in *IEEE Transactions on Industry Applications.*

Brakelmann, H. (1980), "Lateralkühlung von Hochspannungskabeln," *etz-Archiv,* pp. 77–86

Brakelmann, H. (1985), *Belastbarkeiten der Energiekabe* 1, VDE-Verlag, Berlin.

Brakelmann, H. (1989), *Energietechnik programmiert,* VDE-Verlag, Berlin/Offenburg.

Brakelmann, H. (1995), "EMV-Maßnahmen für Drehstrom-Einleiterkabel," *El. Wirtsch.,* pp. 926–930.

Brakelmann, H. (1996), "Magnetfeldreduktion durch Zusatzleiter in Energiekabeltrassen," *El. wirtsch,* pp. 274–279.

Brakelmann, H. (1998) "Kompensationsleiter im Muffenbereich von Energiekabeln," *Bull. d. SEV,* pp. 31–35

Brakelmann H. (1999a), "Hydraulische Kaskadenschaltung für Bündelgekülte Einleiterkabel," *etz,* H15, pp. 48–51.

Brakelmann H. (1999b), "EMC Measures for Underground Power Cables," Paper BPT99-234-51 presented at the IEEE Power Tech '99 Conference, Budapest, Hungary, Aug. 29–Sept. 2, 1999.

Brakelmann, H. (1999c), "Reinforcement of Power Cables crossing Thermally Unfavourable Regions," *ETEP,* no.3, 1999.

Brakelmann, H. (2000), "Autarke Energieversorgung von Verbrauchern in Hochspannungskabeltrassen," *Bull. SEV 93,* H7, pp. 32–36.

Brakelman, H., and Anders, G.J. (2004a), "Increasing Ampacity for Cables Crossing Thermally Unfavorable Regions," accepted for publication in *IEEE Transactions on Power Delivery.*

Bucea, G., and Kent, H. (1998), "Shielding Techniques to Reduce Magnetic Fields Associated with Underground Power Cables," CIGRE-Report 21-201.

Haubrich, H.J. (1973), "Belüftete Kabelkanäle für unterirdische Hochleistungsübertragung," *ETZ-A,* pp. 147–152

Heinhold, L. (1987), *Kabel und Leitungen für Starkstrom,* Siemens AG, Berlin/München.

Holman, J.P. (1990), *Heat Transfer,* McGraw-Hill, New York.

IEC Standard. (1985), "Calculation of the Cyclic and Emergency Current Ratings of Cables. Part 1: Cyclic Rating Factor for Cables up to and Including 18/30 (36) kV," Publication 853-1.

IEC Standard. (1989), "Calculation of the Cyclic and Emergency Current Ratings of Cables. Part 2: Cyclic Rating Factor of Cables Greater than 18/30 (36) kV and Emergency Ratings for Cables of All Voltages," Publication 853-2.

IEC Standard 60287, Part 2-1 (1994), "Calculation of thermal resistances".

Ihle, C. (1977), "*Lüftungs- und Raumheizung,* Bd. 3," Werner-Verlag, Düsseldorf.

Incropera, F.P., and De Witt, D.P. (1990), *Introduction to Heat Transfer,* Wiley, New York.

Morgan, V.T. (1982), "The Thermal Rating of Overhead-Line Conductors, Part I. The Steady-State Thermal Model," Electric Power Systems Research, 5, pp. 119–139.

Ozisik, M.N. (1985), *Heat Transfer. A Basic Approach,* McGraw-Hill, Singapore.

Pommerenke, D., and Shen, J. (1993), "Elektromagnetische Verträglichkeit von Kabelnetzen," Annual report Inst. f.

Weedy, B.M. (1988), *Thermal Design of Underground Systems,* Wiley, Chichester, UK.

Williams, J.A. (1997), "Distributed Fiber Optic Temperature Monitoring Results on 69 kV XLPE Cable in Duct," IEEE PES, Insulated Conductor Committee, Minutes of the 101 Meeting, Scottsdale, Arizona, April 20–23, 1997, Appendix 7-D-1.

CHAPTER 3

Cable Crossings— Derating Considerations

Dangerously high interference temperatures can occur at points where cables cross external heat sources, even when the crossing occurs at 90°. For perpendicular and oblique crossings, these interference temperatures are usually ignored for distribution circuits, whereas for transmission cables, corrective actions in physical installation conditions are sometimes taken. Analytical solutions are almost never used to determine the effect of external heat sources on the ampacity of the rated cable. The main reason no computations are performed is an absence of either derating formulas or derating tables (curves) and not the lack of need. To fill this gap, an analytical solution for the computation of the derating factors has been developed and is presented in this chapter. The approach discussed in this chapter originated with the work of Brakelmann (1985), with further developments by Anders and Brakelmann, (1999a,b), Anders and Dorison (2004), and Dorison (2003). The solution is simple and accurate enough to be suitable for standardization purposes.

3.1 INTRODUCTION

A great variety of infrastructure can be found under the streets of congested urban areas. It is not uncommon to find power cables crossing steam pipes or other cables. When a rated cable is crossed by another heat source, its temperature will be higher than if it was isolated. As will be shown below, this additional conductor temperature rise may exceed 20°C. Therefore, the ampacity of the rated cable should be reduced accordingly. Ten and often 20% reduction may be required, as illustrated in the numerical example presented in this section. This is much larger than the reduction of up to 5% considered by some utilities. Many utilities do not derate cables crossed by other heat sources. This practice may lead to premature cable failures, as reported, for example, in Orton et al. (1996). An informal survey conducted by the author has clearly indicated a need to establish derating factors for cable crossings.

The temperature increase in the rated cable resulting from the presence of another heat source will depend on several parameters. The most important are:

Rating of Electric Power Cables in Unfavorable Thermal Environment. By George J. Anders **121**
ISBN 0-471-67909-7 © 2004 the Institute of Electrical and Electronics Engineers.

- The amount of heat dissipated by the crossing heat source
- The distance between rated cable and the heat source at the point of intersection
- The angle between the heat source and the cable (the more parallel the two are, the larger the influence of the heat source on the rated cable)
- The size of the rated cable

A very limited number of publications have dealt with cable crossings. Abdel Aziz et al. (1986) presented the thermal analysis of temperature distributions along the crossing regions of high-voltage cables. Analytical results for perpendicular crossings were developed and checked against measured values. The hot-spot temperature for the experimental setup of 66 kV and 220 kV cables was measured. This temperature at the 66 kV cable was found to be higher by about 25°C than the cable temperature far away from the crossing region. The corresponding value for the 220 kV cable was only 8°C higher. The authors also considered the effect of applying heating pipes and special backfill for hot-spot elimination. Imajo et al. (1986) described the evaluation of the thermal interference that power cables receive from steam pipes located in close vicinity. Experimental and analytical models were described. 90° crossings were considered.

A similar problem was discussed by Iwata et al. (1992). This paper describes a new type of antithermal-interference system consisting of long heat pipes. A very simple mathematical model for heating of the cable by the interfering source was presented for the 90° crossing. In the example presented in the paper, the cable conductor temperature rise at the intersection with a steam pipe with temperature of 150°C located 1.8 m below the cable reaches 22°C. The effect of the steam pipe disappears about 6 m from the crossing.

A theoretical model involving fourth-order differential equations was presented by Thevenon and Couquet (1999). The proposed solution involved an application of the finite element method. For practical applications, simple formulae are required that would allow calculation of the derating factors for many cable crossing situations that occur in practice.

In this chapter, we will examine the issues related to derating of cables caused by the presence of another heat source. Section 3.2 describes briefly the results of an informal survey whose aim was to find out the practices in various utilities with regard to cable crossings. Section 3.3 discusses modeling issues and outlines the work required to obtain derating factors. We will close this chapter by examining the computation required to obtain derating curves.

3.2 UTILITY PRACTICES

In order to find out how cable crossings are dealt with in electric utilities, an informal survey was conducted among several engineers active in the field of cable ratings. The following utilities in Canada have been contacted: Ontario Hydro, North York Hydro (Toronto), British Columbia Hydro, Manitoba Hydro, and New Brunswick Hydro. In addition, practices in England, Germany, Italy, the Netherlands, France, Belgium, Poland, and Sweden were investigated.

In general, all the contacted utilities are dealing with the subject of cable crossings in a similar way. For distribution installations (low and medium voltage), in a great majority of cases, no derating is considered at all. The most often-quoted reason was that the cables are rated very conservatively (e.g., high soil thermal resistivity or high ground ambient temperature are assumed). In some cases, for example at one European utility, special precautions are taken in the case when a distribution cable (0.4–24 kV) crosses a district heating pipe. In particular, the country's standards state that the cable should always be above the pipe, at a distance not less than 20 cm, and should be as perpendicular as possible. A heat-isolating plate 50 mm thick is required between the cable and the pipe system. In another utility, for cable ratings at the steam pipe crossings, it is assumed that the ambient temperature is higher by 5°C than the values used for isolated cables, which typically results in a 3% decrease in the cable rating. In several countries, a minimal distance is specified between the rated cable and the external heat source. This distance varies between 20 and 50 cm.

The situation is only slightly different when a high-voltage cable is involved. With the exception of France, no computations are performed to determine the actual effect of the cable crossing another heat source (EDF may conduct a three-dimensional finite-element analysis in some important cases). However, most utilities would take special actions to reduce the mutual heating effect. Several approaches have been applied:

- Isolation of the heat source from the cable
- Use of heat-transfer pipes
- Use of thermal backfills in an increased trench cross section
- Use of insulating liquid circulation
- Use of larger cable
- Installation of ventilating manholes at crossing points
- At steam heat crossings, utilities sometimes build a styrofoam (closed cell) box around the steam heat pipe, then fill the box with Gilsulate, an underground insulating material used by steam heating companies
- In one instance, for 90 degree crossings, the utility sometimes would follow a very rough approximation by doing ratings assuming parallel laying, but at twice the actual vertical separation distance

Finally, several engineers indicated that the main reason no computations are performed is an absence of either derating formulas or derating tables (curves) and not the lack of need. All indicated that it would be very beneficial if a simple derating method were available. Such a method is presented in this chapter.

3.3   DERATING FACTOR

A simple model will be developed to account for the effect of the presence of crossing heat sources, taking into account all the important factors mentioned in the Section 3.1.

To illustrate the development of the derating equation, consider two cables crossing as shown in Figure 3-1.

Suppose that cable 2 is to be derated because of the heat W_1 (W/m) generated by cable 1. Let $\Delta\theta_{max} - \Delta\theta_d$ be the maximum permissible conductor temperature rise of the conductor of cable 2 for current-dependent losses. Because of the mutual heating, the maximum allowable heat rate in cable 2 will be reduced as shown below:

$$W_{2r} = \frac{\Delta\theta_{max} - \Delta\theta_d - W_1 T_{12}}{T_t} \tag{3.1}$$

where
W_{2r} = reduced maximum allowable losses in conductor of cable 2 (W/m)
$\Delta\theta_d$ = conductor temperature rise caused by dielectric losses, (K)

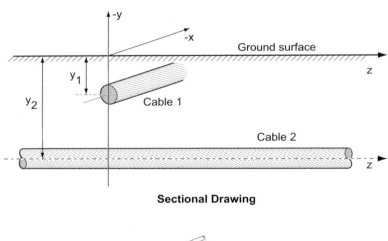

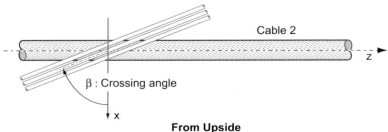

Figure 3-1   Crossing of two cables at an angle $90° - \beta$.

T_{12} = the mutual thermal resistance between cables 1 and 2 (K · m/W)
T_t = total thermal resistance of cable 2 (K · m/W) with

$$T_t = T_1 + n \cdot [(1 + \lambda_1)T_2 + (1 + \lambda_1 + \lambda_2)(T_3 + T_4)] \tag{3.2}$$

The notation is the same as in IEC 60287 (1994) and discussed in Chapter 1. Since, for the isolated cable

$$W_2 = \frac{\Delta\theta_{max} - \Delta\theta_d}{T_t} \tag{3.3}$$

The reduced rating I_{2r} of cable 2 can now be obtained by combining Equations (3.1) and (3.3) as

$$I_{2r} = I_2 \sqrt{1 - \frac{W_1 T_{12}}{\Delta\theta_{max} - \Delta\theta_d}} \tag{3.4}$$

where I_2 is the rating of the isolated cable 2.

The term $W_1 T_{12}$ in Equation (3.4) represents the conductor temperature in cable 2 at point $z = 0$ caused by heat generated in cable 1. Denoting this temperature rise by $\Delta\theta(0)$, the derating factor in Equation (3.4) can be written as

$$DF = \sqrt{1 - \frac{\Delta\theta(0)}{\Delta\theta_{max} - \Delta\theta_d}} \tag{3.5}$$

Thus, the derating factor can be obtained from Equation (3.4) when the mutual thermal resistance T_{12} is known. Computation of this resistance is discussed next.

3.4 TEMPERATURE DISTRIBUTION ALONG THE RATED CABLE AND THE MUTUAL THERMAL RESISTANCE

In this section, we will develop a differential equation describing the temperature distribution along the rated cable, taking into account the presence of the external heat sources. The procedure is based on the discretization of the cable thermal network and is similar to the one employed in Section 2.1.2. Next, we will develop a solution to this equation that will lead to an expression for the mutual thermal resistance between the rated cable and the external heat source [Equation (3.32)].

3.4.1 Single External Heat Source

Let us first assume that there is one external heat source, for example a steam pipe, crossing the route of the rated cable. Referring to Figure 3-1, the temperature rise in the conductor of cable 2 caused by the heat rate W_1 generated by the heat source 1 can be obtained by applying Kennelly's principle [see Equation (9.18) in Anders (1997)], as

$$\Delta\theta_{u12}(z, y = y_2) = \frac{\rho W_1}{4\pi} \ln \frac{(y_1 + y_2)^2 + (z \cdot \cos \beta)^2}{(y_1 - y_2)^2 + (z \cdot \cos \beta)^2} \quad (3.6)$$

where

ρ = thermal resistivity of the soil, (K · m/W)

$\Delta\theta_{u12}$ = the temperature rise at the location of the conductor of cable 2 caused by the heat generated by the heat source 1, uninfluenced (subscript "u") by any effect of cable 2 (K)

To find the temperature at any point of the rated cable, taking into account its longitudinal heat flux and the interfering temperature given by Equation (3.6), we will discretize the length of the rated circuit as shown in Figure 3-2 (Brakelmann, 1985).

In Figure 3-2, the thermal resistances with subscripts 1, 2, 3, and 4 are the characteristic quantities for the cable as defined in the IEC 60287 (1994). ΔT_L [K/ (W-m)] is the longitudinal thermal resistance of the conductor and its value per unit length is computed from

$$T_L = \frac{\rho_c}{A} \quad (3.7)$$

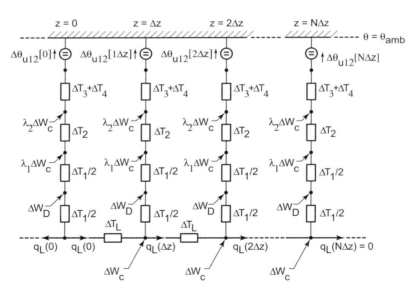

Figure 3-2 Discretized length $N \cdot \Delta z$ of the rated cable with the thermal resistances ΔT corresponding to the length Δz of the discretization interval. W_c are the conductor joule losses and q_L is the conductor longitudinal heat flux (from Brakelmann and Anders, 1999a, with permission from IEEE Press).

where A (m²) is the cross-sectional area of the conductor. ρ_c is the thermal resistivity of the material (K · m/W).

Similar to the case of a cable crossing unfavorable regions, the network representation in Figure 3-2 allows us to consider the heat flow in two directions—radial and longitudinal—in what essentially is a three-dimensional problem.

In the following development leading to the differential Equation (3.20) describing the temperature distribution along the cable route, we will consider again a small element of the cable with length dz, as shown in Figure 3-3 (Dorison, 2003).

In the steady state, from the energy conservation principle we have

$$W_c(z) \cdot \Delta z + W_L \cdot (z - \Delta z) = W_r(z) \cdot \Delta z + W_L(z) \tag{3.8}$$

From Equation (3.8), we obtain

$$(W_r - W_c) + \frac{\partial W_L}{\partial z} = 0 \tag{3.9}$$

The derivative in Equation (3.9) can be computed from the Fourier's Law:

$$W_L = -\frac{1}{T_L}\frac{d\theta}{dz} \tag{3.10}$$

Considering the heat flow in the radial direction, we have

$$\theta = \theta_{amb} + \Delta\theta_{u12} + \Delta\theta_d + W_c \cdot T_t + (W_r - W_c) \cdot T_r \tag{3.11}$$

where

$$T_r = T_1 + n \cdot (T_2 + T_3 + T_4) \tag{3.12}$$

θ_{amb} = ambient temperature (K)
W_c = the heat losses generated by the conductor current, (W/m)

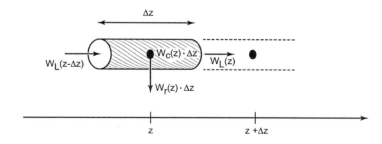

Figure 3-3 An element of the conductor with the length of dz.

The conductor joule losses W_c are temperature dependent. This dependence can be written as

$$W_c = W_{c0}[1 + \alpha_T(\theta - \theta_0)] = W_{ct} + \Delta W \cdot \theta \tag{3.13}$$

with

$$W_{ct} = W_{c0}(1 + \alpha_T\theta_0) \quad \text{and} \quad \Delta W = \alpha_T W_{c0} \tag{3.14}$$

where
α_T = conductor temperature coefficient of resistance (1/K)
W_{c0} = heat rate in the rated cable at the reference temperature θ_0 (usually 20°C) (W/m)

In order to express the losses in terms of the rated current, we will introduce the following notation:

$$\Delta W_0 = \frac{R \cdot \alpha_T \cdot I^2}{1 + \alpha_T \cdot (\theta_{max} - \theta_0)} \tag{3.15}$$

where
R = conductor resistance at maximum operating temperature, ohm/m
I = rated current, A

We observe that $\Delta W_0 = \alpha_T W_{c0} = \Delta W$ for the rated current. Since the reduced current is related to the rated current I by the derating factor in Equation (3.5), we have

$$\Delta W = \Delta W_0 \cdot \left[1 - \frac{\Delta\theta(0)}{\Delta\theta_{max} - \Delta\theta_d} \right] \tag{3.16}$$

The first estimate of $\Delta\theta(0)$ in Equation (3.16) is obtained from Equation (3.6) as

$$\Delta\theta(0) = W_1 \cdot \frac{\rho}{4\pi} \cdot \ln \frac{(y_1 + y_2)^2}{(y_1 - y_2)^2} \tag{3.17}$$

Substituting Equations (3.9), (3.10), and (3.13) into (3.11), we obtain the following differential equation describing conductor temperature:

$$\theta - \frac{1}{\gamma^2} \cdot \frac{\partial^2\theta}{\partial z^2} = \frac{\Delta\theta_{u12} + \theta_{amb} + \Delta\theta_d + W_{ct} \cdot T_t}{(1 - \Delta W \cdot T_t)} = \frac{\theta_{amb} + \Delta\theta_d + W_{ct} \cdot T_t}{1 - \Delta W \cdot T_t} \tag{3.18}$$

$$+ \frac{\Delta\theta_{u12}}{1 - \Delta W \cdot T_t} = \theta_\infty + \frac{\Delta\theta_{u12}}{1 - \Delta W \cdot T_t}$$

where

$$\gamma^2 = (1 - \Delta W \cdot T_t) \cdot \frac{T_L}{T_r} \tag{3.19}$$

Substituting

$$\Delta\theta = (1 - \Delta W \cdot T_t)(\theta - \theta_\infty)$$

$$\Delta\theta_u = \Delta\theta_{u12}$$

we finally obtain

$$\Delta\theta - \frac{1}{\gamma^2}\frac{d^2\Delta\theta}{dz^2} = \Delta\theta_u(z) \qquad (3.20)$$

The solution of this equation yields the mutual thermal resistance given by Equation (3.32). This equation has an identical form to Equation (2.16) but we will employ a different solution procedure from the one used in Section 2.1.3. The applied procedure is described below. To solve Equation (3.20), we will discretize the temperature function $\Delta\theta$ as shown in Figure 3-4 so that in each interval $v\Delta z$, the value of $\Delta\theta_u(z)$ is constant and equal to $\Delta\theta_{uv}$. Under this assumption, the solution in the interval $v\Delta z$ can be written in a form of the following hyperbolic function (Brakelmann, 1985):

$$\Delta\theta^{(v)} = \Delta\theta_{uv} + A_v \sinh(\gamma \cdot z) + B_v \cosh(\gamma \cdot z) \qquad (3.21)$$

The constants A_v and B_v are obtained taking into account the following boundary conditions:

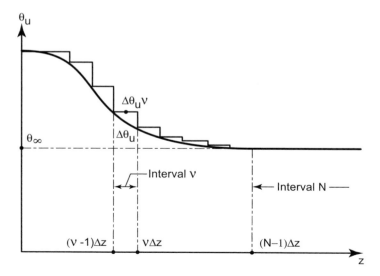

Figure 3-4 Discretization of the temperature curve along the z axis (from Anders and Brakelmann, 1999a, with permission from IEEE Press).

1. The longitudinal heat flux, and hence the heat transfer rate (the heat transfer rate is obtained by multiplying heat flux by the area) at the center of the crossing is equal to zero. Hence, from Equation (3.10)

$$W_L(0) = -A_1 \frac{\gamma}{T_L} = 0 \Rightarrow A_1 = 0 \tag{3.22}$$

2. The temperature is a continuous function when it crosses the interval boundary; that is,

$$\Delta\theta_{uv} + A_v \sinh(\gamma \cdot v \cdot z) + B_v \cosh(\gamma \cdot v \cdot \Delta z)$$
$$= \Delta\theta_{u(v+1)} + A_{v+1} \sinh[\gamma \cdot (v+1) \cdot \Delta z] + B_{v+1} \cosh[\gamma \cdot (v+1) \cdot \Delta z] \tag{3.23}$$

3. The heat flux is a continuous function when it crosses the interval boundary; that is,

$$A_v \cosh(\gamma \cdot v \cdot \Delta z) + B_v \sinh(\gamma \cdot v \cdot \Delta z)$$
$$= A_{v+1} \cosh[\gamma \cdot (v+1) \cdot \Delta z] + B_{v+1} \sinh[\gamma \cdot (v+1) \cdot \Delta z] \tag{3.24}$$

with the solution in the first ($v = 1$) and the last ($v = N$) intervals given by

$$\Delta\theta^{(1)} = \Delta\theta_{u1} + B_1 \cdot \cosh(\gamma \cdot z) \tag{3.25}$$

$$\Delta\theta^{(N)} = B_N \cdot e^{-\gamma z} \tag{3.26}$$

From Equations (3.23)–(3.24), we obtain the following recursive relationships

$$A_v = A_{v-1} - (\Delta\theta_{u(v-1)} - \Delta\theta_{uv}) \cdot \sinh[\gamma \cdot (v-1) \cdot \Delta z] \tag{3.27}$$

$$B_1 = -\Delta\theta_{u1} + \sum_{v=1}^{N} \Delta\theta_{uv}(e^{-(v-1)\cdot\gamma\Delta z} - e^{-v\cdot\gamma\cdot\Delta z}) + \Delta\theta_{uN}e^{-N\cdot\gamma\cdot\Delta z}$$
$$B_v = B_{v-1} + (\Delta\theta_{u(v-1)} - \Delta\theta_{uv}) \cosh[(v-1) \cdot \gamma \cdot \Delta z] \text{ for } v > 1 \tag{3.28}$$

Of particular interest will be the temperature at the crossing point where $z = 0$. From Equations (3.25) and (3.28) we obtain

$$\Delta\theta^{(1)}(z=0) = (e^{\gamma \cdot \Delta z} - 1) \sum_{v=1}^{N} \Delta\theta_{uv}e^{-v\cdot\gamma\cdot\Delta z} \tag{3.29}$$

with

$$\Delta\theta_{uN}e^{-N\cdot\gamma\cdot\Delta z} \approx 0 \tag{3.30}$$

The incremental temperature rise of the rated cable at the point of intersection is equal to

$$\Delta\theta(0) = W_1 T_{12} \tag{3.31}$$

Thus, the mutual thermal resistance at the point of intersection can be obtained from Equation (3.31) by substituting Equation (3.6) into Equation (3.29):

$$T_{12} = \frac{\rho(e^{\gamma \cdot \Delta z} - 1)}{4\pi} \sum_{\upsilon=1}^{N} e^{-\upsilon \cdot \gamma \cdot \Delta z} \ln \frac{(y_1 + y_2)^2 + (\upsilon \cdot \Delta z \cdot \cos \beta)^2}{(y_1 - y_2)^2 + (\upsilon \cdot \Delta z \cdot \cos \beta)^2} \tag{3.32}$$

From Equations (3.32) and (3.19), we can observe that the mutual thermal resistance T_{12} is a function of the heat rate W_2. Therefore, an iterative procedure has to be used to evaluate this resistance in conjunction with Equation (3.4). Alternatively, a conservative estimate of the reduced current in cable 2 can be obtained by substituting the numerator in Equation (3.4) by the unaltered temperature $\Delta\theta_{u12}$ computed from Equation (3.6) at the point of intersection ($z = 0$); that is,

$$I_{2r \min} = I_2 \sqrt{1 - \frac{\Delta\theta_{u12}(0)}{\Delta\theta_{\max} - \Delta\theta_d}} \tag{3.33}$$

If the external heat source is another cable, the most pessimistic result is obtained when the temperature $\Delta\theta_{u12}$ is computed with W_1 corresponding to the ampacity of the external circuit being isolated.

3.4.2 Heating by a Steam Pipe

Equation (3.4) can also be used to derate a cable being crossed by a steam pipe. To illustrate the procedure, consider the installation shown in Figure 3-5.

The heat generated by the pipe is a function of the surface temperature of the pipe:

$$W_p = \frac{\theta_w - \theta_R}{T_R} \tag{3.34}$$

where T_R is the thermal resistance of the pipe insulation and the temperatures are shown in Figure 3-5. The pipe heat rate is influenced by the crossing cable. Therefore, in the first step, the temperature θ_R is determined from

$$\theta_R = W_p T_{4R} + \Delta\theta_{u21} + \Delta\theta_{amb} \tag{3.35}$$

where $\Delta\theta_{u21}$ is the influence temperature of the cable on the pipe, T_{4R} is the pipe external thermal resistance, and W_p is an assumed initial value of the heat generated in the pipe.

Equations (3.34) and (3.35) give the heat generated in the pipe, and this value can be used in Equation (3.4).

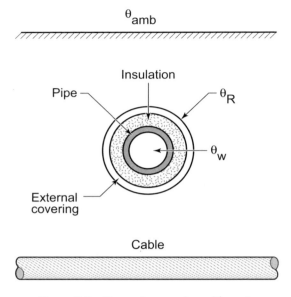

Figure 3-5   Steam pipe crossing cable route.

Example 3.1

Let us consider cable model No. 5. This circuit crosses a steam pipe at $60°$ ($\beta = 30°$) with the following characteristics. The steam temperature is $\theta_w = 150°C$. The thermal resistance of the pipe insulation is equal to 1.0 K $\cdot$ m/W. The outer diameter of the pipe insulation is 0.5 m and the center of the pipe is located 1 m below the ground surface (80 cm above the cables).

The isolated circuit No. 5 has ampacity of 1365 A. The parameters required for the derating calculations (Table 3-1) are taken from Table A1 in the Appendix.

Table 3-1   Values for numerical example 3.1

Cable ampacity	I (A)	1365
Conductor resistance at θ_{max}	R (ohm/km)	0.0126
Concentric/skid wires loss factor	λ_1	0.150
Pipe loss factor	λ_2	0.0
Thermal resistance of insulation	T_1 (K $\cdot$ m/W)	0.579
Thermal resistance of bedding	T_2 (K $\cdot$ m/W)	0.0
Thermal resistance of jacket/pipe covering	T_3 (K $\cdot$ m/W)	0.056
External thermal resistance 100% LF	T_4 (K $\cdot$ m/W)	1.276
Losses in the conductor	W_c (W/m)	23.4
Dielectric losses	W_d (W/m)	6.53
Total joule losses per cable	W_I (W/m)	26.92
Total losses per cable	W_t (W/m)	33.45

We will start the analysis by calculating the external thermal resistance of the pipe:

$$T_{4R} = \frac{\rho}{2\pi} \ln \frac{4L}{D} = \frac{1}{2\pi} \ln \frac{4 \cdot 1}{0.5} = 0.331 \text{ K} \cdot \text{m/W}$$

The influence of a single cable on the pipe is obtained from Equation (3.6) with $z = 0$. Assuming that the influence of all three cables is the same, we obtain

$$\Delta\theta_{u12}(z, y = y_2) = \frac{\rho W_1}{4\pi} \ln \frac{(y_1 + y_2)^2 + (z \cdot \cos \beta)^2}{(y_1 - y_2)^2 + (z \cdot \cos \beta)^2} = \frac{1 \cdot 3 \cdot 33.45}{4\pi} \ln \frac{(1.8 + 1.0)^2}{(1.8 - 1.0)^2}$$

$$= 20.0°C$$

Let us initially assume that the pipe generates heat at the rate of 200 W/m. We can now compute the pipe surface temperature as

$$\theta_R = W_p T_{4R} + \Delta\theta_{u21} + \theta_{amb} = 200 \cdot 0.331 + 20 + 25 = 111.2°C$$

We can now verify the amount of heat generated by the pipe from Equation (3.34):

$$W_p = \frac{\theta_w - \theta_R}{T_R} = \frac{150 - 111.2}{1} = 38.8 \text{ W/m}$$

Substituting this value again to Equation (3.35), we obtain the revised value of the pipe temperature equal to 57.8°C. The process converges after four iterations with the final values of $W_p = 78.9$ W/m and $\theta_R = 71.1°C$.

Next, we will compute the mutual thermal resistance of the pipe and the middle cable. The longitudinal thermal resistance is obtained from Equation (3.7) as

$$T_L = \frac{\rho}{A} = \frac{0.0026}{0.002} = 1.3 \text{ K/m-W}$$

The combined thermal resistances are obtained from Equations (3.2) and (3.12):

$$T_t = T_1 + n(1 + \lambda_1)T_2 + n(1 + \lambda_1 + \lambda_2)(T_3 + T_4) = 0.579 + 1 \cdot 1.15 \cdot (0.056 + 1.276)$$

$$= 2.11 \text{ K} \cdot \text{m/W}$$

$$T_r = T_1 + n(T_2 + T_3 + T_4) = 0.579 + 1 \cdot (0.056 + 1.276) = 1.91 \text{ K} \cdot \text{m/W}$$

The loss values are now computed from Equations (3.14), (3.15), and (3.16). The value of ΔW_0 is obtained from

$$\Delta W_0 = \frac{\alpha_T W_c}{1 + \alpha_T(\theta_{max} - 20)} = \frac{0.00393 \cdot 23.4}{1 + 0.00393 \cdot 65} = 0.073 \text{ W/m}$$

The first estimate of $\Delta\theta(0)$ is obtained from Equation (3.17) as

$$\Delta\theta(0) = W_1 \cdot \frac{\rho}{4\pi} \cdot \ln\frac{(y_1 + y_2)^2}{(y_1 - y_2)^2} = 78.9 \cdot \frac{1.0}{4\pi} \ln\frac{(1.8 + 1)^2}{(1.8 - 1)^2} = 15.7°C$$

With the temperature reduction caused by the dielectric losses equal in this case to 10.6°C, from Equation (3.16), we have

$$\Delta W = \Delta W_0 \cdot \left[1 - \frac{\Delta\theta(0)}{\Delta\theta_{max} - \Delta\theta_d}\right] = 0.073 \cdot \left[1 - \frac{15.7}{85 - 25 - 10.6}\right] = 0.050$$

From Equation (3.19), we have

$$\gamma = \sqrt{(1 - \Delta W \cdot T_I) \cdot \frac{T_L}{T_r}} = \sqrt{(1 - 0.05 \cdot 2.11) \cdot \frac{1.3}{1.91}} = 0.78 \ 1/m$$

Assuming $\Delta z = 0.01$ and $z_{max} = 6$ m, the derating factor equals to 0.88, with maximum current of 1205 A. If the cable is derated to 1205 A, the amount of heat generated by the cable circuit is decreased and the whole computational procedure should be repeated. ∎

3.4.3 Multiple Crossing Heat Sources

Equation (3.32) can be generalized for several cables crossing the rated cable by applying a superposition principle. In order to make this generalization, we must assume that the point $z = 0$ is the place where the temperature of the rated cable reaches its maximum.

Let the rated cable have the designation r and let z_r be the z-coordinate of the hottest point in cable r. Then, for any other cable k ($z = z_k$) away from the point $z = z_r$ we have

$$T_{kr} = \frac{\rho \cdot (e^{\gamma_r \cdot \Delta z} - 1)}{4\pi} + \sum_{v=1}^{N} e^{-v \cdot \gamma_r \cdot \Delta z} \ln\left[\frac{(y_k + y_r)^2 + [(z_r - z_k + v \cdot \Delta z) \cdot \cos\beta]^2}{(y_k - y_r)^2 + [(z_r - z_k + v \cdot \Delta z) \cdot \cos\beta]^2}\right] \quad (3.36)$$

If we consider only two cables, the maximum temperature may be reached somewhere between them and not at the two crossing points. In this case, Equation (3.36) is not really valid but the temperatures at various values of z can still be computed by taking a general solution of Equation (3.18).

Example 3.2

We will consider a crossing of a 10 kV XLPE cable circuit (cable No. 1) and a 145 kV three-core armored cable (cable No. 2). The cables and their characteristics are described in the Appendix and the summary of the computed parameters are given in Table 3-2. However, in this example, we will adopt slightly different

Table 3-2 Values for numerical example 3.2

		Computed parameters (hottest cable)	
Cable ampacity	I (A)	631	871
Conductor resistance at θ_{max}	R (ohm/km)	0.0781	0.033
Concentric/skid wires loss factor	λ_1	0.090	0.121
Armor loss factor	λ_2		0.206
Thermal resistance of insulation	T_1 (K · m/W)	0.214	0.754
Thermal resistance between sheath and armor	T_2 (K · m/W)		0.040
Thermal resistance of jacket	T_3 (K · m/W)	0.104	0.047
External thermal resistance 100% LF	T_4 (K · m/W)	1.621	0.385
Losses in the conductor	W_c (W/m)	31.07	24.92
Dielectric losses	W_d (W/m)	0	0
Total joule losses per cable	W_I (W/m)	33.82	33.11
Total losses per cable	W_t (W/m)	33.82	33.11

laying conditions for both cables. The soil thermal resistivity is equal to 0.8 K · m/W and ambient soil temperature is 25°C. Some parameters of both cables had to be recomputed because of the laying conditions are different than those used in Appendix A.

We will compute derating factors for both circuits. At the start of the process, when calculating the derating factor for the three-core cable, we will assume that the 10 kV circuit operates at its maximum power, corresponding to the case when this circuit is isolated. Similarly, when computing the derating factor for the 10 kV circuit, we will assume that the three-core cable is fully loaded, as in the isolated case.

Perpendicular Crossing. We will consider a 90° crossing; that is, ($\beta = 0$). We will assume $\Delta z = 0.01$ and $z_{max} = 6$ m, which results in $N = z_{max}/\Delta z = 600$.

We will investigate the effect of the three-core cable on the center (the hottest) cable of the 10 kV circuit. The effect of the 10 kV circuit on the three-core cable has to include the effect of all three cables. The coordinates are as shown in Figure 3-6, with the y-axis passing through the center cable. Three values of T_{12} will be required from the three cables of the 10 kV circuit, as described below.

We will start with the derating calculations for the 10 kV circuit by assuming that the three-core cable is loaded to 871 A. The loss values are now computed from Equations (3.14), (3.15), and (3.16). The value of ΔW_0 is obtained from

$$\Delta W_0 = \frac{\alpha_T W_c}{1 + \alpha_T(\theta_{max} - 20)} = \frac{0.00393 \cdot 0.0781 \cdot 10^{-3} \cdot 631^2}{1 + 0.00393 \cdot (90 - 20)} = 9.58 \cdot 10^{-2} \text{ W/m}$$

The first estimate of $\Delta\theta(0)$ is obtained from Equation (3.17) as

$$\Delta\theta(0) = W_1 \cdot \frac{\rho}{4\pi} \cdot \ln\frac{(y_1 + y_2)^2}{(y_1 - y_2)^2} = \frac{0.8 \cdot 3 \cdot 33.11}{4 \cdot \pi} \cdot \ln\left[\frac{(1.20 + 0.9)^2}{(1.20 - 0.9)^2}\right] = 24.6°C$$

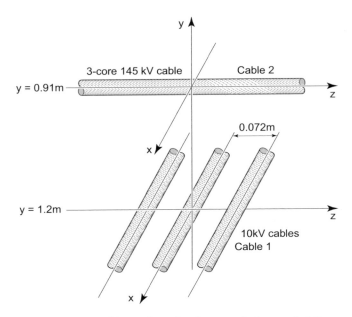

Figure 3-6 Cable configuration for numerical example 3.2.

The total thermal resistances are obtained from Equations (3.2) and (3.12) as

$$T_t = T_1 + n(1 + \lambda_1)T_2 + n(1 + \lambda_1 + \lambda_2)(T_3 + T_4) = 0.214 + 1.09 \cdot (0.104 + 1.621)$$
$$= 2.091 \text{ K} \cdot \text{m/W}$$

$$T_r = T_1 + n(T_2 + T_3 + T_4) = 0.214 + 0.104 + 1.621 = 1.939 \text{ K} \cdot \text{m/W}$$

The longitudinal thermal resistance of the conductor is (other, parallel paths for the longitudinal heat transfer are neglected here; see, however, Section 3.5)

$$T_{L1} = \frac{\rho}{A} = \frac{0.0026}{0.0003} = 8.67 \text{ K/(m-W)}$$

From Equation (3.16), we have

$$\Delta W = \Delta W_0 \cdot \left[1 - \frac{\Delta \theta(0)}{\Delta \theta_{\max} - \Delta \theta_d} \right] = 9.58 \cdot 10^{-2} \cdot \left(1 - \frac{24.6}{65} \right) = 0.060 \text{ W/m}$$

The attenuation factor is obtained from Equation (3.19):

$$\gamma = \sqrt{(1 - \Delta W \cdot T_t) \cdot \frac{T_L}{T_r}} = \sqrt{\frac{(1 - 0.060 \cdot 2.091) \cdot 8.67}{1.939}} = 1.98 \text{ 1/m}$$

We can now compute the mutual thermal resistance from Equation (3.32):

$$T_{12} = \frac{\rho(e^{\gamma \Delta z} - 1)}{4\pi} \sum_{v=1}^{N} e^{-v \cdot \gamma \cdot \Delta z} \ln \frac{(y_1 + y_2)^2 + (v \cdot \Delta z \cdot \cos \beta)^2}{(y_1 - y_2)^2 + (v \cdot \Delta z \cdot \cos \beta)^2}$$

$$= \frac{0.8(e^{1.98 \cdot 0.01} - 1)}{4\pi} \cdot \left[\sum_{v=1}^{600} e^{-v \cdot 1.98 \cdot 0.01} \ln \frac{(0.9 + 1.2)^2 + (0.01v)^2}{(0.9 - 1.2)^2 + (0.01v)^2} \right] = 0.179 \text{ K} \cdot \text{m/W}$$

The derating factor after the first iteration is obtained from Equations (3.5) and (3.31):

$$DF = \sqrt{1 - \frac{\Delta \theta(0)}{\Delta \theta_{max} - \Delta \theta_d}} = \sqrt{1 - \frac{99.33 \cdot 0.179}{90 - 25}} = 0.85$$

Similar calculations are now repeated for the three-core cable with the current in the 10 kV cable fixed at 631 A. The calculations are summarized in Table 3-3.

Table 3-3 Calculation of the derating factor for the three-core cable (cable type: 800 mm², 145 kV)

Characteristics		
Longitudinal thermal resistances of the conductors T_L	K/(m-W)	$0.0026/800 \cdot 10^{-6} = 3.25$
T_r	K · m/W/W	$0.754 + 3 \cdot [0.04 + 0.047 + 0.385] = 2.17$
T_t	K · m /W	$0.754 + 3 \cdot [1.121 \cdot 0.04 + 1.327 \cdot (0.047 + 0.385)] = 2.61$
$\Delta \theta_{max}$	°C	$90 - 25 = 65$
$\Delta \theta_d$	°C	0
ΔW_0	W/m	$\dfrac{0.033 \cdot 10^{-3} \cdot 0.00393 \cdot 870^2}{1 + 0.00393 \cdot (90 - 20)} = 7.68 \cdot 10^{-2}$
Computing rating factor		With : $\Delta z = 0.01$ m, $N - N > 500$
First estimate of $\Delta \theta$	°C	$\dfrac{0.8 \cdot 3 \cdot 33.82}{4 \cdot \pi} \cdot \left\{ \ln \left[\dfrac{(1.20 + 0.9)^2}{(1.20 - 0.9)^2} \right] + 2 \right.$ $\left. \cdot \ln \left[\dfrac{(1.20 + 0.9)^2 + 0.072^2}{(1.20 - 0.9)^2 + 0.072^2} \right] \right\} = 24.9$
First estimate of ΔW		$7.68 \cdot 10^{-2} \cdot \left(1 - \dfrac{24.9}{65} \right) = 0.047$
First estimate of γ		$\sqrt{\dfrac{(1 - 0.047 \cdot 2.61) \cdot 3.25}{2.17}} = 1.15$
Final estimate of $\Delta \theta$ (2nd iteration)	°C	14.9
Derating factor DF		$\sqrt{1 - \dfrac{14.9}{65}} = 0.88$

Simultaneous Rating of the Two Links. Six iterations are necessary to get the derating factors of the 2 links, when taking into account mutual thermal effects. The final result is as follows:

Rating of the two links	
Cable type	Rating factor
300 mm^2 XLPE 10 kV	0.88
800 mm^2 145 kV	0.91

The Influence of the Crossing Angle and the Distance between Circuits. The derating factors will, of course, be affected by the value of the crossing angle. Figure 3-7 shows this relation for both cables. We recall that $\beta = 90°$ represents the parallel route of both circuits.

From Figure 3-7, we can observe that the decrease of the derating factor is more pronounced for the high-voltage cable as the circuits become more parallel. For the 145 kV cable, the derating factor decreases from 0.91 for a perpendicular crossing to 0.85 when the circuits are parallel. A conservative practice, used by many utilities in cases like this, would be to assume that the cables are parallel. For parallel laying, the ampacity of the high-voltage cable is 736 A. Thus, when this conservative assumption is used, the conventional approach would unnecessarily decrease the ampacity of this cable by about 52 A or about 7%.

The temperature influence of the external heat source on the rated cable will depend on both the crossing angle and the distance between the circuits. Figures 3-8 and 3-9 illustrate the dependence on the crossing angle for both cables.

From Figures 3-8 and 3-9, it is apparent that the influence of the distribution circuit on the high-voltage cable disappears at a distance exceeding 6 m for crossings

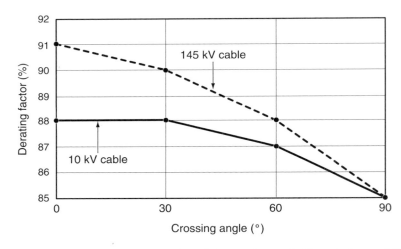

Figure 3-7 Derating factors as a function of the crossing angle (from Anders and Brakelmann, 1999b, with permission from IEEE Press).

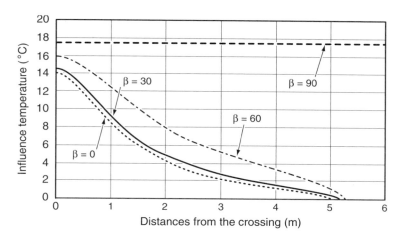

Figure 3-8 Temperature influence on the 10 kV cable as a function of the crossing angle and the distance from the crossing.

with angles smaller than 60°. On the other hand, for the 10 kV cable, this influence extends to about 5 m.

As expected, the more parallel the cables are, the higher will the influence temperature be. The influence temperature can be quite high at the point of intersection, even for a perpendicular crossing, and reaches 14°C in our example for the 10 kV cable (this can be compared with the 17.5°C influence when the circuits are parallel).

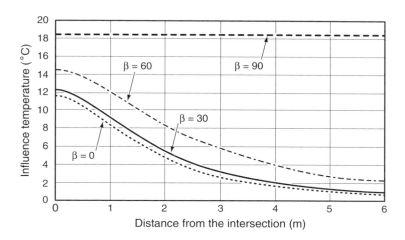

Figure 3-9 Temperature influence on the 145 kV cable as a function of the crossing angle and the distance from the crossing (from Anders and Brakelmann, 1999b, with permission from IEEE Press).

The dependence of the influence temperature on the 10 kV cable on the distance between the cables is shown in Figure 3-10. In this case, we increased the depth of the distribution circuit to 1.5 m, thus doubling the separation of the circuits to 0.6 m with the crossing angle $\beta = 0°$.

The vertical distance between the circuits plays a smaller role as far as the variation in the influence temperature is concerned. It is interesting to note that this influence temperature is almost identical for both separations at a distance greater than about 1 m from the intersection.   ■

3.5   CONSIDERATION OF A SCREEN LONGITUDINAL HEAT FLOW

When the cable has metallic components other than a conductor, additional joule losses are created. In this section, we will illustrate how the longitudinal heat flow in the screen can be taken into account by considering a cable with a metallic screen or sheath but without armor. The development of the equations when armor is present would follow the same procedure as described below (Anders and Dorison, 2004). Our aim is to develop an expression for the conductor temperature rise at the crossing [Equation (3.52)] but since the conductor and screen temperature rises are now dependent on each other, we will first develop a matrix equation linking these two quantities [Equation (3.43)].

In order to take into account the effect of the longitudinal heat flow in the cable screen, we will apply Equations (3.9) and (3.10) to the conductor and screen separately. Thus, we have

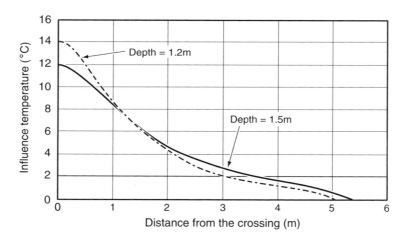

Figure 3-10   Temperature influence on the rated cable as a function of the vertical distance between the circuits and the distance from the crossing. The 145 kV cable is in the fixed position and the 10 kV circuit is shifted downward. Perpendicular crossing is represented. (From Anders and Brakelmann, 1999b, with permission from IEEE Press.)

$$(W_{cr} - W_c) - \frac{1}{T_{Lc}} \cdot \frac{\partial^2 \theta_c}{\partial z^2} = 0 \qquad (3.37)$$

$$(W_{sr} - W_s) - \frac{1}{T_{Ls}} \cdot \frac{\partial^2 \theta_s}{\partial z^2} = 0 \qquad (3.38)$$

Applying the energy conservation principle to both the conductor and the screen, and using relationships between the temperatures and radial heat flux, the following equations are obtained.

For the temperature drop between conductor and the sheath,

$$\theta_c - \theta_s = W_{cr} \cdot T_1 + W_d \cdot T_1/2 = (W_{cr} - W_c) \cdot T_1 + W_c \cdot T_1 + W_d \cdot T_1/2 \qquad (3.39)$$

Taking into account relation (3.13), Equation (3.39) can be written as

$$\theta_c \cdot [1 - \Delta W \cdot T_1] - \theta_s = (W_{cr} - W_c) \cdot T_1 + W_{ct} \cdot T_1 + W_d \cdot T_1/2$$
$$= \frac{T_1}{T_{Lc}} \cdot \frac{\partial^2 \theta_c}{\partial z^2} + W_{ct} \cdot T_1 + W_d \cdot T_1/2 \qquad (3.40)$$

The temperature drop between sheath and ambient is equal to

$$\theta_s - \theta_{amb} = \Delta\theta_{u12} + W_d \cdot \left(T_r - \frac{T_1}{2} \right) + W_{cr} \cdot (T_r - T_1) + W_{sr} \cdot (T_r - T_1)$$
$$= \Delta\theta_{u12} + W_d \cdot (T_r - T_1/2) + W_c \cdot (T_r - T_1) + W_s \cdot (T_r - T_1) \qquad (3.41)$$
$$+ [(W_{cr} - W_c) + (W_{sr} - W_s)] \cdot (T_r - T_1)$$

or, remembering that $W_c + W_s = (1 + \lambda_1)W_c$ and $(1 + \lambda_1) \cdot (T_r - T_1) = (T_t - T_1)$,

$$\theta_s - \theta_{amb} = \Delta\theta_{u12} + W_d \cdot (T_r - T_1/2) + W_{ct} \cdot (T_t - T_1) + \Delta W \cdot (T_t - T_1) \cdot \theta_c$$
$$+ [(W_{cr} - W_c) + (W_{sr} - W_s)] \cdot (T_r - T_1)$$
$$= \Delta\theta_{u12} + W_d \cdot (T_r - T_1/2) + W_{ct} \cdot (T_t - T_1) + \Delta W \cdot (T_t - T_1) \cdot \theta_c \qquad (3.42)$$
$$+ \left[\frac{1}{T_{Lc}} \cdot \frac{\partial^2 \theta_c}{\partial z^2} + \frac{1}{T_{Ls}} \cdot \frac{\partial^2 \theta_s}{\partial z^2} \right] \cdot (T_r - T_1)$$

where
W_s, W_c = conductor and screen losses per unit length (W/m)
W_{cr}, W_{sr} = radial heat flowing from the conductor and the screen per unit length (W/m)
θ_c, θ_r = conductor and screen temperatures (°C)
T_{Lc}, T_{Lr} = longitudinal thermal resistances of conductor and screen (K/W-m)
λ_1 = screen loss factor (assumed to be independent of the temperature)

and T_r and T_t are defined in Equations (3.12) and (3.2), respectively.

Using a matrix formulation, Equations (3.40) and (3.42) can be written as

$$\begin{bmatrix} \Delta\theta_c \\ \Delta\theta_s \end{bmatrix} - \begin{bmatrix} M_{11} & M_{12} \\ M_{21} & M_{22} \end{bmatrix} \cdot \frac{\partial^2}{\partial z_2} \begin{bmatrix} \Delta\theta_c \\ \Delta\theta_s \end{bmatrix} = \begin{bmatrix} \Delta\theta_u \\ \Delta\theta_u \end{bmatrix} \tag{3.43}$$

where

$$\Delta\theta_c = [1 - \Delta W \cdot T_t] \cdot \theta_c - \Delta\theta_{c0}$$

$$\Delta\theta_s = \frac{[1 - \Delta W \cdot T_t]}{[1 - \Delta W \cdot T_1]} \cdot (\theta_s + \Delta\theta_x) - \Delta\theta_{c0} \tag{3.44}$$

and

$$\Delta\theta_x = W_{ct} \cdot T_1 + \frac{T_1}{2} \cdot W_d$$

$$\Delta\theta_{c0} = \theta_{amb} + W_d \cdot (T_r - T_1) + W_{ct} \cdot (T_t - T_1) + \Delta\theta_x = \theta_{amb} + W_d \cdot \left(T_r - \frac{T_1}{2}\right) + W_{ct} \cdot T_t \tag{3.45}$$

$$M_{11} = \frac{1}{[1 - \Delta W \cdot T_t]} \cdot \frac{T_r}{T_{Lc}}$$

$$M_{12} = \frac{1}{[1 - \Delta W \cdot T_t]} \cdot \frac{(T_r - T_1)[1 - \Delta W \cdot T_t]}{T_{Ls}}$$

$$M_{21} = \frac{1}{[1 - \Delta W \cdot T_t]} \cdot \frac{1}{T_{Lc}} \cdot \left[T_r - T_1 \cdot \frac{[1 - \Delta W \cdot T_t]}{[1 - \Delta W \cdot T_1]} \right] \tag{3.46}$$

$$M_{22} = \frac{1}{[1 - \Delta W \cdot T_t]} \cdot \frac{(T_r - T_1)[1 - \Delta W \cdot T_1]}{T_{Ls}}$$

and, to simplify the notation, $\Delta\theta_u = \Delta\theta_{u12}$.

We will now solve the differential Equation (3.43). Let e_1 and e_2 be the eigenvalues of matrix **M**. Denoting

$$\eta = \frac{1}{\sqrt{e_1}} \quad \text{and} \quad \mu = \frac{1}{\sqrt{e_2}} \tag{3.47}$$

the general form of the conductor temperature rise is given by

$$\Delta\theta_c = \Delta\theta_u + A \cdot e^{\eta z} + B \cdot e^{-\eta z} + C \cdot e^{\mu z} + D \cdot e^{-\mu z} \tag{3.48}$$

The expression of the screen temperature is obtained from Equation (3.40) as

$$\theta_s = \Delta\theta_c - \frac{T_1}{T_{Lc}} \cdot \frac{1}{1 - \Delta W \cdot T_t} \cdot \frac{d^2\Delta\theta_c}{dz^2} + \Delta\theta_{c0} - \Delta\theta_x$$

or

$$\theta_s = \Delta\theta_u - \frac{T_1}{T_{Lc}} \cdot \frac{1}{1 - \Delta W \cdot T_t} \frac{d^2\Delta\theta_u}{dz^2} + \Delta\theta_{c0} - \Delta\theta_s + \alpha_1 \cdot (A \cdot e^{\eta z} + B \cdot e^{-\eta z})$$
$$+ \alpha_2 \cdot (C \cdot e^{\mu z} + D \cdot e^{-\mu z}) \tag{3.49}$$

$$\alpha_1 = 1 - \frac{T_1}{T_{Lc} \cdot [1 - \Delta W \cdot T_t]} \cdot \eta^2$$
$$\alpha_2 = 1 - \frac{T_1}{T_{Lc} \cdot [1 - \Delta W \cdot T_t]} \cdot \mu^2 \tag{3.50}$$

Since $\Delta\theta_{c0}$ and $\Delta\theta_x$ are not dependent on z, they vanish when expressing screen temperature or flux continuity at discretization points. Also, $\Delta\theta_u$ is constant in each interval; hence, its derivative also vanishes. Thus, when expressing the equality of the temperature and flux at points z_v for both the conductor and the screen, recursive relationships are derived for coefficients A, B, C, and D:

$$A_{v-1} = A_v + \frac{\alpha_2 - 1}{\alpha_2 - \alpha_1} \cdot \frac{\Delta\theta_{uv} - \Delta\theta_{u(v-1)}}{2} \cdot e^{-\eta \cdot v \cdot \Delta z}$$

$$B_{v-1} = B_v + \frac{\alpha_2 - 1}{\alpha_2 - \alpha_1} \cdot \frac{\Delta\theta_{uv} - \Delta\theta_{u(v-1)}}{2} \cdot e^{\eta \cdot v \cdot \Delta z}$$

$$C_{v-1} = C_v - \frac{\alpha_1 - 1}{\alpha_2 - \alpha_1} \cdot \frac{\Delta\theta_{uv} - \Delta\theta_{u(v-1)}}{2} \cdot e^{-\mu \cdot v \cdot \Delta z} \tag{3.51}$$

$$D_{v-1} = D_v - \frac{\alpha_1 - 1}{\alpha_2 - \alpha_1} \cdot \frac{\Delta\theta_{uv} - \Delta\theta_{u(v-1)}}{2} \cdot e^{\mu \cdot v \cdot \Delta z}$$

If $\Delta\theta_{cp}(\eta, 0)$ and $\Delta\theta_{cp}(\mu, 0)$ are the conductor and the screen temperature rises at the crossing location derived as in Section 3.4.1, with propagation coefficients η and μ, respectively, instead of γ, then

$$\Delta\theta_c(0) = u \cdot \Delta\theta_{cp}(\eta, 0) + (1 + u) \cdot \Delta\theta_{cp}(\mu, 0) \tag{3.52}$$

with

$$u = \frac{\alpha_2 - 1}{\alpha_2 - \alpha_1} = \frac{\mu^2}{\mu^2 - \eta^2} \tag{3.53}$$

Example 3.3
We will reconsider Example 3.1, but this time we will include the effect of the longitudinal heat transfer in the screen. We will show calculations for one iteration only.

The longitudinal resistance of the screen is obtained from Equation (3.7) as

$$T_{Ls} = \frac{\rho}{A} = \frac{0.00495}{\pi 0.5 \cdot (0.102 + 0.113) \cdot 0.006} = 1.28 \text{ K/W-m}$$

Since all other cable parameters were computed in Example 3.1, we can proceed to calculation of the components of the matrix $\mathbf{M}$ from Equation (3.46):

$$M_{11} = \frac{T_r}{T_{Lc} \cdot [1 - \Delta W \cdot T_t]} = \frac{1.91}{1.3 \cdot [1 - 0.05 \cdot 2.11]} = 1.64$$

$$M_{12} = \frac{(T_r - T_1)[1 - \Delta W \cdot T_1]}{T_{Ls} \cdot [1 - \Delta W \cdot T_t]} = \frac{(1.91 - 0.579) \cdot [1 - 0.05 \cdot 0.579]}{2.59 \cdot [1 - 0.05 \cdot 2.11]} = 0.559$$

$$M_{22} = M_{12} = 0.559$$

$$M_{21} = \frac{1}{T_{Lc} \cdot [1 - \Delta W \cdot T_t]} \cdot \left[T_r - T_1 \cdot \frac{[1 - \Delta W \cdot T_t]}{[1 - \Delta W \cdot T_1]} \right]$$

$$= \frac{1}{1.3 \cdot [1 - 0.05 \cdot 2.11]} \cdot \left[1.91 - 0.579 \cdot \frac{[1 - 0.05 \cdot 2.11]}{[1 - 0.05 \cdot 0.579]} \right] = 1.18$$

The eigenvalues of matrix $\mathbf{M}$ are equal to $e_1 = 0.123$ and $e_2 = 2.08$, hence $\eta = 2.85$ and $\mu = 0.694$. The parameter u in Equation (3.52) is computed next from Equation (3.53):

$$u = \frac{\mu^2}{\mu^2 - \eta^2} = \frac{0.694^2}{0.694^2 - 2.85^2} = -0.063$$

Since our goal is to evaluate the conductor temperature rise at the intersection, we will need the quantities $\Delta\theta_{cp}(\eta, 0)$ and $\Delta\theta_{cp}(\mu, 0)$. These are obtained from Equation (3.31) and are equal to $\Delta\theta_{cp}(\eta, 0) = 14.6°C$ and $\Delta\theta_{cp}(\mu, 0) = 10.5°C$, respectively.

The conductor temperature rise at the intersection is then equal to

$$\Delta\theta_c(0) = u \cdot \Delta\theta_{cp}(\eta, 0) + (1 - u) \cdot \Delta\theta_{cp}(\mu, 0) = -0.063 \cdot 14.6 + 1.063 \cdot 10.5 = 10.2°C$$

The derating factor is now obtained from Equation (3.5) as

$$DF = \sqrt{1 - \frac{\Delta\theta_{cp}(0)}{\Delta\theta_{max} - \Delta\theta_d}} = \sqrt{1 - \frac{10.2}{85 - 25 - 10.6}} = 0.89$$

This corresponds to the derated current of 1217 A. This value can be compared with the current of 1205 A obtained when the screen longitudinal heat flow was not considered. The new derated current is 12 A larger than the one computed in Example 3.1.

We will conclude this example with the sensitivity analysis showing the changes of the derating factor with respect to the crossing angle. Figure 3-11 illustrates this dependence for this example.

The more parallel the heat source is to the cable, not only is the ampacity of the cable smaller, but so is the effect of the screen longitudinal heat flow. The crossing angle of 0 degrees represents perpendicular crossing. When the cable is parallel to the heat source, there is no longitudinal heat flow in the conductor or the screen, and the rating of the cable is the lowest.　■

3.6 TRANSIENT TEMPERATURE RISE OF CABLE CROSSINGS

3.6.1 Transient External Thermal Resistance

The developments so far have dealt with the derating considerations when the intervening heat source emitted a constant heat flux. In practice, the external heat source is very often another cable circuit and its loading changes with time. A simple modification of Equation (3.32) will allow us to include the effect of transient loading in cable 1 (considered to be the external heat source) on the temperature rise in cable 2 (Brakelmann and Sahin, 1999).

First, we observe that Equation (3.6) takes the form (Anders, 1997)

$$\Delta\theta_{u12}(t; y = y_2, z) = \frac{\rho W_1}{4\pi}\left\{-Ei\left[-\frac{d_p^2}{4\delta \cdot t}\right] + Ei\left[-\frac{d_p'^2}{4\delta \cdot t}\right]\right\} \tag{3.54}$$

with (geometric variables as in Figure 3-1)

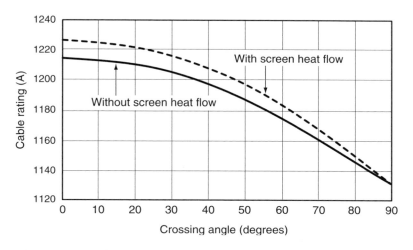

Figure 3-11 Derating factor as a function of the crossing angle (from Anders and Brakelmann, 1999b, with permission from IEEE Press).

$$d_p = \sqrt{(y_1 - y_2)^2 + (z \cdot \cos \beta)^2} \quad \text{and} \quad d_p' = \sqrt{(y_1 + y_2)^2 + (z \cdot \cos \beta)^2} \quad (3.55)$$

where

t = time (s)

δ = soil thermal diffusivity, m^2/s.

Substituting Equation (3.54) and (3.55) into Equation (3.29), the transient external thermal resistance is obtained as

$$T_{12}(t) = \frac{\rho(e^{\gamma \cdot \Delta z} - 1)}{4\pi} \sum_{v=1}^{N} e^{-v \cdot \gamma \cdot \Delta z}$$

$$\left\{ -Ei\left[-\frac{(y_1 - y_2)^2 + (v \cdot \Delta z \cdot \cos \beta)^2}{4\delta \cdot t} \right] + Ei\left[-\frac{(y_1 + y_2)^2 + (v \cdot \Delta z \cdot \cos \beta)^2}{4\delta \cdot t} \right] \right\} \quad (3.56)$$

Example 3.4

Let us revisit Example 3.2, in which we investigated the effect of a three-core 145 kV cable (cable model No. 2) crossing a circuit composed of three 10 kV cables (cable model No. 1).

We will now consider the following situation. The three-core cable is initially loaded with the current of 740 A, corresponding to 85% of its isolated rating. The 10 kV cables initially carry 50% of the continuous load while isolated; that is, 315 A. At $t = 0$, the current in the distribution cable increases to its full rating when isolated; that is, to 631 A. We are interested in the change of the conductor temperature of the three-core cable. The isolated cable parameters corresponding to these loading conditions are listed in Table 3-4.

The conductor temperature of this cable following a step change of current in the distribution circuit is shown in Figure 3-12 for two crossing angles, $\beta = 0$ and $\beta =$

Table 3-4   Parameters of two crossing circuits before time step change of current

		Computed parameters (hottest cable)	
Cable		10 kV	145 kV
Cable current	I (A)	315	740
Conductor temperature,	θ_{init} (°C)	40.5	76.2
Conductor resistance at θ_{init}	R (ohm/km)	0.0672	0.033
Concentric/skid wires loss factor	λ_1	0.173	0.175
Pipe/armor loss factor	λ_2		0.295
Thermal resistance of insulation	T_1 (K · m /W)	0.214	0.754
Thermal resistance of armor bedding	T_2 (K · m /W)		0.040
Thermal resistance of jacket	T_3 (K · m /W)	0.104	0.047
External thermal resistance 100% LF	T_4 (K · m /W)	1.621	0.385
Losses in the conductor	W_c (W/m)	6.67	18.31
Dielectric losses	W_d (W/m)	0	0
Total joule losses per cable	W_I (W/m)	7.82	26.93
Total losses per cable	W_t (W/m)	7.82	26.93

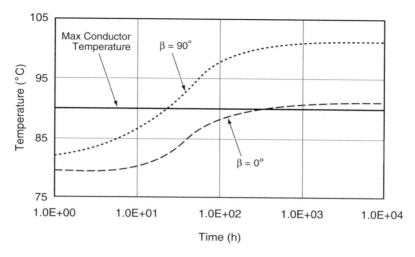

Figure 3-12 Conductor temperature of the three-core cable after a step change of current in the distribution cable at time $t = 0$ (from Anders and Brakelmann, 1999b, with permission from IEEE Press).

$90°$, corresponding to perpendicular and parallel crossings, respectively. When the effect of mutual heating is taken into account, the initial conductor temperatures for the three-core cable are $79.6°C$ and $82°C$, whereas the steady-state temperatures reached after doubling the load in the 10 kV circuit are $91.2°C$ and $101.3°C$ for the perpendicular and parallel crossings, respectively.

The maximum permissible conductor temperature is first exceeded after a time of about 743 h (31 days) if a perpendicular crossing is considered. However, for the parallel crossing, the cable will already be overheated after 36 h. This underlines again the importance of longitudinal heat flow in the metallic parts of the cable for oblique crossings. ■

3.6.2 Cyclic Rating

We will consider two different situations that can arise in practice. In one case, the external heat source exhibits cyclic loading; in the other, the rated cable is a subject to a cyclic current. When both situations occur, the results are simply additive.

3.6.2.1 *Cyclic Loading of the External Heat Source.* To obtain an expression for the rated cable conductor temperature rise caused by cyclic variation of losses in the external heat source at an arbitrary time τ, we will make an assumption that the heat dissipated in the external cable is directly proportional to the square of the applied current. This simplifying assumption gives quite accurate results, as confirmed by the author by performing finite-element studies. The required temper-

ature is then obtained as an algebraic sum of the hourly temperature rises as given by expression

$$\Delta\theta_{u12}(\tau, z) - \Delta\theta_{u12}(\tau - 1, z)$$ (3.57)

Let the rated cable conductor temperature rise at time i after the application of a step function of losses corresponding to the rated current I_R in the external heat source be $\Delta\theta_{u12R}(i, z)$. The temperature rise corresponding to the cyclic current with a maximum value I_{max} will be denoted by $\Delta\theta_{u12}(i, z)$. Since the temperature rise is assumed to be proportional to the square of the current, the temperature rise at time $t = 0$ is equal to

$$\Delta\theta_{u12}(0, z) = \frac{I_{max}^2}{I_R^2} \, \mu\Delta\theta_{u12R}(\infty, z)$$ (3.58)

Equation (3.58) assumes that a uniform current $\sqrt{\mu}I_{max}$ has been applied for a long time before $t = 0$. At the end of time $t = \tau$, the temperature rise caused by this average current is equal to

$$\frac{I_{max}^2}{I_R^2} \, \mu[\Delta\theta_{u12R}(\infty, z) - \Delta\theta_{u12R}(\tau, z)]$$ (3.59)

If at $t = 0$, a current pulse with magnitude I_0 is applied, the maximum conductor temperature rise at time $t = 1$ due to the external heat source is obtained from Equations (3.57) and (3.59) as

$$\Delta\theta_{u12max} = \frac{I_{max}^2}{I_R^2} \{\mu[\Delta\theta_{u12R}(\infty, z) - \Delta\theta_{u12R}(1, z)] + Y_0[\Delta\theta_{u12R}(1, z) - \Delta\theta_{u12R}(0, z)]\}$$ (3.60)

where Y_0 is the ratio I_0^2/I_{max}^2 and $\Delta\theta_{u12R}(0, z) = 0$.

Similarly, when a cyclic load is applied to the conductor of the external source, at the end of τ h, the maximum influence temperature rise is equal to

$$\Delta\theta_{u12max}(\tau, z) = \frac{I_{max}^2}{I_R^2} \left\{ \mu[\Delta\theta_{u12R}(\infty, z) - \Delta\theta_{u12R}(\tau, z)] \right.$$

$$\left. + \sum_{i=0}^{\tau-1} Y_i[\Delta\theta_{u12R}(i + 1, z) - \Delta\theta_{u12R}(i, z)] \right\}$$ (3.61)

$$= \frac{I_{max}^2 \Delta\theta_{u12R}(\infty, z)}{I_R^2} \left\{ \mu\left[1 - \frac{\Delta\theta_{u12R}(\tau, z)}{\Delta\theta_{u12R}(\infty, z)}\right] \right.$$

$$\left. + \sum_{i=0}^{\tau-1} Y_i\left[\frac{\Delta\theta_{u12R}(i + 1, z)}{\Delta\theta_{u12R}(\infty, z)} - \frac{\Delta\theta_{u12R}(i, z)}{\Delta\theta_{u12R}(\infty, z)}\right] \right\}$$

where each of the Y's is expressed as a fraction of their maximum value, and Y_i is a measure of the equivalent square current between i and $(i + 1)$ h prior to the expected time of maximum influence temperature.

The ratio $\Delta\theta_{u12R}(t, z)/\Delta\theta_{u12R}(\infty, z)$ is obtained as follows. The expression for $\Delta\theta_{u12R}(t, z)$ is given by Equation (3.54). On the other hand, the value $\Delta\theta_{u12R}(\infty, z)$ is given by Equation (3.6). In both equations, W_1 is assumed to correspond to the rated current. Denoting this ratio by $\beta(t)$, we obtain

$$\beta(t, z) = \frac{\alpha(t)\left\{Ei\left(-\dfrac{d_p^2}{4 \cdot \delta \cdot t}\right) - Ei\left(-\dfrac{d_p'^2}{4 \cdot \delta \cdot t}\right)\right\}}{\ln\left[\left(\dfrac{d_p'}{d_p}\right)^2\right]} \tag{3.62}$$

with d_p and d_p' defined in Equation (3.55) and $\alpha(t)$ the cable surface attainment factor (Anders, 1997).

Remembering that

$$\Delta\theta_{u12R}(\infty, z) = \frac{\rho}{4\pi} \cdot I_R^2 \cdot R \cdot \ln\frac{(y_1 + y_2)^2 + (z \cdot \cos\beta)^2}{(y_1 - y_2)^2 + (z \cdot \cos\beta)^2} = I_R^2 \cdot R \cdot T_{u12}(z) \tag{3.63}$$

and assuming the attainment factor close to 1, Equation (3.61) can be written for $\tau = 6$ as

$$\Delta\theta_{u12max}(z) = \mu I_{max}^2 \cdot R \cdot T_{u12}(z) \cdot \left\{[1 - \beta(6, z)] + \sum_{i=0}^{5} \frac{Y_i}{\mu}[\beta(i+1, z) - \beta(i, z)]\right\} \tag{3.64}$$

Equation (3.64) can be simplified as follows. For $d_p \geq 0.24\sqrt{n}$ and $t = n$ hours, $Ei[(-d_p^2/4 \cdot \delta \cdot t)] \approx 0$. Also, for $d_p \geq 0.2$ and $d_p' \geq 1$, $\beta(6, z) \leq 0.057$ and $\beta(i + 1, z) - \beta(i, z) \leq 0.02$ for $i = 0$ to 5.

Therefore, generally, the following approximation will be on the safe side and sufficiently accurate:

$$\Delta\theta_{u12max}(z) \approx \mu \cdot I_{max}^2 \cdot R \cdot T_{u12}(z) = \mu \cdot W_1 \cdot T_{u12}(z) \tag{3.65}$$

Dorison (2003) examined the accuracy of this approximation. He observed that for the usual cable spacing (larger than 0.2 m) and depth of laying larger than 1 m, the second component of Equation (3.64) that was omitted in Equation (3.65) will not be larger than 1.06. Only under very unfavorable conditions (very close proximity of the heat source to the rated cable) should the full Equation (3.64) be used.

3.6.2.2 Rated Cable with Cyclic Load.
We will start by recalling that in the steady state the following two equations describe the dynamics of the heat transfer in the rated cable [see Equations (3.9) to (3.11)]:

$$(W_r - W_c) - \frac{1}{T_L} \cdot \frac{\partial^2 \theta}{\partial z^2} = 0 \tag{3.66}$$

$$\theta = \theta_{amb} + \Delta\theta_{u12} + \Delta\theta_d + W_c \cdot T_t + (W_r - W_c) \cdot T_r \tag{3.67}$$

where T_t and T_r are defined in Equations (3.2) and (3.12), respectively.

As a result, the conductor temperature may be expressed in a simple way in which the effect of the crossing heat source is localized in the term

$$\Delta\theta = \Delta\theta_{u12} + (W_r - W_c) \cdot T_r \tag{3.68}$$

which clearly includes the longitudinal heat dissipation, when considered. Substituting the loss difference from Equation (3.66), and taking into account the variation of conductor resistance with temperature, we can write Equations (3.67) and (3.68) as [see also Equation (3.20)]

$$\theta = \theta_{amb} + \Delta\theta_d + W_c \cdot T_t + \Delta\theta \tag{3.69}$$

$$\Delta\theta = \Delta\theta_{u12} + \frac{1}{\gamma^2} \cdot \frac{\partial^2 \Delta\theta}{\partial z^2} \tag{3.70}$$

where γ is defined in Equation (3.19).

The following, somewhat lengthy developments, will result in an expression for the temperature of the conductor in a steady-state cyclic operation [Equation (3.90)], which can be used to compute the derating factor, as illustrated in Example 3.5 below.

To deal with a cyclic load, we need to consider first the transient temperature resulting from a current step function. Equations (3.69) and (3.70) will now take the form

$$\theta(t) = \theta_{amb} + \Delta\theta_d + W_c \cdot T_t(t) + \Delta\theta(t) \tag{3.71}$$

$$\Delta\theta(t) = \Delta\theta_{u12} + \frac{1}{\gamma^2(t)} \cdot \frac{\partial^2 \Delta\theta(t)}{\partial z^2} \tag{3.72}$$

with

$$\gamma^2(t) = [1 - \Delta W \cdot T_t(t)] \cdot \frac{T_L}{T_r(t)} \tag{3.73}$$

The thermal resistances in Equations (3.71) and (3.72) are time dependent since they now take into account the cable thermal capacitances. They are defined by

$$T_t(t) = \alpha(t) \cdot [T_1 + (1 + \lambda_1) \cdot T_2 + (1 + \lambda_1 + \lambda_2) \cdot ([T_3 + \beta(t) \cdot T_4])]$$
$$T_r(t) = \alpha(t) \cdot [T_1 + T_2 + T_3 + \beta(t) \cdot T_4] \tag{3.74}$$

The variables $\alpha(t)$ and $\beta(t)$ are the conductor and external surface attainment factors, and are defined by (Anders, 1997)

$$\alpha(t) = \frac{\theta(t)}{\theta(\infty)} \tag{3.75}$$

$$\beta(t) = \frac{-Ei\left(-\frac{D_e^{*2}}{16 \cdot t \cdot \delta}\right) + Ei\left(-\frac{L^{*2}}{t \cdot \delta}\right)}{2 \ln\left(\frac{4L^*}{D_e^*}\right)} \tag{3.76}$$

where
D_e^* = external diameter of the cable or duct, m
L^* = depth of laying, m.

Following the IEC standard approach, only the six hourly steps before the time when the temperature reaches its highest value are involved, the remaining hourly steps being represented by a representative load, as illustrated in Figure 3-13.

The conductor temperature may be expressed from individual steps responses as

$$\theta(t_j) = \theta_{amb} + \Delta\theta_d + \sum_{j=0}^{6} W_c(t_j) \cdot [T_r(t_{j+1}) - T_r(t_j)] + \sum_{j=0}^{6} (W_r(t_j) - W_c(t_j))[T_r(t_{j+1}) - T_r(t_j)] \tag{3.77}$$

The last term of Equation (3.77) can be expressed as

$$\Delta\theta_6(t_7) - \Delta\theta_{u12} + \sum_{j=1}^{6} [\Delta\theta_{j-1}(t_j) - \Delta\theta_j(t_j)] \tag{3.78}$$

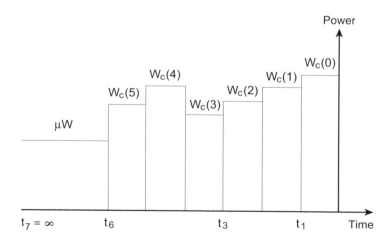

Figure 3-13 Counting of hourly steps in cyclic calculations.

To evaluate the expression under the summation sign in Equation (3.78), we will write Equation (3.68) for time t_j and the step with magnitude W_k. We have

$$\Delta\theta_k(t_j) = \Delta\theta_{u12} + (W_{rk} - W_{ck}) \cdot T_r(t_j) \tag{3.79}$$

with

$$\Delta\theta_k(t_j) = \Delta\theta_{u12} + \frac{1}{\gamma_k^2(t_j)} \cdot \frac{\partial^2 \Delta\theta_k(t_j)}{\partial z^2} \tag{3.80}$$

and

$$\gamma_k^2(t_j) = [1 - \Delta W_k \cdot T_r(t_j)] \cdot \frac{T_L}{T_r(t_j)} \tag{3.81}$$

$\Delta W_k \cdot T_r(t_j)$ can be expressed as follows

$$\Delta W_k \cdot T_r(t_j) = u\frac{T_r(t_j)}{T_t} Y_k \tag{3.82}$$

where

$$Y_k = \frac{I_k^2}{I_{max}^2} \qquad u = \frac{\alpha_{20} \cdot R_{max} \cdot I_{max}^2 \cdot T_t}{1 + \alpha_{20} \cdot (\theta_{max} - 20)} \approx \frac{\alpha_{20} \cdot (\theta_{max} - \theta_{amb})}{1 + \alpha_{20} \cdot (\theta_{max} - 20)} \tag{3.83}$$

We will try to simplify Equation (3.77) by finding the bounds on the variations of the attenuation coefficient $\gamma_k(t_j)$. Taking derivates of both sides of Equation (3.81), we obtain

$$2\left|\frac{d\gamma_k(t_j)}{\gamma_k(t_j)}\right| < \frac{u\dfrac{T_r(t_j)}{T_t}Y_k}{1 - u\dfrac{T_r(t_j)}{T_t}Y_k} \cdot \frac{dY_k}{Y_k} \tag{3.84}$$

Denoting

$$x = u\frac{T_r(t_j)}{T_t} Y_k \tag{3.85}$$

we observe that $x \ll 1$ since

$$u \ll 1, \ T_r(t_j) < T_t \ \text{and} \ Y_k < 1$$

Hence, we obtain from Equation (3.84)

$$\left|\frac{d\gamma_k(t_j)}{\gamma_k(t_j)}\right| < \frac{x}{2}(1 + x) \cdot \frac{dY_k}{Y_k} \tag{3.86}$$

Finally, since $x < u$, we obtain the following inequality (Dorison, 2003):

$$\left| \frac{d\gamma_k(t_j)}{\gamma_k} \right| < \frac{u}{2}(1+u) \cdot \frac{dY_k}{Y_k} \tag{3.87}$$

From the last inequality, for usual cables ($\alpha_{20} \approx 0.004$ K^{-1} and $\theta_{max} = 90°$C) and load curves, where the magnitudes of hourly steps do not show large deviations, the variations of $\gamma_k(t_j)$ are small. Since in the expression [Equation (3.80)] for $\theta_k(t_j)$ only the attenuation factor depends on k, with a small loss of accuracy, we have

$$\Delta\theta_k(t_j) \approx \Delta\theta_{k-1}(t_j) \tag{3.88}$$

Substituting Equation (3.78) to (3.77) and taking into account Equation (3.88), we obtain

$$\theta(t) = \theta_{amb} + \Delta\theta_d + \sum_{j=0}^{6} W_c(t_j) \cdot [T_j(t_{j+1}) - T_t(t_j)] + \Delta\theta_6(t_7) \tag{3.89}$$

As t_7 is infinite, the expression for $\Delta\theta_6(t_7)$ is the equation for the steady-state operation. Introducing the load loss factor, the final result is given by

$$\theta = \theta_{amb} + \Delta\theta_d + \frac{W_{c,max}}{M^2} \cdot T_t + \Delta\theta \tag{3.90}$$

where $\Delta\theta$ is determined from a steady-state operation, with the average losses equal to $\mu W_{c,max}$.

To calculate the derating factor, we proceed in the same way as for the steady-state operation; but we need to replace the steady-state current I in Equation (3.15) by the representative current of the load cycle $\mu M^2 I^2$.

Example 3.5
To study the effect of cyclic loading, we will revisit Example 3.2. Let us assume that the cables are perpendicular and the 10 kV cable circuit is loaded with a cyclic current shown in Figure 2-8, with the loss factor $\mu = 0.504$. The numerical values for this curve are summarized in Table 2-5.

We want to determine the effect of a cyclic load in the 10 kV cable on the ratings of both circuits. We will assume the losses as in the isolated cable study. First we need to determine the cyclic rating factor for the low-voltage cable. The calculations of this factor were already performed in Example 2.2 but the cables were located at the depth of 1 m, whereas now their depth of burial is 1.2 m. This difference has an effect on two components of Equation (1.77),[1] namely $\beta_1(t)$ and k_1. As shown in Example 2.2, the value of $\beta_1(t)$ can be assumed to be independent of the depth of burial for the durations considered here. Therefore, we need to recalculate

---

[1]We use this simplified equation for the cyclic rating factor together with Equation (1.8.5) because the attainment factor for this cable is equal to 1 for all six time points, as shown in Example 2.2.

only the value of k_1, which is given by Equation (1.84). The external thermal resistance of the middle cable, including the effect of mutual heating, is now equal to 1.621 K · m/W (see Example 3.2). Thus,

$$k_1 = \frac{W_f(T_4 + \Delta T_4)}{W_c T + W_f(T_4 + \Delta T_4)} = \frac{33.82 \cdot 1.621}{31.09 \cdot (0.214 + 0.104) + 33.82 \cdot 1.621} = 0.847$$

The cyclic rating factor for the 10 kV cable is equal to

$$M = \frac{1}{\sqrt{(1 - k_1)Y_0 + k_1\{B + \mu[1 - \beta(6)]\}}}$$

$$= \frac{1}{\sqrt{(1 - 0.847) \cdot 0.992 + 0.847 \cdot [0.274 + 0.504(1 - 0.337)]}} = 1.22$$

We will assume that the three-core cable is loaded to 871 A. We will recompute only those quantities that are affected by the change in the current in the 10 kV circuit. The loss values are computed from Equations (3.14), (3.15), and (3.16). The value of ΔW_0 is obtained from a revised Equation (3.15):

$$\Delta W_0 = \frac{\alpha_T R \cdot \mu \cdot M^2 \cdot I^2}{1 + \alpha_T(\alpha_{max} - 20)} = \frac{0.00393 \cdot 0.0781 \cdot 10^{-3} \cdot 0.504 \cdot 1.22^2 \cdot 631^2}{1 + 0.00393 \cdot (90 - 20)}$$

$$= 7.23 \cdot 10^{-2} \text{ W/m}$$

From Equation (3.16), we have

$$\Delta W = \Delta W_0 \cdot \left[1 - \frac{\Delta \theta(0)}{\Delta \theta_{max} - \Delta \theta_d}\right] = 7.23 \cdot 10^{-2} \cdot \left(1 - \frac{24.6}{65}\right) = 0.045 \text{ W/m}$$

The attenuation factor is obtained from Equation (3.19):

$$\gamma = \sqrt{(1 - \Delta W \cdot T_t) \cdot \frac{T_L}{T_r}} = \sqrt{\frac{(1 - 0.045 \cdot 2.091) \cdot 8.67}{1.939}} = 2.01 \text{ 1/m}$$

We can now compute the mutual thermal resistance from Equation (3.32):

$$T_{12} = \frac{\rho(e^{\gamma \cdot \Delta z} - 1)}{4\pi} \sum_{v=1}^{N} e^{-v \cdot \gamma \Delta z} \ln \frac{(y_1 + y_2)^2 + (v \cdot \Delta z \cdot \cos \beta)^2}{(y_1 - y_2)^2 + (v \cdot \Delta z \cdot \cos \beta)^2}$$

$$= \frac{0.8 \cdot (e^{2.01 \cdot 0.01} - 1)}{4\pi} \cdot \left[\sum_{v=1}^{600} e^{-v \cdot 2.01 \cdot 0.01} \ln \frac{(0.9 + 1.2)^2 + (0.01v)^2}{(0.9 - 1.2)^2 + (0.01v)^2}\right] = 0.180 \text{ K · m/W}$$

The derating factor after the first iteration is obtained from Equations (3.5) and (3.31):

$$DF = \sqrt{1 - \frac{\Delta\theta(0)}{\Delta\theta_{max} - \Delta\theta_d}} = \sqrt{1 - \frac{99.33 \cdot 0.180}{90 - 25}} = 0.85$$

Similar calculations are now repeated for the three-core cable with the current in the 10 kV cable fixed at 631 A.

The first estimate of $\Delta\theta$ is equal to

$$\Delta\theta(0) = \frac{0.8 \cdot 0.504 \cdot 1.22^2 \cdot 33.82}{4 \cdot \pi}$$

$$\cdot \left\{ \ln\left[\frac{(1.20 + 0.9)^2}{(1.20 - 0.9)^2} \right] + 2 \cdot \ln\left[\frac{(1.20 + 0.9)^2 + 0.072^2}{(1.20 - 0.9)^2 + 0.072^2} \right] \right\} = 18.7°C$$

Since the value of ΔW_0 for the 145 kV cable is the same as in Example 3.2, from Equation (3.16), we have

$$\Delta W = \Delta W_0 \cdot \left[1 - \frac{\Delta\theta(0)}{\Delta\theta_{max} - \Delta\theta_d} \right] = 7.68 \cdot 10^{-2} \cdot \left(1 - \frac{24.6}{65} \right) = 0.055 \text{ W/m}$$

The attenuation factor is obtained from (3.19)

$$\gamma = \sqrt{(1 - \Delta W \cdot T_t) \cdot \frac{T_L}{T_r}} = \sqrt{\frac{(1 - 0.055 \cdot 2.61) \cdot 3.25}{2.17}} = 1.13 \text{ 1/m}$$

We can now compute the mutual thermal resistance from Equation (3.36):

$$T_{21} = \frac{\rho \cdot (e^{\gamma_r \cdot \Delta z} - 1)}{4\pi} + \sum_{v=1}^{N} e^{-v \cdot \gamma_r \cdot \Delta z} \ln\left[\frac{(y_k + y_r)^2 + [(z_r - z_k + v \cdot \Delta z) \cdot \cos\beta]^2}{(y_k - y_r)^2 + [(z_r - z_k + v \cdot \Delta z) \cdot \cos\beta]^2} \right]$$

$$= \frac{0.8 \cdot (e^{1.13 \cdot 0.01} - 1)}{4\pi} \cdot \left[\sum_{v=1}^{600} e^{-v \cdot 1.13 \cdot 0.01} \ln\frac{(0.9 + 1.2)^2 + (0.072 + 0.01v)^2}{(0.9 - 1.2)^2 + (0.072 + 0.01v)^2} \right]$$

$$= 0.137 \text{ K} \cdot \text{m/W}$$

The derating factor after the first iteration is obtained from Equations (3.5) and (3.31):

$$DF = \sqrt{1 - \frac{\Delta\theta(0)}{\Delta\theta_{max} - \Delta\theta_d}} = \sqrt{1 - \frac{33.82 \cdot 0.137}{90 - 25}} = 0.96$$

After five iterations, the final results are as follows:

Rating of the 2 links

Cable type	Rating factor
300 mm² XLPE, 10 kV	0.86
800 mm², 145 kV	0.97

Considerably higher current can be imposed on the 145 kV circuit in comparison with the results obtained in Example 3.2.   ■

3.7   SOIL DRYOUT CAUSED BY MOISTURE MIGRATION

In this section, we will show how the heat-transfer differential equation can be modified to include the effect of the soil drying out around the rated cable. The resulting expression [see Equation (3.103)] has the identical form as Equation (3.20), but some of the coefficients are modified to include the effect of the soil dry out. The concept of the dry zones forming around the rated cable and around the heat source is illustrated in Figure 3-14.

The conductor temperature is the sum of the temperature rise inside the cable and the external temperature. The internal temperature rise is given by the last term of Equation (3.11) plus the expression

$$W_c \cdot [T_1 + n(1 + \lambda_1) \cdot T_2 + n(1 + \lambda_1 + \lambda_2) \cdot T_3] + W_d \cdot [T_1 + n \cdot T_2 + n \cdot T_3]$$

On the other hand, the temperature at the surface of the cable is obtained from a slight modification of Equation (2.50) as

$$\theta'_e - \theta_x = \frac{\rho_2}{\rho_1}[\Delta\theta_{u12} + W_t \cdot T_4 - (\theta_x - \theta_{amb})] = v \cdot [\Delta\theta_{u12} + W_t \cdot T_4 - (\theta_x - \theta_{amb})]$$

Thus, Equation (3.11) takes the form

$$\theta = \theta_{amb} + v \cdot \Delta\theta_{u12} + \Delta\theta'_d - (v - 1) \cdot \Delta\theta_x + W_c \cdot T'_t + (W_r - W_c) \cdot T'_r \qquad (3.91)$$

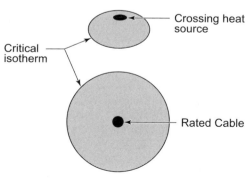

Figure 3-14   Crossing cable dry zone formation.

where T_i' and T_r' are the modifications of the definitions given in Equations (3.2) and (3.12), respectively, obtained by replacing T_4 by $v \cdot T_4$; that is,

$$T_r' = T_1 + n \cdot (1 + \lambda_1) \cdot T_2 + n \cdot (1 + \lambda_1 + \lambda_2) \cdot (T_3 + v' \cdot T_4)$$
$$\text{and} \quad T_r' = T_1 + n \cdot (T_2 + T_3 + v' \cdot T_4) \tag{3.92}$$

The temperature rise $\Delta\theta_d'$ caused by the dielectric losses is obtained in a similar way by replacing T_4 by $v \cdot T_4$.

The conductor joule losses W_c are temperature dependent. This dependence has been explicitly written in Equations (3.13) to (3.15) and is repeated here:

$$W_c = W_{c0}[1 + \alpha_T(\theta - \theta_0)] = W_{ct} + \Delta W \cdot \theta \tag{3.93}$$

with

$$W_c = W_{c0}(1 - \alpha_T\theta_0) \quad \text{and} \quad \Delta W = \alpha_T W_{c0} \tag{3.94}$$

where
α_T = conductor temperature coefficient of resistance (1/K)
W_{c0} = heat rate in the rated cable at the reference temperature θ_0 (usually 20°C) (W/m)

In order to express the losses in terms of the rated current, we introduced the following notation in Equation (3.15)

$$\Delta W_0 = \frac{R \cdot \alpha_T \cdot I^2}{1 + \alpha_T \cdot (\theta_{max} - \theta_0)} \tag{3.95}$$

where
R = conductor resistance at maximum operating temperature, ohm/m
I = the rated current when the soil is moist, A

Since the reduced current is related to the rated current I by the derating factor in Equation (3.5), we have

$$\Delta W' = \Delta W_0 \cdot \left[1 - \frac{\Delta\theta'(0)}{\Delta\theta_{max} - \Delta\theta_d' + (v - 1) \cdot \Delta\theta_x}\right] \tag{3.96}$$

The first estimate of $\Delta\theta'(0)$ in Equation (3.96) is obtained from Equation (3.6) as

$$\Delta\theta'(0) = W_1 \cdot \frac{v \cdot \rho}{4\pi} \cdot \ln\frac{(y_1 + y_2)^2}{(y_1 - y_2)^2} \tag{3.97}$$

From Equations (3.8) and (3.9), we obtain

$$(W_r - W_c) = \frac{1}{T_L}\frac{d^2\theta}{dz^2} \tag{3.98}$$

Substituting Equations (3.93) and (3.98) into Equation (3.91), we obtain the following differential equation describing conductor temperature:

$$\theta - \frac{1}{\gamma'^2} \cdot \frac{\partial^2 \theta}{\partial z^2} = \frac{v \cdot \Delta\theta_{u12} + \theta_{amb} + \Delta\theta'_d - (v-1) \cdot \Delta\theta_x + W_{ct} \cdot T'_t}{(1 - \Delta W' \cdot T'_t)}$$

$$= \frac{\theta_{amb} + \Delta\theta'_d - (v-1) \cdot \Delta\theta_x + W_{ct} \cdot T'_t}{1 - \Delta W' \cdot T'_t} + \frac{v \cdot \Delta\theta_{u12}}{1 - \Delta W' \cdot T'_t} = \theta'_\infty + \frac{v \cdot \Delta\theta_{u12}}{1 - \Delta W' \cdot T'_t} \qquad (3.99)$$

where

$$\gamma'^2 = (1 - \Delta W' \cdot T'_t) \cdot \frac{T_L}{T'_r} \qquad (3.100)$$

Substituting

$$\Delta\theta = (1 - \Delta W' \cdot T'_t)(\theta - \theta'_\infty) \qquad (3.101)$$

$$\Delta\theta'_u = v \cdot \Delta\theta_{u12} \qquad (3.102)$$

we obtain an equation of exactly the same form as Equation (3.20); that is,

$$\Delta\theta - \frac{1}{\gamma'^2} \cdot \frac{d^2 \Delta\theta}{dz^2} = \Delta\theta'_u(z) \qquad (3.103)$$

The solution of Equation (3.103) was derived in Section 3.4.1. The revised mutual thermal resistance is obtained by modifying Equation (3.32); that is,

$$T_{12} = \frac{v \cdot \rho_1 (e^{\gamma' \cdot \Delta z} - 1)}{4\pi} \sum_{j=1}^{N} e^{-j \cdot \gamma' \cdot \Delta z} \ln \frac{(y_1 + y_2)^2 + (j \cdot \Delta z \cdot \cos \beta)^2}{(y_1 - y_2)^2 + (j \cdot \Delta z \cdot \cos \beta)^2} \qquad (3.104)$$

Example 3.6

We will return to Example 3.1 in which we investigated the case where the route of the cable model No. 5 was crossed by a steam pipe with the steam temperature of θ_w = 150°C. The thermal resistance of the pipe insulation is equal to 1.0 K · m/W. The outer diameter of the pipe insulation is 0.5 m and the center of the pipe is located 1 m below the ground surface (80 cm above the cables) as shown in Figure 3-15.

Let us assume that the soil around the cable dries out and the thermal resistivity of the dry soil is ρ_2 = 2.0 K · m/W, that is, $v = 2$, and that $\Delta\theta_x$ = 35°C. The parameters required for the derating calculations (Table 3-5) are taken from Table A1 in the Appendix.

The isolated circuit No. 5 has ampacity of 1365 A when the soil is wet. When the drying of the soil around the cable occurs, the ampacity of this circuit becomes [see Equation 4.13 in Anders (1997) and Equation (1.51)]

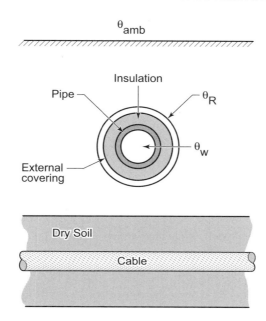

Figure 3-15 Steam pipe crossing cable route.

$$I' = \left[\frac{\Delta\theta - W_d[0.5T_1 + n(T_2 + T_3 + vT_4)] + (v - 1)\,\Delta\theta_x}{RT_1 + nR(1 + \lambda_1)T_2 + nR(1 + \lambda_1 + \lambda_2)(T_3 + vT_4)} \right]^{0.5}$$

$$= \left[\frac{60 - 6.53[0.5 \cdot 0.579 + 0.056 + 2 \cdot 1.276] + (2 - 1) \cdot 35}{1.26E - 5 \cdot [0.579 + (1 + 0.15) \cdot 0.056 + (1 + 0.15) \cdot (0 + 2 \cdot 1.276)]} \right]^{0.5} = 1299 \text{ A}$$

In Example 3.1, we calculated $W_p = 78.9$ W/m. We will assume that the same value applies in this example. Also, $T_L = 1.3$ K/m-W. The combined thermal resistances are now obtained from Equations (3.92):

Table 3-5 Values for numerical example 3-6

Cable ampacity in moist soil	I (A)	1365
Conductor resistance at θ_{max}	R (ohm/km)	0.0126
Concentric/skid wires loss factor	λ_1	0.150
Armor loss factor	λ_2	0.0
Thermal resistance of insulation	T_1 (K · m/W)	0.579
Thermal resistance of armor bedding	T_2 (K · m/W)	0.0
Thermal resistance of jacket	T_3 (K · m/W)	0.056
External thermal resistance, 100% LF	T_4 (K · m/W)	1.276
Dielectric losses	W_d (W/m)	6.53

$$T_t = T_1 + n(1 + \lambda_1)T_2 + n(1 + \lambda_1 + \lambda_2)(T_3 + vT_4)$$
$$= 0.579 + 1 \cdot 1.15 \cdot (0.056 + 2 \cdot 1.276) = 3.58 \text{ K} \cdot \text{m/W}$$

$$T_r = T_1 + n(T_2 + T_3 + vT_4) = 0.579 + 1 \cdot (0.056 + 2 \cdot 1.276) = 3.19 \text{ K} \cdot \text{m/W}$$

The loss values are computed now from Equations (3.95), (3.94), and (3.96):

$$\Delta W_0 = \frac{\alpha_T I^2 R}{1 + \alpha_T(\theta_{max} - 20)} = \frac{0.00393 \cdot 1365^2 \cdot 1.26 E - 5}{1 + 0.00393 \cdot 65} = 0.073 \text{ W/m}$$

$$W_{ct} = W_{c0}(1 - \alpha_T \theta_0) = \frac{1365^2 \cdot 1.26 E - 5}{1 + 0.00393 \cdot (85 - 20)} \cdot (1 - 0.00393 \cdot 20) = 17.2 \text{ W/m}$$

The first estimate of $\Delta\theta(0)$ is obtained from Equation (3.97) as

$$\Delta\theta'(0) = W_1 \cdot \frac{v \cdot \rho}{4\pi} \cdot \ln\frac{(y_1 + y_2)^2}{(y_1 - y_2)^2} = 78.9 \cdot \frac{2 \cdot 1.0}{4\pi} \ln\frac{(1.8 + 1)^2}{(1.8 - 1)^2} = 31.5°\text{C}$$

The temperature reduction caused by the dielectric losses equals, in this case,

$$\Delta\theta'_d = W_d \cdot (T_1/2 + T_2 + vT_4) = 6.53 \cdot (0.579 + 0.056 + 2 \cdot 1.276) = 18.9°\text{C}$$

From Equation (3.96), we have

$$\Delta W' = \Delta W_0 \cdot [1 - \frac{\Delta\theta'(0)}{\Delta\theta_{max} - \Delta\theta'_d + (v - 1) \cdot \Delta\theta_x}]$$

$$= 0.073 \cdot \left[1 - \frac{31.5}{85 - 25 - 18.9 + 1 \cdot 35}\right] = 0.043$$

The attenuation factor is obtained from Equation (3.100) as

$$\gamma' = \sqrt{(1 - \Delta W' \cdot T'_t) \cdot \frac{T_L}{T'_r}} = \sqrt{(1 - 0.043 \cdot 3.58) \cdot \frac{1.3}{3.19}} = 0.59 \text{ 1/m}$$

Assuming $\Delta z = 0.01$ and $z_{max} = 6$ m, the mutual thermal resistance is obtained from Equation (3.104):

$$T_{12} = \frac{v \cdot \rho_1 \cdot (e^{\gamma' \cdot \Delta z} - 1)}{4\pi} \sum_{j=1}^{N} e^{-j \cdot \gamma' \cdot \Delta z} \ln\frac{(y_1 + y_2)^2 + (j \cdot \Delta z \cdot \cos\beta)^2}{(y_1 - y_2)^2 + (j \cdot \Delta z \cdot \cos\beta)^2}$$

$$= \frac{2 \cdot 1.0 \cdot (e^{0.59 \cdot 0.01} - 1)}{4\pi} \sum_{j=1}^{600} e^{-j0.59 \cdot 0.01} \ln\frac{(1.8 + 1.0)^2 + (j \cdot 0.01 \cdot \cos 30)^2}{(1.8 - 1.0)^2 + (j \cdot 0.01 \cdot \cos 30)^2}$$

$$= 0.247 \text{ K} \cdot \text{m/W}$$

The temperature rise $\Delta\theta'(0)$ is now obtained from Equation (3.31):

$$\Delta\theta'(0) = W_1 T_{12} = 78.9 \cdot 0.247 = 19.5°C$$

Hence, the cable conductor temperature computed from Equation (3.101) becomes

$$\theta = \frac{\Delta\theta}{1 - \Delta W' \cdot T_t'} + \frac{\theta_{amb} + \Delta\theta_d' - (v - 1) \cdot \Delta\theta_x + W_{ct} \cdot T_t'}{1 - \Delta W \cdot T_t'}$$

$$= \frac{19.5 + 25 + 18.9 - (2 - 1) \cdot 35 + 17.2 \cdot 3.58}{1 - 0.043 \cdot 3.58} = 106.3°C$$

After two iterations, the derating factor equals 0.86 with the maximum current of 1176 A. This value can be compared with the rating of 1205 A when moist soil conditions are assumed. ∎

3.8 CONCLUDING REMARKS

From the information reported in this chapter, it is quite clear that the presence of external heat sources may have a significant effect on the rating of power cables. In particular, the derating of 3 to 5% used by some utilities may be insufficient, especially for cables with smaller conductors. Hence, the need for either derating curves or computational procedures, or both, seems to be quite evident.

Similarly, as in the case of cables crossing unfavorable regions, using Equations (3.4) or (3.5) and (3.34)–(3.35), derating curves similar to those in Figure 3-7, where the derating is a function of the crossing angle and the size of the conductor, can be developed. In order to derive such curves, some standard installation conditions have to be assumed and the information about the external heat source has to be used. A set of derating graphs could be developed, depending on the distance between the cable and the heat source, the crossing angle, and also on the depth of burial. The amount of work involved would depend on the number of cases to be examined.

In addition, diagrams similar to those in Figures 3-9 and 3-10, showing the influence temperature as a function of the distance from the crossing point, could be developed to give guidance on the length of the area in which isolating boards, heat pipes, or special backfills should be applied to reduce the mutual heating effect.

3.9 CHAPTER SUMMARY

The most important development presented in this chapter relates to the calculation of the derating factor when an external heat source (possibly another cable) crosses the cable for which the rating is to be determined. This derating factor is obtained from

For a single crossing,

$$DF = \sqrt{\frac{\Delta\theta(0)}{\Delta\theta_{max} - \Delta\theta_d}}$$

$$D\theta(0) = \frac{W_h \cdot \rho \cdot (e^{\gamma \cdot \Delta z} - 1)}{4\pi} \sum_{v=1}^{N} e^{-v \cdot \Delta z} \ln \frac{(y + y_h)^2 + (v \cdot \Delta z \cdot \cos \beta)^2}{(y - y_h)^2 + (v \cdot \Delta z \cdot \cos \beta)^2}$$

where:
ρ is the soil thermal resistivity (K $\cdot$ m/W)
W_h is the heat generated by the external heat source (W/m)
β is the crossing angle
y is the laying depth of the rated cable (m)
y_h is the laying depth of the heat source (m)

The attenuation factor γ is expressed as

$$\gamma = \sqrt{(1 - \Delta W \cdot T) \cdot \frac{T_L}{T_r}}$$

with

$$T_L = \frac{\rho_c}{A}$$

$$T_r = T_1 + n \cdot (T_2 + T_3 + T_4)$$

$$T = T_1 + n \cdot [(1 + \lambda_1) \cdot T_2 + (1 + \lambda_1 + \lambda_2) \cdot (T_3 + T_4)]$$

$$\Delta\theta_d = W_d \cdot \left[\frac{T_1}{2} + n \cdot (T_2 + T_3 + T_4) \right]$$

$$\Delta W = \Delta W_0 \cdot \left[1 - \frac{\Delta\theta(0)}{\Delta\theta_{max} - \Delta\theta_d} \right]$$

$$\Delta W_0 = \frac{R \cdot \alpha_{20} \cdot I^2}{1 + \alpha_{20} \cdot (\theta_{max} - 20)}$$

where
ρ_c is the conductor thermal resistivity (K $\cdot$ m/W)
A is the conductor cross-sectional area (m^2)
α_{20} is the temperature coefficient of electrical resistivity for the conductor material
I is the current flowing in the conductor of the rated cable (A)

The remaining variables are the classical thermal resistances of cable components and the sheath and armor loss factors.

Typically, a value of $\Delta z = 0.01$ m may be used; it must be checked that

$$\gamma \cdot \Delta z < \varepsilon$$

N is determined from

$$\Delta \theta_{uh}(N \cdot \Delta z) < \varepsilon$$

where ε is a small value, typically 0.01.

$\Delta \theta_{uh}(z)$ represents the temperature rise in the conductor as a function of the distance z from the crossing, caused by the crossing heat source. This temperature can be obtained by applying Kennelly's principle:

$$\Delta \theta_{uh}(z) = \frac{\rho}{4\pi} \cdot W_h \cdot \ln \frac{(y + y_h)^2 + z^2 \cdot \cos^2 \beta}{(y - y_h)^2 + z^2 \cdot \cos^2 \beta}$$

As γ depends upon the current in the rated cable, which is to be determined, an iterative solution is necessary, using as a first estimation of this current, the rated current when the heat source is assumed to be parallel to the rated cable.

The first estimate of $\Delta \theta(0)$ is

$$\Delta \theta(0) = W_h \cdot \frac{\rho}{4\pi} \cdot \ln \frac{(y + y_h)^2}{(y - y_h)^2}$$

For several crossings, the rating factor has the same expression as above but

$$\Delta \theta(0) = \sum_{k=1}^{n} T_{kr} \cdot W_k$$

Let the rated cable have the designation r and z_r be the z-coordinate of the hottest point in cable r. Then, for any other heat source k at a distance z_k away from the point z_r, we have

$$T_{kr} = \frac{\rho \cdot (e^{\gamma_r \cdot \Delta z} - 1)}{4\pi} + \sum_{v=1}^{N} e^{-v \cdot \gamma_r \cdot \Delta z} \ln \left[\frac{(y_k + y_r)^2 + [(z_r - z_k + v \cdot \Delta z) \cdot \cos \beta]^2}{(y_k - y_r)^2 + [(z_r - z_k + v \cdot \Delta z) \cdot \cos \beta]^2} \right]$$

where

n is the number of heat sources crossing the rated cable

y_k is the laying depth of heat source k (m)

The attenuation factor γ has to be calculated as above with, as a first estimate,

$$\Delta \theta(0) = \frac{\rho}{4 \cdot \pi} \cdot \sum_{k=1}^{n} W_h \cdot \ln \frac{(y + y_k)^2 + (z_r - z_k)^2}{(y - y_k)^2 + (z_r - z_k)^2}$$

REFERENCES

Abdel Aziz, E.M., Zahab, A., and Kassem, A.A.E. (1986), "Prediction and Elimination of Hot Spots at the Crossing Regions of H.V. Cables," *IEEE Transactions on Power Delivery,* Vol. 3 No. 3, July 1988, pp. 845–849. Also presented at the 2nd International Conference on Power Cables and Accessories, London, UK, Nov. 26–28, pp. 227–231.

Anders, G.J. (1997), *Rating of Electric Power Cables: Ampacity Computations for Transmission, Distribution and Industrial Applications,* IEEE Press, New York.

Anders, G.J., and Brakelmann, H. (1999a), "Cable Crossings—Derating Considerations. Part I—Derivation of the Derating Equations," *IEEE Trans. On Power Delivery,* Vol. PWRD-13, No.4, July 1999, pp. 709–714.

Anders, G.J., and Brakelmann, H. (1999b), "Cable Crossings—Derating Considerations. Part II—Example of derivation of the derating curves," *IEEE Trans. on Power Delivery,* Vol. PWRD-13, No.4, July 1999, pp. 715–720.

Anders, G.J., and Dorison, E. (2004), "Derating Factors for Cable Crossings with Considerations of Longitudinal Heat Flow in Cable Screen," *IEEE Transactions on Power Delivery,* Vol. 19, No. 3, July 2004, pp. 926–932.

Brakelmann, H. (1985), *Belastbarkeiten der Energiekabe, 1,* VDE-Verlag, Berlin.

Brakelmann, H., and Sahin, M. (1999), "Transient temperature rise of cable crossings," *ETAP,* Vol. 8, No. 3, pp. 217–220.

Dorison, E. (2003), "Cable Crossings," In Proceedings of Jicable '03 meeting, Versailles, France, June 2003.

IEC Standard 228 (1978) "Conductors of Insulated Cables," Second edition. First Supplement (1982).

IEC Standard 60287, Part 2-1 (1994), "Calculation of thermal resistances".

IEC Standard, 60853 (1989), "Calculation of the Cyclic and Emergency Current Ratings of Cables. Part 2: Cyclic Rating Factor of Cables Greater than 18/30 (36) kV and Emergency Ratings for Cables of All Voltages," Publication 853-2.

Imajo, T., Fukagawa H., and Itoh, T. (1986), "Thermal Interference in Underground Cable Heat Dissipation by Steam Pipes and its Prevention," paper presented at the 2nd International Conference on Power Cables and Accessories, London, UK, Nov. 26–28, 1986, pp. 221–226.

Iwata, Z., Sakuma, S., Dam-Andersen M., and Jacobsen E. (1992), "Heat Pipe Local Cooling System Applied for 145 kV Transmission Lines in Copenhagen," *IEEE Transactions on Power Delivery,* Vol. 7, No. 2, April 1992, pp. 767–775. A companion paper "Installation and Supervision of 145 kV Cables in Copenhagen City," was presented at the 1990 CIGRE session, paper No. 21-204.

Orton, H., MacPhail, A., and Buchholtz, V. (1996), "Elevated Temperature Operation—A Precautionary Note," IEEE PES, Insulated Conductor Committee, Minutes of the 99th Meeting, Houston Texas, April 14–17, 1996, Appendix 5-K-1.

Thevenon, H., and Couqet, J.M. (1999), "Potential Overheating in the Event of a Cable Crossing," Proceedings of Jicable'99 meeting in Versailles, France, June 1999, pp. 230–234.

Application of Thermal Backfills for Cables Crossing Unfavorable Thermal Environments

The main focus of this book is the calculation of cable ampacities when the cable route crosses an unfavorable thermal environment. The most common method used in practice to mitigate bad thermal conditions is an application of thermal backfills. These are used to modify the external cable resistance of the cable.

The earth's thermal resistance is the largest resistance to heat dissipation for underground cables and is by far the most variable of the cable thermal network parameters. Accurate knowledge of this resistance and its modification can provide substantial ampacity increases, with confidence that the cable will remain within allowable operating temperatures for all designed loading conditions.

The heat generated by an electric-power-transmitting conductor has to dissipate through the cable insulation and the surrounding soil. In order to facilitate heat dissipation, the cable is normally backfilled in a controlled manner. The purpose of this "corrective backfill" is to provide a stable thermal environment for the cables that mitigates the effects of seasonal weather factors and variability of natural soil conditions. In order to reach optimal transmission efficiency, the thermal resistance of the backfill should be as low and as stable as possible. This thermal resistance is affected not only by the thermal resistivity of the backfill material, but also by the dimensions of the thermal envelope.

There are two competing factors affecting backfilling. One is the substantial cost involved, and the other is the need to increase cable ampacity. In this chapter, we will present an optimization algorithm that tries to balance these two competing factors. The first attempt to address this problem can be found in El-Kady (1982).

Soil thermal resistivity, together with soil ambient temperature and cable load, are the most important factors determining the cable operating temperature. All these parameters are random in nature and their most probable values have to be predicted if the cable operating temperature is to be computed with reasonable accuracy. One way to account for the variability of these parameters is to apply probabilistic analysis to cable ampacity calculations. This topic will also be discussed in this chapter.

Rating of Electric Power Cables in Unfavorable Thermal Environment. By George J. Anders
ISBN 0-471-67909-7 © 2004 the Institute of Electrical and Electronics Engineers.

We will start our discussions with a brief review of the history of soil thermal resistivity measurements.

4.1 A BRIEF HISTORY OF SOIL THERMAL RESISTIVITY MEASUREMENTS

Measurement of heat transfer in soils has been of interest to utility and manufacturing engineers, research scientists, and others for many years. A significant amount of research work has been devoted to physical and thermal properties of soils in North America and Europe. This work relating to buried cables has been carried out by universities, independent research laboratories, and electric utilities since the early 1940s. Since then, numerous contributions have been made and reported in a host of technical journals. A variety of measurement techniques have been employed over the years and measurements have generated data with a large spectrum of results for soil thermal properties.

Soil thermal resistivity has been measured over the years with numerous types of devices, including spheres and line heat sources, that is, probes, guarded hot plates, buried cylinders, and heat flow meters. These measurements were made with both steady-state and transient-type devices, yielding unreliable data much of the time. In comparison, transient measurements employing a line heat source were found to be the most reliable because they were quick and avoided the attainment problems for satisfactory equilibrium that are characteristic of steady-state measurements.

Transient thermal resistivity measurements were suggested as early as 1888 (Wiedemann, 1888). Development work continued and successful thermal resistivity measurement on liquids and other materials were made during the 1930s and 1940s.

The first successful work on soil thermal resistivity measurements in North America[1] was reported in 1952 by Mason and Kurtz (1952). Their work demonstrated the practical use of a line heat source for the transient measurement of soil thermal resistivity. Concurrently, the Insulated Conductor Committee of IEEE, formed in 1947, began a special project on soil thermal resistivity in 1951. This task force, chaired by W. Burrell, eventually reached the point where the lack of knowledge became too profound and more research work was needed to continue the project. A special subcommittee was set up in the ICC headed by Professor H.F. Winterkorn of Princeton University. This project continued for 10 years and culminated in an extensive committee report whose summary can be found in IEEE (1966).

In 1962, the Winterkorn subcommittee was disbanded and various utilities continued to use thermal probes and other devices to measure soil thermal resistivity in the field and on recompacted soil samples in the laboratory.

---

[1]Material presented in this section on North American history of soil thermal resistivity measurements is based on the presentation of M.A. Martin (Chu et al., 1981) at the cable workshop in Toronto.

During the late 1960s and early 1970s, a large number of underground cable systems were installed in North America. At the same time, reports started coming to the ICC indicating inconsistent results with soil thermal resistivity measurements. Thermal resistivity values were being measured for uniform mixtures of concrete between 0.4 K·m/W to 200 K·m/W. Therefore, Subcommittee 12 of the ICC organized a project to prepare an IEEE standard on the measurement of soil thermal resistivity. This project was started in 1971. Measurement techniques employed a variety of methods that included the following: the Shannon and Wells method (1947), rhometer (Stolpe, 1969), rapid K measurement (ASTM C518-67, 1967), as well as the classical thermal probe technique. As expected, the results from this work were variable and extensive studies were undertaken before the differences in the measurements could be resolved. After some refinements in the measuring techniques, IEEE prepared a much-needed guide for measurements employing a thermal needle for both in-field and laboratory use. The guide was issued as Standard P442 in 1981.

Commercial equipment for the measurement of soil thermal resistivity was developed in the mid-1970s by the Ontario Hydro Research Division under the sponsorship of the Electric Power Research Institute. The instrument was named the Thermal Property Analyzer (TPA) and, with continuous improvements carried out mainly by the Geotherm company in Canada, it is currently used as a standard measuring device of soil thermal resistivity.

Concurrently with the development of the measuring devices, an extensive research effort was conducted on the development of suitable cable-trench backfilling materials. Early work involved selecting naturally occurring sands that had the proper grain size distribution to give good thermal resistivities for a very a wide range of moisture contents. Later, utilities employed manufactured thermal sands (MTS) by developing sieve analysis specifications that assured the proper material gradation. Other utilities employed limestone screenings. Significant progress in measurement and analysis techniques were described in the proceedings of a 1981 conference held in Toronto (Chu et al., 1981).

In the 1970s and 1980s, utilities in the United States and Canada undertook the task of developing special thermal backfills for underground power cable installations. These materials were to provide very good thermal properties over a wide range of moisture contents, while simplifying the installation techniques. Much of the work was performed under the auspices of organizations such as The Canadian Electrical Association (CEA), EPRI, and Ontario Hydro's Research Division (Radhakrishna, 1981). One of the identified backfills was a fluidized thermal backfill (FTB) containing sand, aggregate, cement, and fly ash. It has not only very low thermal resistivity (see Figure 4-1) but also very long thermal dryout times (it is thermally very stable). Although granular-type backfills (thermal sand and limestone screenings) are still being used by many utilities, the use of FTB over the past several years has become very widespread.

FTB is, however, fairly expensive; hence, there is a need to balance the cost of its installation with the substantial ampacity improvements that can be achieved when it is used. An optimization algorithm, developed by El-Kady (1982) to achieve the desired balance, is presented in the next section.

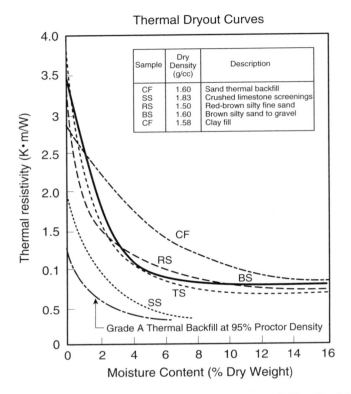

Figure 4-1 Thermal resistivity versus moisture content for some fluidized backfills. (From Anders et al., 1988a, with permission from IEEE Press.)

4.2 OPTIMIZATION OF POWER CABLE AND THERMAL BACKFILL CONFIGURATIONS

The thermal analysis of power cables in backfill represents a basic part in the cable system design and it has been well established (Neher & McGrath, 1957; Fink, 1954; Fink & Smerke, 1958, Schmill, 1967; Saleeby et al., 1979; Anders, 1997) that a relatively small amount of backfill material around the cable leads to an appreciable increase in cable ampacity. Thermal backfill design involves the determination of several parameters, for example, thickness, thermal resistivities, width, and so on. These parameters establish both the overall cost of a cable system and its thermal performance. Efficient cable design requires the selection of optimal values of these parameters, taking into account the costs criteria for different alternatives as well as other constraints. The cable backfill optimization problem has been addressed in works of Saleeby et al. (1979) and El-Kady (1982). The first paper investigated an improved solution based on varying one parameter at a time to determine the resulting effect on other variables. In the following section, we will use a similar approach employing a tornado chart tool. The work of El-Kady looks at several design parameters at the same time, including different cable configurations in a particular backfill envelope. We will continue this chapter by formulating the optimization problem and conclude with some numerical examples.

A discussion of the incremental worth of proper design and installation of cable backfill is included in the conclusion of a paper by Fink (1965). In view of the critical influence of backfill characteristics on cable operating temperatures, it is worth noting that work has been reported in the literature regarding the most effective design geometries for backfill (Fink and Smerke, 1958).

4.2.1 Analysis of the Effect of Parameter Variations

Sensitivity analysis forms an important part of engineering studies. In order to facilitate the collection of the input data, which is required to perform either the selection of backfill parameters or the steady-state/transient ampacity computations, one should first determine which parameters play an important role in the analysis. A useful tool for displaying the results of sensitivity analysis when one parameter is varied at the time is a tornado chart. Example 4.1 illustrates this concept.

There are basically five dimensional and physical variables associated with the cable thermal backfill. These are backfill thickness h, width w, depth of the bank center L_G, cable spacing s_1, and the thermal resistivity ρ_c. There may by some other design variables that we may wish to include in the optimization process, but in order to illustrate the concept, we will start with the backfill characteristics.

Example 4.1

We will consider cable No. 4 laid in a backfill as shown in Figure 4-2.

First, we need to define the ranges in which the design parameters: L_G, w, h, and s_1 will change. The lower limits of the ranges of the backfill design parameters are

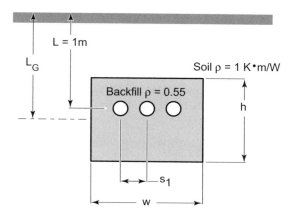

Figure 4-2 Cable model No. 4 in a backfill.

defined by the assumed depth of laying of the cables and their diameters. On the one hand, the cables could be touching (we assume a horizontal formation), and on the other, they could be far apart. The upper limit on the trench width is arbitrarily set to 5 m. The same upper limit is set for the height of the backfill. Since we assumed that there is a need for at least 50 cm of the backfill material on either side of the outer cables as well as on the top and the bottom of the circuit, the lower and upper bounds of the depth of the center of the backfill are set accordingly. The nominal values are those specified in Appendix A. Table 4-1 summarize the bounds imposed on the design variables.

We will now check which of the four design parameters will have the greatest influence on the objective function; that is, on the value of the conductor current. A useful tool for this analysis is a tornado chart. This tool measures the impact of each model variable one at a time on a conductor current. While analyzing one variable, the tool freezes the other variables at their base values. This measures the effect each variable on the computed current while removing the effects of the other variables. This method is also known as "one-at-a-time perturbation" or "parametric analysis." It tests the range of each variable at specified percentiles and then calculates the value of the current at each point. The tornado chart illustrates the swing between the maximum and minimum current values for each variable, placing the variable that causes the largest swing at the top and the variable that causes the smallest swing at the bottom.

The tornado chart with the full range of each variable[2] for the optimization problem in this example is shown in Figure 4-3.

The bars next to each variable represent the ampacity value range across the variable tested. The values are within the bounds specified in Table 4-1 with the middle line corresponding to the specified nominal values. Next to the bars are the values of the variables that produced the greatest swing in the ampacity values.

In our case, increasing the backfill depth causes a decrease in the ampacity of the cable, whereas increasing the height and the width results in an increased rating. We can observe that the depth has the greatest influence on the current, whereas the width has the smallest. Changing the cable spacing has a nonmonotonic effect on the rating. A decreased spacing increases the mutual heating effect but reduces the circulating current. Another interesting observation that can be drawn from Figure 4-3 is that we can expect the current to vary in a narrow range between 700 and 850 A when one of the parameters is changed.

The next step is to perform a selection of the backfill design parameters to either minimize the costs or to maximize cable ampacity.   ■

4.2.2   Formulation of the Optimization Problem

In practice, the thermal backfill might be covered by a protective layer of a different material; however, in order to simplify the formulation and the illustration of the optimization aspects, we will assume that the protection layer, if it exists, has ther-

---

[2]The actual range is from 1 to 99% as allowed by the Crystal Ball software.

Table 4-1   Bounds on the design variables

Variable	Lower bound	Nominal value	Upper bound
L_G	0.5	1.0	3.5
w	1.21	1.5	5
h	1.11	1	5
s	0.105	0.5	2

mal resistivity equal to that of the native soil, so that only two different media exist in the region surrounding the cables. We will use the Neher/McGrath modification of the external thermal resistance to account for the presence of the backfill.

Since our goal is to minimize the cost and maximize cable ampacity, we will formulate the cable rating equation in such a way as to make it explicitly dependent on the backfill characteristics. The basic rating equation is given by Equation (4.1),

$$I = \left[\frac{\Delta\theta - W_d[0.5T_1 + n(T_2 + T_3 + T_4)]}{RT_1 + nR(1 + \lambda_1)T_2 + nR(1 + \lambda_1 + \lambda_2)(T_3 + T_4)} \right]^{0.5} = \left[\frac{a + b \cdot T_4}{c + d \cdot T_4} \right]^{0.5} \quad (4.1)$$

with the following definitions of the constants a, b, c, and d and an assumption of a unity load factor:

$$a = \Delta\theta - W_d \cdot (0.5T_1 + nT_2 + nT_3) \qquad b = -n \cdot W_d$$

$$c = R \cdot T_1 + n \cdot R \cdot [(1 + \lambda_1) \cdot T_2 + (1 + \lambda_1 + \lambda_2) \cdot T_3] \quad (4.2)$$

$$d = n \cdot R \cdot (1 + \lambda_1 + \lambda_2)$$

We adopted the above notation since only parameter T_4 in Equation (4.1) depends on the backfill characteristics. When the cable system is contained within an envelope of thermal resistivity ρ_c, the effect of thermal resistivity of the backfill en-

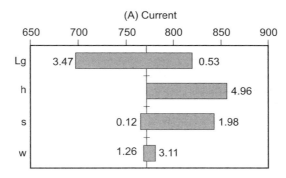

Figure 4-3   Tornado chart for backfill optimization example.

velope being different from that of the surrounding soil ρ_e is handled by first assuming that the thermal resistivity of the medium is ρ_c throughout. A correction is then added algebraically to account for the difference in the thermal resistivities of the envelope and the native soil. The correction to the thermal resistance is given by (Neher & McGrath, 1957)

$$T_4^{corr} = \frac{N}{2\pi}(\rho_e - \rho_c) \cdot \ln(u_b + \sqrt{u_b^2 - 1}) = \frac{N}{2\pi} \cdot (\rho_e - \rho_c) \cdot G_b \qquad (4.3)$$

where
N = number of loaded cables in the envelope
$u_b = \dfrac{L_G}{r_b}$
L_G = depth of laying to the center of the backfill, mm
r_b = equivalent radius of the envelope, mm
$G_b = \ln(u_b + \sqrt{u_b^2 - 1}) \approx \ln \dfrac{2L_G}{r_b}$

The geometric factor G_b contains all the design parameters through the value of L_G and the equivalent radius. The equivalent radius of the thermal envelope is equal to (Neher & McGrath, 1957)

$$r_b = \exp\left[\frac{1}{2} \frac{x}{y}\left(\frac{4}{\pi} - \frac{x}{y} \right) \cdot \ln\left(1 + \frac{y^2}{x^2} \right) + \ln \frac{x}{2} \right] \qquad (4.4)$$

where
$x = \min(w, h)$
$y = \max(w, h)$

Equation (4.4) is only valid for ratios of y/x less than 3 and, since we need an analytical solution to our optimization problem, this is a restriction that will be enforced in all backfill designs considered here.

The external thermal resistance is thus equal to

$$T_4 = \frac{\rho_c}{2\pi} \cdot \ln(u + \sqrt{u^2 + 1}) + \frac{N}{2\pi} \cdot (\rho_e - \rho_c) \cdot G_b \qquad (4.5)$$

where $u = 2L/D_e 2$ with D_e being the cable external diameter (mm). When the depth of laying is much greater than the cable external diameter (this is the usual situation and it is assumed in what follows), we can write

$$T_4 = \frac{\rho_c}{2\pi} \cdot \ln\left(\frac{4L}{D_e} \right) + \frac{N}{2\pi} \cdot (\rho_e - \rho_c) \cdot G_b \qquad (4.6)$$

Substituting Equation (4.6) into Equation (4.1), we obtain the following expression for the cable current as a function of backfill parameters:

$$I = \sqrt{\frac{a + b \cdot \left[\frac{\rho_c}{2\pi} \cdot \ln\left(\frac{4L}{D_e}\right) + \frac{N}{2\pi} \cdot (\rho_e - \rho_c) \cdot G_b\right]}{c + d \cdot \left[\frac{\rho_c}{2\pi} \cdot \ln\left(\frac{4L}{D_e}\right) + \frac{N}{2\pi} \cdot (\rho_e - \rho_c) \cdot G_b\right]}}$$

$$= \sqrt{\frac{a + b_1 \cdot \rho_c \cdot \ln(L) + b_2 \cdot \rho_c + f_1 \cdot (\rho_e - \rho_c) \cdot G_b}{c + d_1 \cdot \rho_c \cdot \ln(L) + d_2 \cdot \rho_c + f_2 \cdot (\rho_e - \rho_c) \cdot G_b}}$$

(4.7)

where the constants are defined as

$$b_1 = \frac{b}{2\pi}; \qquad b_2 = \frac{b}{2\pi}\ln\left(\frac{4}{D_e}\right); \qquad f_1 = \frac{b \cdot N}{2\pi}$$

$$d_1 = \frac{d}{2\pi}; \qquad d_2 = \frac{d}{2\pi}\ln\left(\frac{4}{D_e}\right); \qquad f_2 = \frac{d \cdot N}{2\pi}$$

(4.8)

We can now formulate the backfill optimization problem. The design variables h, w, L_G, and ρ_c will be denoted by a vector $\mathbf{x}$ (defined with n variables for general considerations):

$$\mathbf{x} = \begin{bmatrix} x_1 \\ x_2 \\ x_1 \\ \vdots \\ x_n \end{bmatrix} = \begin{bmatrix} h \\ w \\ L_G \\ \vdots \end{bmatrix}$$

(4.9)

Now we are ready to select the design objective function. In the optimization problem, the objective function to be minimized (or maximized) subject to some constraints is selected according to the design formulation. Two cases may be considered in the cable backfill design problem. First, the total cost of producing the cable trench and laying the backfill material is to be minimized subject to a lower bound on cable ampacity. Alternatively, the cable ampacity itself may be maximized subject to an upper bound on the cost budget. The cost function will be denoted by $C(\mathbf{x}, \boldsymbol{\alpha}, \mathbf{p})$ expressed in terms of the design variables $\mathbf{x}$ of Equation (4.9), the nominal parameters $\boldsymbol{\alpha}$ that include the constants defined by Equations (4.2) and (4.8), as well as some cost indices represented by the vector $\mathbf{p}$, which depends on the type of cost function under consideration. For example, when the cost of cable trench production and backfill is optimized, the components of the vector $\mathbf{p}$ may represent the individual costs of excavation, earth disposal, and so on.

The values assigned to the design variable are usually restricted by several inequality constraints. Some of these constraints represent physical lower and upper bounds on variables. For example, the constraint $\rho_c \geq \rho_c^l$ represents a lower bound on the backfill thermal resistivity associated with the available backfill material. Also, the constraint $w \leq w^u$ represents an upper bound on the width dictated by the available right-of-way. On the other hand, inequality constraints may represent interrelationships between design variables. An important functional constraint that may be considered in a particular problem formulation is the cable ampacity constraint in the form $I \geq I^l$, where I denotes the cable current producing the maximum conductor temperature associated with certain values of backfill design parameters, and I^l denotes the required cable ampacity. In general, we shall denote the set of q inequality constraints considered in the optimization by the vector

$$\mathbf{g(x)} \geq \mathbf{0} \qquad (4.10)$$

where a component $g_i(\mathbf{x})$ may represent any cable inequality constraints described above.

In the cable design problem, some equality constraints of the form

$$\mathbf{h(x)} = \mathbf{0} \qquad (4.11)$$

may also be defined. The vector $\mathbf{h(x)}$ in Equation (4.11) has m components—$h_k(\mathbf{x})$; $k = 1, \ldots, m$—which may denote a specific relationship between design parameters to be satisfied. For example, in a particular application, the cable may be centered vertically in the backfill, hence the equality constraint $L = L_G$.

We will illustrate how the optimization problem is formulated in the following example.

Example 4.2

In this example, we will minimize the cost of backfilling the cable trench with the constraints imposed on the minimal current that the cable should carry.

The cost of backfilling the trench includes the cost of removing the asphalt pavement, excavation, disposal of earth, installation of the backfill, and repaving with asphalt. Table 4-2 (based on El-Kady, 1982) shows the assumed civil engineering costs for cable trenches in suburban Northeastern United States. The total costs are expressed in \$/linear meter of trench right-of-way. The values in Table 4-2 may not reflect contemporary costs but are sufficient for the illustrative purpose of the present example.

Based on the values in Table 4-2, the cost function can be formulated as

$$C(h, w, L_G) = 31.3 \cdot (w + 0.6) + 47 \cdot w \cdot L_G + 23.5 \cdot w \cdot h$$

Table 4-2   Assumed cable trench costs

Task	Cost basis	Cost term
Remove asphalt pavement	\$13.5/m^2	$w + 0.6$
Excavation	\$36.6/m^3	$w \cdot L_G + w \cdot h/2$
Dispose of earth	\$10.4/m^3	$w \cdot L_G + w \cdot h/2$
Backfill with thermal sand	\$23.5/m^3	$w \cdot h$
Repave with asphalt	\$17.8/m^2	$w + 0.6$

where an extra 0.3 m is added to w on each side of the right-of-way for removing and repaving with asphalt. Constants 31.3, 47, and 23.5 were computed from Table 4-2 as follows: asphalt removal and repavement = 13.5 + 17.8 = \$31.3; excavation and disposal of earth = 36.6 + 10.4 = \$47; cost of backfill = \$23.5.

In the above expression, the cost associated with the backfill material is not included and all dimensional variables are in meters.

The cost function is to be minimized subject to a lower bound on cable ampacity and other parameter constraints. We will consider cable model No. 2 in a backfill. The installation is shown in Figure 4-4.

We will first consider a case of two-dimensional optimization in with the trench width w and the depth L_G are the only design variables. The other parameters are set to fixed values, namely, $h = 0.6$ m, native soil thermal resistivity $\rho_e = 1.2$ K $\cdot$ m/W, and backfill resistivity $\rho_c = 0.5$ K $\cdot$ m/W. We also assume that the remaining parameters are the same as specified in Table A1; that is, $\theta_c = 90°C$, $\theta_{amb} = 15°C$, $W_d = 0.52$ W/m, $T_1 = 0.754$ K $\cdot$ m/W, $T_2 = 0.040$ K $\cdot$ m/W, $T_3 = 0.047$ K $\cdot$ m/W, $\lambda_1 = 0.123$, $\lambda_2 = 0.210$, and the conductor resistance at 90°C $R = 0.000033$ ohm/m. Even though the laying conditions are different from those in Appendix A, the above parameters remain unchanged.

The constraints on the design variables are as follows:

$$0.2 \leq w \leq 0.6$$
$$0.8 \leq L_G \tag{4.13}$$

We will also assume that the maximum current must be greater than 0.82 kA. We observe that with the above assumptions, the width of the trench is always

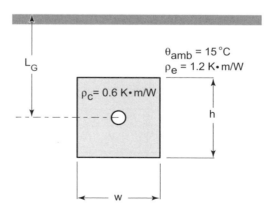

Figure 4-4   Single cable in backfill.

shorter than its height. In view of this, the value of the equivalent radius in Equation (4.4) becomes

$$r_b = \exp\left[\frac{w}{1.2}\left(\frac{4}{\pi} - \frac{w}{0.6} \right)\ln\left(1 + \frac{0.36}{w^2}\right) + \ln\frac{w}{2} \right] \tag{4.14}$$

and the geometric factor becomes

$$G_b = \ln\frac{2L_G}{r_b} = \ln\left(\frac{4L_G}{w}\right) - \frac{w}{1.2}\left(\frac{4}{\pi} - \frac{w}{0.6}\right) \cdot \ln\left(1 + \frac{0.36}{w^2}\right) \tag{4.15}$$

The constants defined by Equations (4.2) and (4.8) are computed next:

$$a = \Delta\theta - W_d \cdot (0.5T_1 + nT_2 + nT_3) = (90 - 15) - 0.52$$
$$\cdot (0.5 \cdot 0.754 + 3 \cdot 0.040 + 3 \cdot 0.047) = 74.67$$

$$b = -n \cdot W_d = -3 \cdot 0.52 = -1.56$$

$$c = R \cdot T_1 + n \cdot R \cdot [(1 + \lambda_1) \cdot T_2 + (1 + \lambda_1 + \lambda_2) \cdot T_3] \tag{4.16}$$
$$= 0.000033 \cdot [\, 0.754 + 3 \cdot (1 + 0.123) \cdot 0.04 + 3 \cdot (1 + 0.123 + 0.210) \cdot 0.047]$$
$$= 3.55\text{E-}05$$

$$d = n \cdot R \cdot (1 + \lambda_1 + \lambda_2) = 3 \cdot 0.000033 \cdot (1 + 0.123 + 0.210) = 0.000132$$

$$b_1 = \frac{b}{2\pi} = \frac{-1.56}{2\pi} = -0.248; \qquad b_2 = \frac{b}{2\pi}\ln\left(\frac{4}{D_e}\right) = \frac{-1.56}{2\pi}\ln\frac{4}{0.175} = -0.777;$$

$$d_1 = \frac{d}{2\pi} = \frac{0.000132}{2\pi} = 2.10\text{E-}05; \qquad d_2 = \frac{d}{2\pi}\ln\left(\frac{4}{D_e}\right) = \frac{0.000132}{2\pi}\cdot\ln\frac{4}{0.175}$$
$$= 6.57\text{E-}05 \tag{4.17}$$

$$f_1 = \frac{b \cdot N}{2\pi} = \frac{-1.56 \cdot 1}{2\pi} = -0.248; \qquad f_2 = \frac{d \cdot N}{2\pi} = \frac{0.000132 \cdot 1}{2\pi} = 2.10\text{E-}05$$

Substituting Equations (4.15), (4.16), and (4.17) into Equation (4.7) and noting that, in this example, $L = L_G$, we obtain the desired constraint on the minimum current as a function of the two design parameters:

$$I = \sqrt{\frac{a + b_1 \cdot \rho_c \cdot \ln(L) + b_2 \cdot \rho_c + f_1 \cdot (\rho_e - \rho_c) \cdot G_b}{c + d_1 \cdot \rho_c \cdot \ln(L) + d_2 \cdot \rho_c + f_2 \cdot (\rho_e - \rho_c) \cdot G_b}}$$
$$= \sqrt{\frac{74.28 - 0.124 \cdot \ln(L_G) - 0.124 \cdot G_b}{0.684 + 0.105 \cdot \ln(L_G) + 0.105 \cdot G_b}} \geq 0.82 \tag{4.18}$$

The solution is $w = 0.2$ m and $L_G = 0.8$ m, for which the cost function has the minimum value of \$35.4/linear m.

Figure 4-5 illustrates the optimization aspects of this problem. From this figure, we can see that the optimal point is located at the intersection of the trench depth and the trench width lower bounds (the active constraints) where other constraints are already satisfied.

The ampacity constraint is not active at the constrained minimum of Figure 4-5. In other words, the optimal solution obtained provides a cable capability more than the minimum required of 0.82 kA. In fact, the cable ampacity at the optimal solution is 0.853 A.

To assess the gain in the cable ampacity achieved by the application of the backfill, we now calculate the cable current in a uniform soil with the resistivity $\rho_e = 1.2$ K · m/W. The external thermal resistance of the cable laid at $L = 0.8$ m is equal to

$$T_4 = \frac{\rho_e}{2\pi} \cdot \ln\frac{4L}{D_e} = \frac{1.2}{2\pi} \cdot \ln\frac{4 \cdot 0.8}{0.175} = 0.555 \text{ K} \cdot \text{m/W}$$

Hence,

$$I = \left[\frac{\Delta\theta - W_d[0.5T_1 + n(T_2 + T_3 + T_4)]}{RT_1 + nR(1 + \lambda_1)T_2 + nR(1 + \lambda_1 + \lambda_2)(T_3 + T_4)} \right]^{0.5}$$

$$= 100 \left[\frac{75 - 0.52 \cdot [0.5 \cdot 0.754 + 3 \cdot (0.04 + 0.047 + 0.555)]}{0.33 \cdot \{0.754 + 3 \cdot [(1 + 0.123) \cdot 0.04 + (1 + 0.123 + 0.210) \cdot (0.047 + 0.555)]\}} \right]^{0.5}$$

$$= 824 \text{ A}$$

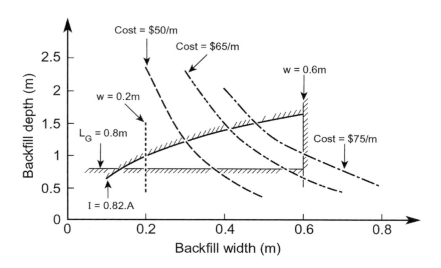

Figure 4-5 Cost minimization subject to ampacity and parameter.

Thus, we gained about 30 A in ampacity at a cost of \$35.4/m of trench. We note that if a higher cable rating is required, say 900 A, the solution will be bounded by the ampacity constraint with variables $w = 0.48$ m and $L_G = 0.8$ m, and with minimal cost of \$58.4/m.

When more design variables are considered in the optimization, more flexibility in reducing the cost function is usually gained. However, this flexibility is restricted by the inclusion of more boundary constraints on the extra variables. In the case considered in this example, there are four design variables that can be included in the optimization process, namely, h, w, L_G, and ρ_c. We will now include all of these variables in the optimization with the following additional bounds [note that bounds on w and L_G are given by Equation (4.13)]. The additional bounds on the backfill thickness are

$$0 \leq h \leq 0.8 \qquad (4.19)$$

Five different types of backfill material are considered: sea sand, crushed limestone, river sand, loamy sand, and crushed dolomite. The corresponding thermal resistivities are bounded as

$$0.4 \leq \rho_c \leq 0.8 \qquad (4.20)$$

Since the lowest current for this system is 824 A, we will seek an optimal solution requiring a current value of at least 900 A. The optimization procedure yields a solution constrained at the lower bound of the backfill thermal resistivity, the lower bound of the backfill depth, and the lower bound of the current, with the optimal cost of \$49.8/m. We can observe that the cost is now reduced by about \$10 per linear meter compared with the case when only two variables were considered.   ■

When a cost budget is available for cable trench production, the flexibility of assessing the trench cost may be exploited to maximize the cable ampacity. Obviously, it is assumed that the upper bound assigned for the trench cost is at least sufficient to provide the minimum ampacity required. Ampacity maximization involves the selection of a set of optimal backfill design parameters that supports the cable against service fluctuations. In other words, a cable design with optimally selected backfill parameters may withstand higher overloading for longer periods than other arbitrary designs of the same trench cost. The formulation of the optimization problem is quite similar to the one previously defined, with the objective function being the current maximization.

4.2.3 Assessment of Sensitivities to Ambient Fluctuations

The thermal design of buried cable systems is usually based on assumed constant values of the soil thermal resistivity and the ambient temperature. Although relatively high values of these variables are normally assumed, based on the designer's estimates and the available records, both the soil resistivity and ambient temperature may experience unexpected variations during future cable service. In order to assess the effect of these changes on cable ampacity, a pair of dimensionless sensitivity measures are defined as follows (El-Kady, 1982):

$$S_\rho = \frac{\partial I}{\partial \rho_e} \cdot \frac{\rho_e}{I} \tag{4.21}$$

$$S_\theta = \frac{\partial I}{\partial \theta_{amb}} \cdot \frac{\theta_{amb}}{I} \tag{4.22}$$

The sensitivity parameters given by these definitions indicate the sensitivity of the cable ampacity with respect to respective parameters at the nominal point. In order to compute the sensitivity coefficients, we will rewrite Equation (4.7) as

$$I = \sqrt{\frac{a + b_1 \cdot \rho_c \cdot \ln(L) + b_2 \cdot \rho_c + f_1 \cdot (\rho_e - \rho_c) \cdot G_b}{c + d_1 \cdot \rho_c \cdot \ln(L) + d_2 \cdot \rho_c + f_2 \cdot (\rho_e - \rho_c) \cdot G_b}} = \sqrt{\frac{E}{Z}} \tag{4.23}$$

Taking partial derivatives from this equation [θ_{amb} is contained in the parameter a defined in Equation (4.2)], we obtain

$$S_\rho = \frac{\rho_e}{2I^2} \cdot \frac{(f_1 \cdot Z - f_2 \cdot E) \cdot G_b}{Z^2} \tag{4.24}$$

$$S_\theta = \frac{\theta_{amb}}{2I^2} \cdot \frac{-1}{Z} \tag{4.25}$$

In practical cable designs, both S_ρ and S_θ are negative, indicating that the cable ampacity associated with maximum conductor temperature and a certain cable configuration decreases as the soil thermal resistivity or the ambient temperature increase.

Sensitivity coefficients S_ρ and S_θ can be used in the optimization procedure with appropriately selected upper bounds; that is,

$$|S_\rho| \leq S_\rho^u$$
$$|S_\theta| \leq S_\theta^u \tag{4.26}$$

Application of the sensitivity coefficients is illustrated in the following example.

Example 4.3

We will consider a circuit composed of cable model No. 4 arranged as shown in Figure 4-6. Our goal is to maximize cable ampacity subject to a cost constraint. In

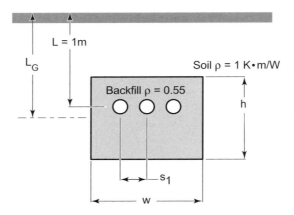

Figure 4-6 Cable model No. 4 in a backfill.

maximizing cable ampacity I, the proximity effect tends to increase the cable spacing s_1, whereas, for a given backfill width w, the circulating current losses tend to decrease so that the cables are more deeply embedded in the thermal backfill. The optimal value of s that maximizes the ampacity is obtained when the two effects balance each other.

We will assume that the cable is located in a fluidized backfill of thermal resistivity $\rho_c = 0.6$ K · m/W. The cost indices are specified in Table 4-3.

The cable depth is fixed at $L = 1.0$ m and the optimization will be carried out with respect to cable spacing s_1, the backfill depth L_G, width w, and height h.

The parameters of this cable, extracted from Appendix A, are summarized in Table 4-4.

In addition to the design parameters of the backfill, the objective function should now contain the cable spacing s_1 as an explicit variable. The spacing appears in two places in the rating equation; namely, in the calculation of the circulating current loss factor and in the external thermal resistance.

Cable No. 4 has a lead sheath with copper reinforcing tapes and copper armor wires. For the purpose of the calculation of the loss factors, these three components are first combined and the equivalent circulating current loss factor computed. Then, this loss factor is split into two components, one corresponding to the sheath and the other to the armor on the basis of their resistances.

Table 4-3 Cost parameters for backfill optimization for cable No. 4

Task	Cost basis	Cost term
Excavation	\$16.5 /m³	$w \cdot L_G + w \cdot h/2$
Disposal of earth	\$13.5 /m³	$w \cdot L_G + w \cdot h/2$
Backfill with thermal sand	\$28.5 /m³	$w \cdot h$

Table 4-4 Parameters of cable model No. 4

Cable ampacity	I (A)	771
Conductor resistance at θ_{max}	R (ohm/km)	0.0356
Sheath loss factor	λ_1	0.229
Armor loss factor	λ_2	0.990
Thermal resistance of insulation	T_1 (K · m/W)	0.568
Thermal resistance of armor bedding	T_2 (K · m/W)	0.082
Thermal resistance of armor serving	T_3 (K · m/W)	0.066
External thermal resistance, 100% LF	T_4 (K · m/W)	0.814
Losses in the conductor	W_c (W/m)	22.11
Dielectric losses	W_d (W/m)	5.46
Total joule losses per cable	W_I (W/m)	49.08
Total losses per cable	W (W/m)	54.53

The circulating current loss factor for the middle cable, which is usually the hottest, is obtained from (see Section 1.6.4.2)[3]

$$\lambda_{1m} = \frac{R_s}{R} \frac{Q^2}{R_s^2 + Q^2} \qquad Q = X - \frac{X_m}{3} \qquad (4.27)$$

where
R = conductor resistance at the maximum operating temperature, ohm/m
R_s = sheath resistance at the maximum operating temperature, ohm/m

The reactance X is obtained as described in Section 1.6.4.1:

$$X = \omega \cdot 2 \cdot 10^{-7} \ln \frac{2s_1}{d} \qquad (4.28)$$

with
s_1 = distance between conductor axis in the electrical section being considered, mm
d = mean diameter of the sheath, mm
X_m = mutual reactance between the sheath of the outer cable and the conductors of the other two, Ω/m = $2\omega10^{-7} \cdot \ln(2)$, Ω/m

In our example, the conductor resistance at the operating temperatures of 85°C is equal to $R = 0.356 \cdot 10^{-4}$ Ω/m. Assuming, that the sheath and armor temperatures are 70 and 68°C, respectively, the equivalent resistance of the sheath and the reinforcing tape is $R_s = 0.203 \cdot 10^{-3}$ Ω/m. The rms diameter of the equivalent sheath is equal to $d* = 0.0767$ m. The armor resistance is $R_a = 0.0692 \cdot 10^{-3}$ Ω/m; thus, the combined resistance of sheath, tape, and armor is $R_{sa} = 0.516 \cdot 10^{-4}$ Ω/m. This val-

---

[3]For cables in a close, flat formation, the outer cable may be the hottest. A formula that is appropriate for this case should then be used. Since the purpose of this example is to illustrate the approach, we will assume that Equation (4.27) is applicable in this case as well.

ue is used in Equation (4.27) for the resistance R_s. The loss factors λ_1 and λ_2 are then obtained as

$$\lambda_1 = \eta \cdot \lambda_{1m} \qquad \lambda_2 = (1 - \eta) \cdot \lambda_{1m} \qquad (4.29)$$

with

$$\eta = \frac{R_a}{R_s + R_a} = \frac{0.0692}{0.203 + 0.0692} = 0.254 \qquad (4.30)$$

The external thermal resistance of the middle cable is obtained from a modified Equation (4.6) (see Section 1.6.6.5):

$$T_4 = \frac{\rho_c}{2\pi} \cdot \ln\left\{\left(\frac{4L}{D_e}\right) \cdot \left[1 + \left(\frac{2L}{s_1}\right)^2\right]\right\} + \frac{N}{2\pi} \cdot (\rho_e - \rho_c) \cdot G_b \qquad (4.31)$$

The constants c and d in Equation (4.2) will now also depend on the cable spacing; hence, they will be rewritten as

$$c = c_1 + c_2 \cdot \lambda_{1m} \qquad d = d_3 + d_3 \cdot \lambda_{1m} \qquad (4.32)$$

with

$$c_1 = R \cdot T_1 + n \cdot R \cdot (T_2 + T_3) \qquad c_2 = n \cdot R \cdot (\eta \cdot T_2 + T_3) \qquad d_3 = n \cdot R \quad (4.33)$$

Substituting Equations (4.31) and (4.32) into Equation (4.1), we obtain the following expression for the cable current as a function of cable spacing and backfill parameters:

$$I = \sqrt{\frac{a + b \cdot \left[\frac{\rho_c}{2\pi} \cdot \ln\left\{\left(\frac{4L}{D_e}\right)\left[1 + \left(\frac{2L}{s_1}\right)^2\right]\right\} + \frac{N}{2\pi} \cdot (\rho_e - \rho_c) \cdot G_b\right]}{c + d \cdot \left[\frac{\rho_c}{2\pi} \cdot \ln\left\{\left(\frac{4L}{D_e}\right)\left[1 + \left(\frac{2L}{s_1}\right)^2\right]\right\} + \frac{N}{2\pi} \cdot (\rho_e - \rho_c) \cdot G_b\right]}}$$

$$= \sqrt{\frac{a_1 + b_3 \cdot \ln\left[1 + \left(\frac{2L}{s_1}\right)^2\right] + f_3 \cdot G_b}{c_3 + c_4 \cdot \lambda_{1m} + (1 + \lambda_{1m}) \cdot \left\{d_4 \cdot \ln\left[1 + \left(\frac{2L}{s_1}\right)^2\right] + f_3 \cdot G_b\right\}}} \qquad (4.34)$$

where the constants are defined as

$$a_1 = a + b_3 \ln\left(\frac{4L}{D_e}\right); \qquad b_3 = b_1\rho_c; \qquad f_3 = f_1(\rho_e - \rho_c); \qquad c_3 = c_1 + d_5;$$

$$\qquad (4.35)$$

$$c_4 = c_2 + d_5; \qquad d_4 = \frac{d_3 \cdot \rho_c}{2\pi}; \qquad d_5 = d_4 \ln\left(\frac{4L}{D_e}\right); \qquad f_4 = d_3 \frac{N}{2\pi}(\rho_e - \rho_c)$$

The constraints on the design variables are defined as follows:

1. Cost. Based on the values in Table 4-3, the constraint on the cost function can be formulated as

$$C(h, w, L_G) = 30 \cdot w \cdot L_G + 43.5 \cdot w \cdot h \leq C_1 \tag{4.36}$$

The constant C_1 will be a parameter in the analysis.

2. Depth of backfill center. The center of the cable is already fixed at 1.0 m. Assuming that we need at least 0.5 m of the backfill above and below the cable center, we have the following constraints:

$$0 \leq L_G - h/2 \leq 0.5$$

$$L_G + h/2 \geq 1.5$$

3. Width of the backfill. We need at least 0.5 m of the backfill on each side of the center of the outer cables. Therefore,

$$w \geq 1.0 + 2s_1$$

4. Height of the backfill. We will assume that the backfill is at least 0.6 m high. That is,

$$h \geq 0.6$$

5. Height-to-width restriction. Since the cables are laid fairly far apart, we need to restrict the height of the backfill to limit the costs. We will require that the height be smaller than the width. This constraint is not really required in the optimization program but is added here to facilitate the construction of the equivalent radius function:

$$h \leq w$$

6. Cable spacing. The spacing of the must not be smaller than the cable diameter, equal to 0.105 m; that is,

$$s_1 \geq 0.105$$

We observe that with the above assumptions, the width of the trench is always greater than its height. In view of this, the value of the equivalent radius in Equation (4.4) becomes

$$r_b = \exp\left[\frac{1}{2} \frac{h}{w} \left(\frac{4}{\pi} - \frac{h}{w} \right) \cdot \ln\left(1 + \frac{w^2}{h^2} \right) + \ln\frac{h}{2} \right] \tag{4.37}$$

and the geometric factor is equal to

$$G_b = \ln\frac{2L_G}{r_b} = \ln\left(\frac{4L_G}{h}\right) - \frac{h}{2w}\left(\frac{4}{\pi} - \frac{h}{w}\right) \cdot \ln\left(1 + \frac{w^2}{h^2}\right) \qquad (4.38)$$

The constants defined by Equations (4.2), (4.8) and (4.35) are computed next.

$$a = \Delta\theta - W_d \cdot (0.5T_1 + nT_2 + nT_3) = (85 - 20) - 5.46 \cdot (0.5 \cdot 0.568 + 0.083 + 0.066)$$
$$= 62.64$$

$$b = -n \cdot W_d = -5.46 \qquad (4.39)$$

$$c_1 = R \cdot T_1 + n \cdot R(T_2 + T_3) = 0.356E - 4 \cdot (0.568 + 0.083 + 0.066) = 2.55E-5$$

$$c_2 = n \cdot R \cdot (\eta \cdot T_2 + T_3) = 0.356E - 4 \cdot (0.254 \cdot 0.083 + 0.066) = 3.10E-06$$

$$b_1 = \frac{b}{2\pi} = \frac{-5.46}{2\pi} = -0.869 \qquad b_2 = \frac{b}{2\pi}\ln\left(\frac{4}{D_e}\right) = \frac{-5.46}{2\pi}\ln\frac{4}{0.105} = -3.16$$

$$d_3 = n \cdot R = 1 \cdot 0.356E - 4 = 0.356E - 4$$

$$(4.40)$$

$$d_4 = \frac{d_3 \cdot \rho_c}{2\pi} = \frac{0.0000356 \cdot 0.6}{2\pi} = 3.40E-06$$

$$d_5 = d_4 \cdot \ln\left(\frac{4L}{D_e}\right) = 3.4E - 6 \cdot \ln\frac{4 \cdot 1}{0.105} = 1.24E-05$$

$$a_1 = a + b_3\ln\left(\frac{4L}{D_e}\right) = 62.64 - 0.521 \cdot \ln\left(\frac{4 \cdot 1.0}{0.105}\right) = 60.74$$

$$b_3 = b_1\rho_c = -0.869 \cdot 0.6 = -0.521$$

$$c_4 = c_2 + d_5 = 3.10E - 6 + 1.24E - 5 = 1.55E - 5 \qquad (4.41)$$

$$f_1 = \frac{b \cdot N}{2\pi} = \frac{-5.46 \cdot 3}{2\pi} = -2.61 \qquad f_3 = f_1(\rho_e - \rho_c) = -2.61 \cdot (1 - 0.6) = -1.04$$

$$f_4 = d_3\frac{N}{2\pi}(\rho_e - \rho_c) = \frac{0.0000356 \cdot 3 \cdot (1 - 0.6)}{2\pi} = 6.80E-06$$

Substituting Equations (4.39) to (4.41) into Equation (4.34), we obtain the desired objective function for the minimum current as a function of the design parameters:

$$I = \sqrt{\frac{a_1 + b_3 \cdot \ln\left[1 + \left(\frac{2L}{s_1}\right)^2\right] + f_3 \cdot G_b}{c_3 + c_4 \cdot \lambda_{1m} + (1 + \lambda_{1m}) \cdot \left\{d_4 \cdot \ln\left[1 + \left(\frac{2L}{s_1}\right)^2\right] + f_4 \cdot G_b\right\}}}$$

(4.42)

$$= 100 \cdot \sqrt{\frac{60.74 - 0.521 \cdot \ln\left[1 + \left(\frac{2}{s_1}\right)^2\right] - 1.04 \cdot G_b}{0.38 + 0.155 \cdot \lambda_{1m} + (1 + \lambda_{1m}) \cdot \left\{0.34 \cdot \ln\left[1 + \left(\frac{2}{s_1}\right)^2\right] + 0.068 \cdot G_b\right\}}}$$

For C_1 = \$300, the maximum current is 890 A, with w = 3.4 m, L_G = 0.75 m, h = 1.5 m, and s_1 = 1.2 m, for which the cost function has the value of \$298/linear m.

In Figures 4-7 and 4-8, contours of S_ρ and S_θ are shown together with the cost contours in the two-dimensional space of the backfill width w and the thickness h. The curves are drawn at values of trench depth L_G = 1 m, backfill thermal resistivity ρ_c = 0.6 K · m/W, cable depth L = 1 m, and spacing s_1 = 0.25 m. In Figure 4-7, the

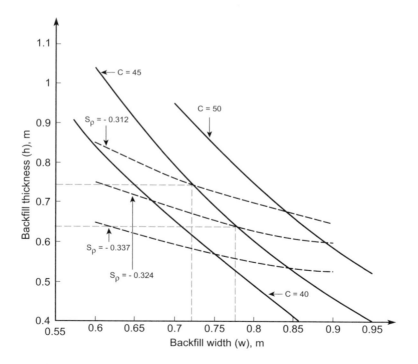

Figure 4-7 Cost and $I - \rho_e$ sensitivities for three cables in a backfill.

sensitivity contours S_ρ are drawn at a nominal value of soil thermal resistivity $\rho_e = 1.0 \text{ K} \cdot \text{m/W}$ and the cable ampacity $I = 0.72 \text{ kA}$.

In Figure 4-8, the sensitivity contours S_θ are drawn at a nominal value of ambient temperature $\theta_{amb} = 20°\text{C}$, cable ampacity $I = 0.72 \text{ kA}$, and soil thermal resistivity $\rho_e = 1.0 \text{ K} \cdot \text{m/W}$.

The information supplied by sensitivity charts similar to those of Figures 4-7 and 4-8 is useful in predicting variations in the allowable cable ampacity associated with ambient fluctuations. Let us consider, for example, the design values of $w = 0.75 \text{ m}$ and $h = 0.65 \text{ m}$. The sensitivity of cable ampacity to the soil thermal resistivity for these values of the design parameters is read from Figure 4-7 as $S_\rho = -0.324$. That is, an increase in soil thermal resistivity by 0.5 K·m/W from the nominal value of 1.0 K·m/W will cause a change ΔI_ρ from the nominal value of 0.72 kA, where

$$\Delta I_\rho = \frac{I}{\rho_e} \cdot S_\rho \cdot \Delta \rho_e = \frac{0.72}{1.0} \cdot (-0.324) \cdot 0.5 = -0.11 \text{ kA}$$

giving a steady-state maximum current of $0.72 - 0.11 = 0.61 \text{ kA}$.

On the other hand, from Figure 4-8, the same cable design provides ampacity sensitivity to ambient temperature of $S_\theta = -0.176$. That is, an increase in ambient

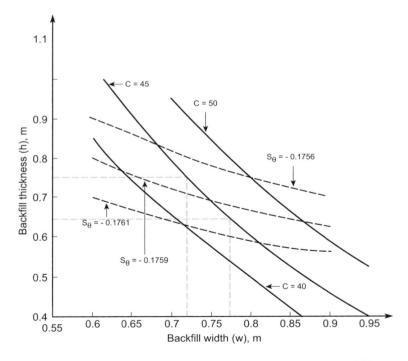

Figure 4-8 Cost and $I - \theta_{amb}$ sensitivities for three cables in a backfill.

temperature by 15°C from the nominal value of 25°C will cause a change in the allowable cable current equal to

$$\Delta I_\theta = \frac{I}{\theta_{amb}} \cdot S_\theta \cdot \Delta\theta_{amb} = \frac{0.72}{20} \cdot (-0.176) \cdot 15 = -0.10 \text{ kA}$$

resulting in a maximum steady-state cable current of only $0.72 - 0.10 = 0.62$ kA.

When the above cable design experiences, simultaneously, changes in soil thermal resistivity and ambient temperature, the total change in the maximum cable current is simply calculated as

$$\Delta I = \Delta I_\rho + \Delta I_\theta = -0.11 - 0.10 = -0.21 \text{ kA}$$

resulting in a maximum steady-state cable current of only 0.51 kA.

The inclusion of the cost curves in Figures 4-7 and 4-8 facilitates the trade-off between different design alternatives since it reduces the task to a simple comparison between sensitivity profiles associated with a given cost. For example, we can observe from Figure 4-7 that a cable design with $w = 0.78$ m and $h = 0.75$ m may be considered better than a design with $w = 0.78$ m and $h = 0.65$ m. Although both designs provide the same cable ampacity of 0.72 kA at the same cost of \$45/linear m, the first design possesses ampacity sensitivity to soil resistivity of only $S_\rho = -0.312$ compared with $S_\rho = -0.324$ associated with the second design. Also, from Figure 4-8, the first design is better in terms of sensitivity to the ambient temperature. Depending on the local conditions of cable installation (soil and weather variations), the designer may decide which design alternative is preferable. ∎

4.3 PARAMETER UNCERTAINTY IN RATING ANALYSIS OF CABLES CROSSING UNFAVORABLE THERMAL ENVIRONMENTS

Underground power cable installations have traditionally been designed using deterministic criteria. A typical design process involves specification of required current-carrying capability of the cable circuit by the system planning unit of the utility, followed by the selection of the cable type and method of burial established by the design department. Once the installation environment is determined, rating computations can be performed to enable the selection of a cable that satisfies the ampacity requirements. The major parameters entering ampacity calculations are the cable dimensions and materials, soil and backfill/duct bank thermal resistivity values, backfill/duct bank dimensions, cable locations, and the ambient temperature.

The most important factors determining the cable operating temperature are the soil thermal resistivity, the ambient temperature, and the cable load. All these parameters are random in nature and their most probable values have to be predicted if the cable operating temperature is to be computed with reasonable accuracy. This is particularly important in the case of cables crossing an unfavorable thermal environment since a large reduction of cable ampacity might be avoided if the stochastic nature of the design parameters is taken into account.

Since there are many uncertainties encountered in cable rating, a probabilistic approach should be considered in some applications, especially those dealing with future investment options. A probabilistic approach to cable rating offers an opportunity to consider variations in the most important parameters affecting the rating of a cable. Such an approach yields the probability distribution function of the cable temperatures. Probabilistic techniques take advantage of the fact that not all variables take on their "worst" values at the same time (as deterministic analysis would assume); therefore, the "worst conditions," as per deterministic design, rarely, if ever occur.

We will introduce in this Chapter a probabilistic method for thermal analysis of power cables in the system planning process (Anders et al., 1988a). We will apply a Monte Carlo simulation for calculating the temperature rise of power cables with provision for statistical variations of various soil, boundary, and loading conditions. The probability distribution of cable operating temperature can be used in several ways in the system design process. We will illustrate several such applications. In one of them, we will convolve the probability distribution of the cable conductor with the probability distribution of maximum permissible operating temperatures. As a result of this convolution, a probability of exceeding permissible operating temperatures is obtained. In another application, we will look at a decision concerning cable uprating. Finally, we will show how a probabilistic approach can be used to help to make decisions about backfill selection. We will start our discussion by selecting a sample cable system that will be used throughout this chapter to illustrate the various concepts.

4.3.1 Sample Cable System

To illustrate various concepts applied in the probabilistic analysis of underground power cables, a realistic system composed of four 115 kV circuits supplying two stations in the city of Toronto, Canada will be used as an illustrative example throughout discussions in this chapter. The system is shown in Figure 4-9.

The cable arrangement is shown in Figure 4-10. Since this is a radial system, the total load at stations A and B will give total current flowing in the cables.

The cables under consideration are 1250 kcmil, hollow-core, low-pressure, liquid-filled cables (LPLF) with copper segmental construction, lead sheath, and rubber bitumen sandwich.

Typical environmental parameters for the area of the city of Toronto would be ambient soil temperature of 20°C, native soil thermal resistivity of 1.2 K · m/W, and backfill thermal resistivity of 2.0 K · m/W.[4]

Under the above conditions, each cable can carry a maximum steady-state current of 800 A. The system is designed and operated in such a way that a loss of a

---

[4]A river sand of high thermal resistivity was used during this cable system construction.

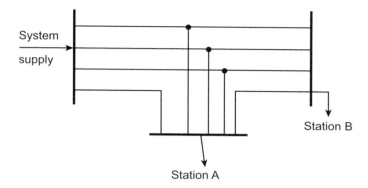

Figure 4-9   Two stations supplied by a four-circuit cable system (from Anders et al., 1988b, with permission from CIGRE).

single circuit should not cause an overload of the remaining circuits under steady-state conditions.

The peak load forecast for stations A and B, translated into current flowing in one cable with one circuit out of service, is given in Table 4-5.

Comparing the computed cable ampacity with the values in Table 4-5, we conclude that the steady-state limits will be exceeded in year 2. Thus, a new circuit should be installed by this year to meet expected load growth.

The following section describes the statistical analysis of the most important parameters in cable rating calculations.

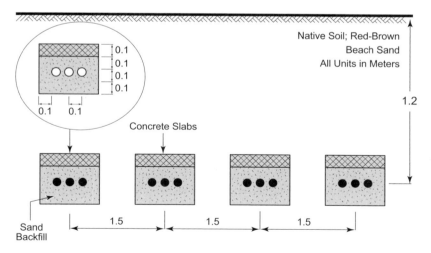

Figure 4-10   Cable arrangement in the backfill for probabilistic analysis (from Anders et al., 1988b, with permission from CIGRE).

Table 4-5 Loading in the study period

Year	1	2	4	7	9
Current (A)	787	818	892	980	1053

4.3.2 Statistical Variations of Cable Circuit Parameters

For a given cable type and size, the temperature θ is, in general, a function of the cable loading as well as other soil and ambient parameters that constitute the thermal circuit of the cable. The following parameters exhibit statistical variations:

ρ_e, native soil thermal resistivity (K·m/W)

ρ_c, backfill thermal resistivity (K·m/W)

I, cable loading (A)

θ_{amb}, ambient temperature (°C)

Using conventional techniques, the cable temperature is evaluated on the basis of known or assumed values of cable circuit parameters (base-case or nominal values). In practice, however, these parameters are subject to random operating, geographical, and seasonal variations that affect the cable temperatures and, hence, the allowable loading level of any particular cable. The random parameter variations are described by associated probability distributions, which are evaluated on the basis of statistical and historical records. The following sections describe how these distributions are obtained from historical records. We will use an actual cable circuit operating in Toronto. Hence, all the environmental data will refer to the conditions in that city.

4.3.2.1 Load Probability Distribution.

The probability distribution of the cable loadings is normally obtained from the historical records for circuits under consideration. A histogram showing the number of occurrences of a particular MVA daily peak demand divided by the monthly peak was obtained for the month of August in three consecutive years and is shown in Figure 4-11. The month of August was chosen because this is the most critical month from cable operating temperature point of view for the circuit chosen to illustrate the concepts.

4.3.2.2 Ambient Temperature Probability Distribution.

To obtain the probability density function of the air ambient mean temperature, the weather records for downtown Toronto provided by Environment Canada for a 66-year period were analyzed. Figure 4-12 shows the histogram of the mean daily temperature in the month of August based on the 66 years of data, consisting of 2046 observations.

4.3.2.3 Native Soil and Backfill Probability Distributions.

The thermal resistivity of the thermal backfill and the surrounding native soil are governed by

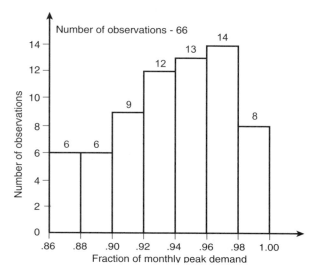

Figure 4-11 Histogram of the daily peak load for the month of August (from Anders et al., 1988a, with permission from IEEE Press).

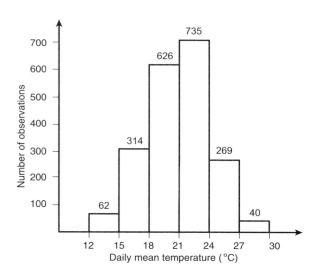

Figure 4-12 Histogram of the ambient temperature in downtown Toronto for the month of August (from Anders et al., 1988a, with permission from IEEE Press).

their composition, density, and moisture content. For a given backfill/soil type, the moisture content is the most important factor influencing the thermal resistivity. For the sands in the Toronto area, the curves relating variation of thermal resistivity with changes in the soil moisture content were established as shown in Figure 4-13. In order to determine these relationships, several field tests and laboratory measurements should be performed along the cable route. Alternatively, published relationships between these quantities could be used (Groeneveld et al., 1984; Abdel-Hadi, 1978).

The observed high values of thermal resistivity in the low moisture content range is partially due to the poor compaction of the sand backfill. Poor compaction of the sand leads to decreased density and to high dependence of the thermal resistivity on moisture content.

The probability distribution curves for backfill sand and native soil (ρ_e, ρ_c) are constructed from the statistical analysis of weather data for the region, cable loadings, and the established relationship between the thermal resistivity and soil moisture content. To determine the initial amount of water in the soil, the formulas de-

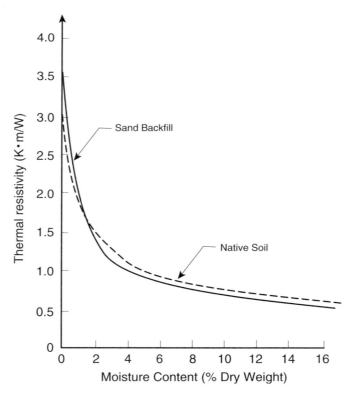

Figure 4-13 Moisture versus thermal resistivity curves for the material in the vicinity of the cables (from Chu et al., 1981, with permission from Pergamon Press).

veloped by Thornthwaite (1948) for agricultural purposes can be adopted. This was done in the case of cable backfills in the Toronto area climate. These formulas build a soil–water balance on the basis of initial moisture content in the soil, recorded precipitation, maximum moisture content that the soil can sustain (so-called field capacity), and recorded temperatures. A brief description of the method is given in Appendix B.

For the approach reported in this chapter, the weekly precipitation values required for the Thornthwaite formulas were obtained from the 330 records for the month of August provided by Environment Canada for the 66-year period. The application of the Thornthwaite method gave 330 values of the initial moisture content in the soil, as shown in Figure 4-14.

It should be pointed out that not only does the water content vary according to climatic conditions, but also due to the thermal gradients caused by the buried loaded cables. The Thornthwaite formulas deal with the moisture fluctuation in the soil caused by the climatic conditions only.

Current flowing in power cables will cause moisture migration away from cables. Therefore, the moisture content of the sand backfill will always be somewhat lower than predicted by the Thornthwaite method. One method to deal with this problem is described in Anders and Radhakrishna (1988). In this work, a series of studies using a heat/moisture finite-element computer program were performed to obtain the actual values of the thermal resistivities of the backfill to be used in the Monte Carlo simulations. The method employed is described in Appendix C.

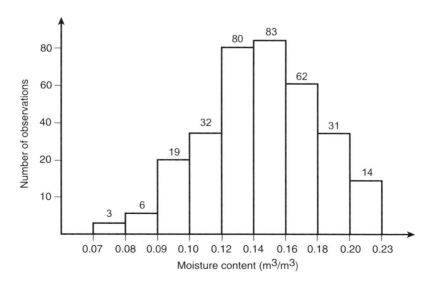

Figure 4-14 Histogram of the initial moisture content in the soil (from Anders and Radhakrishna, 1988, with permission from IEE).

4.3.3 Computation of Temperature Distribution Using Monte Carlo Simulation

4.3.3.1 Steady-State Analysis. The dependence of temperatures around buried power cables on the input parameters can be generally described by a functional relationship as shown below:

$$\theta = \theta(I, \rho_e, \rho_c, \theta_{amb}, \varphi_i, \text{constant parameters}) \tag{4.43}$$

where φ_i is the initial moisture content and, both ρ_e and ρ_c are functions of φ_i.

In the Monte Carlo simulation procedure, the probability density function of θ is obtained as follows. A point is selected from each of the input densities (that is the values of I, φ_i, and θ_{amb} are selected). The value of ρ_e is read from Figure 4-13 and the value of ρ_c is calculated, taking into account the initial moisture content φ_i and the selected value of cable loading I, as described in Appendix C. With these parameter values, the conductor temperature distribution is computed. The procedure is repeated many times, until a convergence is achieved [see Chapter 13 in Anders, (1990) for the details of the Monte Carlo simulation procedure]. This computational procedure is shown in Figure 4-15.

An important consideration in the application of the procedure shown in Figure 4-15 is the interdependence of the random variables representing load, ambient temperature, and moisture content. This interdependence is taken into account in the $\theta - \rho$ transformation shown in the bottom block in Figure 4-15.

For the system under study, the resulting histograms of the thermal resistivities are shown in Figures 4-16 and in 4-17 for the native soil and backfill, respectively.

Next, a rating equation is used to compute resulting temperatures. The selection is performed in such a way that after a sufficient number of selections from a given

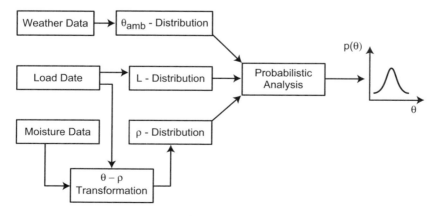

Figure 4-15   General computational procedure for Monte Carlo analysis of power cables.

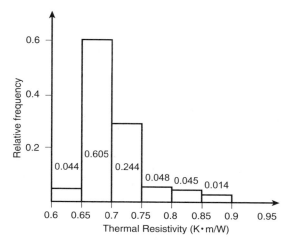

Figure 4-16   Histogram of the native soil thermal resistivity (from Anders et al., 1988a, with permission from IEEE Press).

density function are made and the histogram of the selections is drawn, the resulting shape will closely resemble the input density. The key word in the previous sentence is "sufficient." In practice, this means hundreds and even thousands of Monte Carlo simulations have to be made, depending on the shapes of the original density functions. In the study reported here, almost 4000 simulations were performed for each of the years studied. The probability distribution of the cable surface temerature is shown in Figure 4-19.

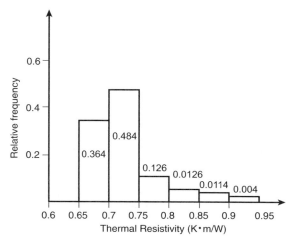

Figure 4-17   Histogram of the backfill thermal resistivity (from Anders et al., 1988a, with permission from IEEE Press).

4.3.3.2 *Transient Analysis.*

The procedure described above gives the probability density function for the steady-state conditions. For the computation of emergency temperature distribution, the following procedure is suggested.

The fist step is to establish the initial conditions. The Monte Carlo selection procedure described above can be used for this purpose. The selected values of ρ_e, ρ_c, and θ_{amb} are then used to compute the precontingency temperature θ_1. Applying this procedure, we assume that the initial conditions lasted long enough for the steady-state conditions to be reached. Next, a circuit is taken out of service and its load is redistributed among the remaining cables. This results in a new value of heat losses in all the cables. Outage duration is sampled from the appropriate probability density function. To obtain this function, the outage data of circuits under consideration should be obtained. In the study reported here, only outages longer than 1 hour were considered because of the long time constants of the soil. There were 72 such outages and the resulting histogram of outage durations is shown in Figure 4-18.

In the transient analysis, it is assumed that the loading and environmental parameters remain unchanged for the selected outage duration. This yields pessimistic values of the temperature because the probability density function for the current flowing in the cables corresponds to the peak load at the customer busses. The temperature at the end of the analyzed period is then recorded in the emergency temperature histograms.

Figure 4-19 shows the probability distribution function of the temperature on the external surface of the fifth cable from the left (see Figure 4-10) based on loading conditions in year No. 2. This is the hottest cable in the group. Both the steady-state and emergency temperatures are shown. The critical cable surface temperature is usually in the range of 50–60°C. Assuming the critical value in this study is 50°C for steady-state and 60°C for transient conditions, respectively, it can be observed that these temperatures are not expected to be reached in that year.

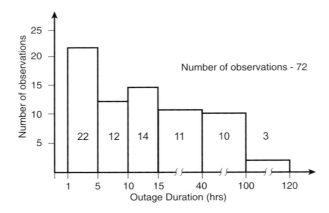

Figure 4-18   Histogram of outage durations > 1 hr (from Anders et al., 1988a, with permission from IEEE Press).

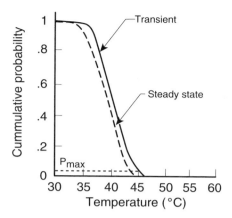

Figure 4-19 Probability distribution of steady-state and emergency cable temperatures (from Anders et al., 1988a, with permission from IEEE Press).

4.3.4 Sample Applications

In this section, we will present several applications of the probabilistic techniques applied to the decision processes involving underground power cables. Two of the examples involve the probability distributions developed in the preceding sections and the third shows an application of the decision tree to the problem of selecting a cable backfill.

Example 4.4

In order to decide when the capacity of existing circuits is no longer sufficient, we will show how the analysis described in the previous section can be used to consider economic aspects of cable uprating (Anders et al., 1988b).

In this example, the probability distribution function of the cable surface temperature was obtained for years 1, 2, and so on. Suppose that an acceptable risk p_{max} of exceeding cable design temperatures is established. Let us assume that, for the purpose of this discussion, the value $p_{max} = 0.05$ is selected. The value of p_{max} could be different for normal and emergency conditions. The corresponding steady-state and emergency temperatures are then read from probability distribution functions for each year as shown in Figure 4-19. These values are then plotted as illustrated in Figure 4-20. From this figure, a critical year is selected in which either the steady-state or emergency cable operating temperature is exceeded. In our case, a new cable has to be installed in year No. 3.

Based on standard deterministic criterion, as discussed in Section 4.3.1, the new cable would have to be installed in the second year. Let us assume an annual discount rate of 10.5%. The installed cost of 1 km of new cable is estimated at $2,500,000. This value was established five years before the analysis for year No.1 is undertaken. The expected cost escalation rate is 6.5%. The three-year postponement gives the savings in the new cable installation shown in Table 4-6.

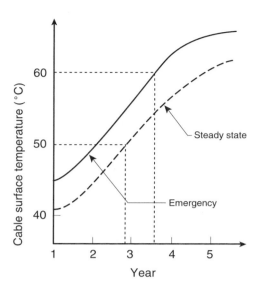

Figure 4-20   Yearly cable surface temperature (from Anders et al., 1988a, with permission from IEEE Press).

The economic value of deferring the installation of 1 km of cable by three years is $197,000. The economic value of deferring the installation of 5 km of cable would be $985,000.

The year chosen for the installation of new facilities will depend on the critical value p_{max} of acceptable risk selected and the cable maximum operating temperatures permitted. The last quantity is determined by the cable manufacturer and is assumed to be constant. The value of p_{max} would have to be established for each cable installation separately, and it would depend on company policies and cable operating conditions.   ∎

Example 4.5
In this example, we will explore the issues related to the computation of the probability of exceeding the permissible cable surface temperature. To obtain numerical values, we will use the cable system shown in Figure 4-10.

Table 4-6   Calculation of saving resulting from postponement in cable investment

Year	Installed cost of cable per km in year shown ($ million)	Present value factor	Present value of expenditure in year shown ($ million)
5	$1.065^5 \cdot 2.5 = 3.4252$	0.5493	1.881
8	$1.065^5 \cdot 2.5 = 4.1375$	0.4071	1.684
Saving			0.197

For simplification, we define in the present analysis the failure probability as the probability of exceeding the permissible cable surface temperature for the given cable design and loading conditions. The calculation of such failure probability requires the applied stress distribution of Figure 4-19 (for one or both of the steady-state and transient cases) as well as the corresponding resistance (withstand) distribution. We will describe next the process of preparing the stress and resistance probability distribution and the final failure probability calculation.

The strength probability distribution represents the spread of cable surface temperature on the basis of the cumulative distribution(s) of Figure 4-19. When both the steady-state and transient cases are to be used, the two distributions have to be combined into one overall stress distribution as follows:

$$P_s(\theta_s) = p_1 \cdot P_{ss}(\theta_{ss}) + p_2 \cdot P_{ts}(\theta_{ts}) \tag{4.44}$$

where $P_{ss}(\theta_{ss})$ and $P_{ts}(\theta_{ts})$ denote the probability distributions of the steady-state and transient-state temperatures, respectively, and p_1 and p_2 are the assigned probabilities of the two cases based, for example, on their relative frequency of occurrences over a given historical period $p_1 + p_2 = 1.0$.

The strength probability distribution, $P_s(\theta_s)$ in Figure 4-21 corresponds to the special case of $p_1 = 1$ and $p_2 = 0$; that is, only the steady-state cable operation is analyzed.

The resistance probability distribution of the cable surface temperature is evaluated on the basis of cable testing/manufacturer data. The relevant data should be screened and assembled as pairs of withstand temperature/condition (or scenario) under consideration. For example, a typical set of withstand temperature data may be assembled as shown in Table 4-7.

An example of a resistance probability distribution based on statistical data for normal cable operation is shown as $P_r(\theta_r)$ in Figure 4-21.

The cable failure probability, that is, the probability that the stress/cable surface temperature exceeds the withstand temperature, is calculated by mathematical convolution of the probability distributions $P_s(\theta_s)$ and $P_r(\theta_r)$ as follows:

$$P_f = \int_0^\infty P_s(\theta_s) \left[\int_0^\infty P_r(\theta_r) d\theta_r \right] d\theta_s \tag{4.45}$$

For the specific example of Figure 4-21, the failure probability calculated using Equation (4.45) is $p_f = 0.02003$; that is, the probability of exceeding the withstand temperature during steady-state operation in year No. 2 is 0.02003. ∎

Example 4.6[5]

In this example, we will investigate a decision problem facing an electric utility encountering a rapidly growing demand in a large urban area. Part of the load in the downtown core is supplied by a four-circuit, 115 kV cable system. According to the

---

[5]This example is extracted from (Anders, 1990).

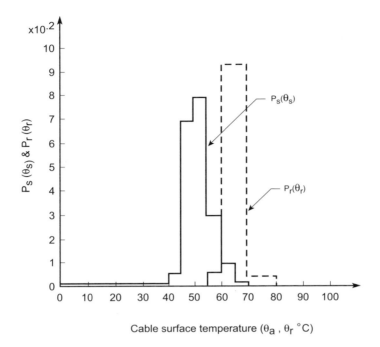

Figure 4-21 Stress (P_s) and resistance (P_r) probability distributions of cable surface temperature (from Anders et al., 1988a, with permission from IEEE Press).

design specifications, the maximum current-carrying capability of each of the 12 cables is 800 A. The cables have been in service for 30 years, and with the present load growth it is expected that a new circuit will be required in 2 years. Construction of an additional circuit is an expensive proposition, and the cost of a 5 km cable route is estimated to be $3 million.

The planning engineer feels that the calculations performed 30 years ago were based on very conservative assumptions and that with the present sophisticated computational tools and better understanding of the influence of various parameters on cable ampacities, the real current-carrying capabilities of these cables would be found to be higher than those provided in the specifications.

Table 4-7 Example of temperature ranges for various operating conditions

Condition/Scenario	Withstand temperature
Normal operation	55–80°C
Emergency long-term	55–120°C
Emergency short-term	110–150°C
Short-circuit	130–210°C

In particular, one of the parameters with large influence on cable ampacity is the value of the soil thermal resistivity, ρ_e. Reviewing the literature on the subject, the planning engineer noticed that most sources seem to suggest that the type of the backfill material used in this particular installation has lower thermal resistivity than assumed in the original design. If the actual soil thermal resistivity is indeed lower than assumed (we will denote this new value by ρ_{ei}), the cable ampacities would increase to 1000 A; furthermore, the construction of a new cable can be postponed an additional 3 years, yielding savings, discounted to present value, of $1 million.

The planning engineer feels that the probability associated with lower value of the soil thermal resistivity is 0.7. If, on the other hand, the resistivity value is equal to or greater than the assumed value (denoted by ρ_{eh}), the utility has three choices: Choice I is to build a new cable circuit; Choice II is to replace the sand in the cable trench with limestone screenings at the cost of $500,000, which will increase cable ampacity to 1000 A and will yield savings, discounted to present value, of $1.0 million; Choice III is to do nothing and hope that the load will not grow at the estimated rate. There is a 40% chance that the load growth will slow down and that the circuits will be adequate for the next 5 years, yielding savings of $500,000. If, on the other hand, the load growth will remain high, the utility faces an additional penalty of $2.0 million if nothing is done to improve cable ratings.

Assuming that each of the contemplated remedial actions will resolve the problem for at least the specified period of time, even in the presence of the highest forecast load, the options facing the planning engineer are as follows:

A_1: Construct a new circuit

A_2: Replace sand backfill with the limestone screenings now

A_3: Do nothing

This is a typical decision analysis problem. A convenient way to solve such problems is to apply the concept of a decision tree (Anders, 1990). The complete decision tree describing this problem is given in Figure 4-22. The monetary loss (in millions of dollars) for each path in the tree is also shown.

The cost of backfilling is obtained by summing (a) the cost of constructing a new cable at the end of the planning period (–$3,000,000) (assumed to be the same as the present cost of construction), (b) the savings resulting from the postponement of the construction ($1,000,000), and (c) the cost of the backfill (–$500,000). Other costs are obtained in a similar manner. The loss (cost) values for alternatives A_1 and A_2 are not affected by the actual value of the soil thermal resistivity since in both cases the money has to be spent within the next 3 years. The cost in the last branch in Figure 4-22 is obtained as the algebraic sum of (a) the construction cost (–$3 million), (b) the saving resulting from postponement of the construction ($1 million), and (c) the penalty (–$2 million).

The expected monetary values (in millions of dollars) for all alternatives are as follows:

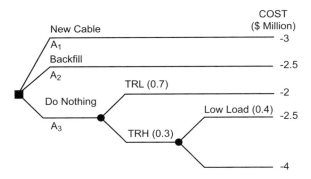

Figure 4-22 Decision tree for cable upgrading (from Anders, 1990, with permission from John Wiley and Sons).

$$E(A_1) = -3.0 \cdot 0.7 - 3.0 \cdot 0.3 = -3.0$$

$$E(A_2) = -2.5 \cdot 0.7 - 2.5 \cdot 0.3 = -2.5$$

$$E(A_3) = -2.0 \cdot 0.7 + 3.0 \cdot (-2.5 \cdot 0.4 - 4.0 \cdot 0.6) = -2.42$$

The decision criterion to be used is called the expected monetary value (EMV) criterion. This criterion states that the alternative with the highest value of EMV should be selected. Therefore, based on the maximum EMV criterion, alternative A_3 should be selected as it yields the smallest monetary losses. ■

4.4 STOCHASTIC OPTIMIZATION

In this section, we will return to the backfill optimization problem, but we will consider the stochastic nature of some of the parameters playing an important role in backfill design. The stochastic optimization problem is an extension of the probabilistic analysis discussed in Section 4.3. In the case of stochastic optimization, we not only take probability distributions of the constituent random variables and compute the probability distribution of the desired variable, but we also find the values of the input variables that lead to the best value of the objective function.

Similar to the deterministic case, in stochastic optimization there are variables that we have control over, such as backfill width or height. These controlled variables are called decision variables. Finding the optimal values for decision variables can make the difference between reaching an important goal and missing that goal.

In a deterministic model, all input data other than the decision variables are constant or assumed to be known with certainty. The decision variables are also constant, except that their values are taken from the specified ranges. In a stochastic model, some of the model data are uncertain and are described with probability distributions. We will use the same terminology as in the Crystal Ball® software that we will apply in the analysis and will call these variables assumptions. The variable whose probability distribution we want to obtain will be called a forecast.

Stochastic models are much more difficult to optimize because they require simulation to compute the objective. The most common method of running these simulations is shown in Figure 4-23.

Once an optimization problem is defined (by selecting decision variables, the objective, and possibly imposing constraints), the mathematical model describing the objective function and the constraints is executed for different sets of decision variable values. The statistical outputs from the simulation model are analyzed and integrated with outputs from previous simulation runs, and a new set of values to evaluate is determined. This is an iterative process that successively generates new sets of values. Not all of these values are feasible or improve the objective, but over time this

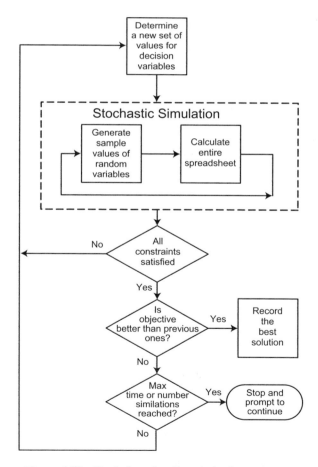

Figure 4-23 Typical stochastic optimization process.

process provides a highly efficient trajectory to the best solutions. The search process continues until some termination criteria, either a limit on the amount of time, or the number of simulations, or the accuracy of the results (Anders, 1990), is reached.

With the probability distributions of the uncertain variables defined in the preceding sections, we can now perform stochastic optimization to obtain not only the best values of the design parameters, but also the probability distribution of the objective function. We will illustrate the procedure in the following example.

Example 4.7
We will return to the optimization problem examined in Example 4.3. The design variables will be the depth of backfill center, backfill width and thickness, and the spacing between the cables. The bounds on these variables are given Table 4-1 and the constraints are defined in Example 4.3. The maximum allowable cost is set at $300/lin m.

The stochastic variables are the thermal resistivities of the native soil and the backfill. Their probability distributions were established in Section 4.3.2.3, taking into account the variability of the ambient air temperature and of the moisture content. The random selection of resistivity values could be based on the histograms in Figure 4-16 and Figure 4-17. However, since commercial software was used for the analysis, and this software requires input in a form of one of the standard continuous distributions, the first step in the analysis was to fit standard probability distributions into resistivity histograms. The resulting density functions are shown in Figures 4-24 and 4-25.

The stochastic optimization process was run with a time limit of 2 hours. For every selection of the design variables, 1000 stochastic simulations were run. For $C_1 = \$300$ and the mean values of the soil and backfill thermal resistivities, the maximum current is 908 A with $w = 3.91$ m, $L_G = 0.95$ m, $h = 1.11$ m, and $s_1 = 1.45$ m, for which the cost function has the value of $300/linear m.

The results of the analysis are also yield the probability density and the cumulative probability distribution functions of the cable current for the optimal backfill

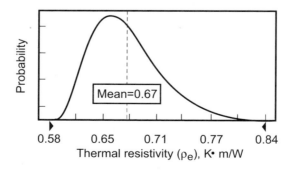

Figure 4-24 Gamma distribution fitted to the native soil histogram. Parameters: location = 0.58, scale = 0.02, shape = 4.60.

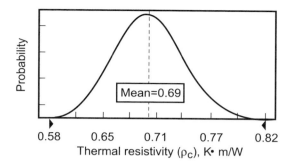

Figure 4-25 Lognormal distribution fitted to the backfill histogram. Parameters: mean = 0.69, standard deviation = 0.04.

dimensions. They are shown in Figures 4-26 and 4-27, respectively, and the corresponding statistics are summarized in Table 4-8.

We can observe that the mean value of the current is equal to 909 A and is the same as the ampacity corresponding to the optimal trench design. However, the graphs in Figures 4-26 and 4-27 and the data in Table 4-8 give us a substantial amount of additional useful information. First, we observe that the frequency chart obtained from Monte Carlo simulations can be fitted with a logistic distribution. It is defined by the following equation:

$$f(x) = \frac{z}{\alpha(1+z)^2} \qquad \text{for } -\infty < x < \infty$$

where

$$z = e^{-\left(\frac{x-\mu}{\alpha}\right)}$$

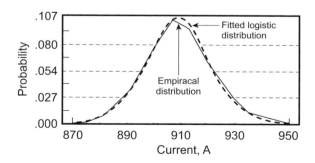

Figure 4-26 Frequency chart of the conductor current from stochastic optimization fitted with a logistic distribution with paramenters mean = 909 A and scale = 7.42.

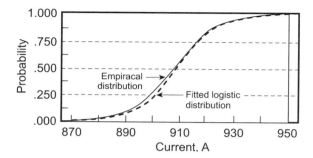

Figure 4-27 Cumulative probability distribution function of the conductor current from stochastic optimization fitted with a logistic distribution.

and

α = scale parameter

μ = mean value

Second, we can see that the current will vary in the range 850 to 990 A. There is about a 10% chance that the rating of this cable will greater than 925 A, and a 10% probability that it will be lower than 893 A. This information could be very useful in deciding when an investment should be made in new transmission facilities, as discussed in Example 4.5. ∎

4.5 CONCLUDING REMARKS

The expectation of continued moderate increases in demand for electricity, the re-structuring of the electric power industry, and the pressure to reduce costs have pro-

Table 4-8 Results of the Monte Carlo simulation

Statistics	Value
Trials	10000
Mean	909.25
Median	909.06
Mode	910
Standard deviation	13.20
Variance	174.25
Skewness	0.12
Kurtosis	3.62
Coefficient of variability	0.01
Range, minimum	848.76
Range, maximum	987.67
Range, width	138.91
Mean standard error	0.13

vided incentive for utilities to look for measures that will increase the capability of existing facilities and thereby defer the construction of new ones. A 10% increase in the ampacity of a transmission circuit in a 3% growth environment, for example, may defer the need for substantial capital expenditures for 3 years. The estimated cost of an underground cable installation in an urban area could be $2.5 million per km. The present value of deferring such a cable installation for a period of three years is estimated to be at least $50,000 per km per year of deferment. To confidently determine the higher cable ampacity that would permit the deferment, reliable, advanced thermal analysis techniques are required.

Two advanced concepts have been introduced in this Chapter to help cable engineers make informed decisions on future cable system investment options. First, we demonstrated the use of a nonlinear programming formulation, in conjunction with trench cost evaluation and cable thermal analysis, to attain an optimal design of cable thermal backfill. Second, we showed how probabilistic analysis techniques could be used to support decisions on cable uprating. With the advent of very fast desktop computers, the numerical challenges involved in these types of analyses can be easily overcome.

In the probabilistic analysis discussed in this chapter, one should remember that the actual loading of underground plant is strongly dependent on the role that a particular line plays within the overall system, and this role may, and often does, change as the system evolves. For instance, in years past when generation was often sited close to load centers, underground lines in metropolitan areas often were installed primarily as a means for providing energy transfers in the event of scheduled or unscheduled unit outages. Such lines were virtually never operated at more than a small fraction of their normal ratings. Even though system conditions have long since changed, it is not surprising that particular underground lines may be operating over long periods at well below their rating. However, as system conditions evolve, the loading patterns of the individual lines will not remain fixed, and ratings based on loading distributions will have to be monitored very closely.

4.6 CHAPTER SUMMARY

The mathematical developments that can be briefly summarized here pertain to the formulation of the backfill optimization problem. Since our goal is to minimize the cost and maximize cable ampacity, we formulated the cable rating equation in such a way as to make it explicitly dependent on the backfill characteristics. The basic rating equation is given by

$$I = \left[\frac{\Delta\theta - W_d[0.5T_1 + n(T_2 + T_3 + T_4)]}{RT_1 + nR(1 + \lambda_1)T_2 + nR(1 + \lambda_1 + \lambda_2)(T_3 + T_4)} \right]^{0.5} = \left[\frac{a + b \cdot T_4}{c + d \cdot T_4} \right]^{0.5}$$

with the following definitions of the constants a, b, c, and d and an assumption of a unity load factor:

$$a = \Delta\theta - W_d \cdot (0.5T_1 + nT_2 + nT_3) \qquad b = -n \cdot W_d$$

$$c = R \cdot T_1 + n \cdot R \cdot [(1 + \lambda_1) \cdot T_2 + (1 + \lambda_1 + \lambda_2) \cdot T_3]$$

$$d = n \cdot R \cdot (1 + \lambda_1 + \lambda_2)$$

We adopted the above notation since only parameter T_4 depends on the backfill characteristics. When the cable system is contained within an envelope of thermal resistivity ρ_c, the effect of thermal resistivity of the backfill envelope being different from that of the surrounding soil ρ_e is handled by first assuming that the thermal resistivity of the medium is ρ_c throughout. A correction is then added algebraically to account for the difference in the thermal resistivities of the envelope and the native soil. Assuming that the depth of cable center is much larger than its diameter, we have

$$T_4 = \frac{\rho_c}{2\pi} \cdot \ln\left(\frac{4L}{D_e}\right) + \frac{N}{2\pi} \cdot (\rho_e - \rho_c) \cdot G_b$$

where
N = number of loaded cables in the envelope
$$G_b = \ln(u_b + \sqrt{u_b^2 - 1}) \approx \ln\frac{2L_G}{r_b}$$
$$u_b = \frac{L_G}{r_b}$$
L_G = depth of laying to the center of the backfill, mm
r_b = equivalent radius of the envelope, mm

The geometric factor G_b contains all the design parameters through the value of L_G and the equivalent radius. The equivalent radius of the thermal envelope is equal to

$$r_b = \exp\left[\frac{1}{2}\frac{x}{y}\left(\frac{4}{\pi} - \frac{x}{y}\right)\ln\left(1 + \frac{y^2}{x^2}\right) + \ln\frac{x}{2}\right]$$

where
$x = \min(w, h)$
$y = \max(w, h)$

The last equation is only valid for ratios of y/x less than 3 and, since we need an analytical solution to our optimization problem, this is a restriction that will be enforced in all backfill designs considered here.

Finally, we obtain the following expression for the cable current as a function of backfill parameters:

$$I = \sqrt{\frac{a + b_1 \cdot \rho_c \cdot \ln(L) + b_2 \cdot \rho_c + f_1 \cdot (\rho_e - \rho_c) \cdot G_b}{c + d_1 \cdot \rho_c \cdot \ln(L) + d_2 \cdot \rho_c + f_2 \cdot (\rho_e - \rho_c) \cdot G_b}}$$

where the constants are defined as

$$b_1 = \frac{b}{2\pi} \qquad b_2 = \frac{b}{2\pi} \ln\left(\frac{4}{D_e}\right) \qquad f_1 = \frac{b \cdot N}{2\pi}$$

$$d_1 = \frac{d}{2\pi} \qquad d_2 = \frac{d}{2\pi} \ln\left(\frac{4}{D_e}\right) \qquad f_2 = \frac{d \cdot N}{2\pi}$$

The last equation for I can be used as an objective function with backfill height, width, and thermal resistivity as parameters. We could also construct an objective function related to the backfill cost, and the equation for the current is then used as a constraint on the minimal value of the cable rating.

REFERENCES

Abdel-Hadi, O.N. (1978), *Flow of Heat and Water around Underground Power Cables,* Ph.D. thesis, University of California at Berkeley.

AIEE (1960), "Soil Thermal Characteristics in Relation to Underground Power Cables," Committee report, AIEE Transactions, Summer Meeting, pp. 792–856.

Anders, G.J. (1990), *Probability Concepts in Electric Power Systems,* Wiley, New York.

Anders, G.J., Foty, M., and Croall, S. (1988a), "Cable Environment Analysis and Probabilistic Approach to Cable Rating," *IEEE Transactions on Power Delivery,* Vol. PWRD-5, No.3, pp. 1628–1633.

Anders, G.J., Krpan, M.M., Roiz, J., and Chaaban, M. (1988b) "Delaying Investments in Transmission Facilities Through Cable Uprating," in *Proceedings of the 1988 CIGRE Meeting,* Paris, France, September 1988, Section 37-08.

Anders, G.J., and Radhakrishna, H.J. (1988), "Computation on the Temperature Field and Moisture Content in the Vicinity of Current Carrying Underground Power Cables," *IEE Proceedings,* Part C, Vol. 153, No.1, pp. 51–62.

ASTM C518-67 (1967), American Society for Testing of Materials, Philadelphia.

Chu, F.Y., Radhakrishna, H.S., and Steinmanis, J.E. (Eds.) (1981), *Underground Cable Thermal Backfill,* Pergamon Press, Toronto.

El-Kady, M. (1982), "Optimization of Power Cable and Thermal Backfill Configurations," *IEEE Transactions on Power Apparatus and Systems,* Vol. PAS-101, No. 12, pp. 4681–4688.

El-Kady, M.A., Chu, F.Y., Radhakrishna, H., Horrocks, D.J., and Ganton, R. (1984), "A Probabilistic Approach to Power Cable Thermal Analysis and Ampacity Calculation," *IEEE Transactions on Power Delivery,* Vol. PAS-103, pp. 2735–2740.

Fink, L.H. (1954), "Control of the Thermal Environment of Buried Cable Systems," *AIEE Transactions on Power Apparatus and Systems,* Vol. 73, pp. 406–412.

Fink, L.H. (1965), "Evaluation of Soil Thermal Characteristics," *IEEE Transactions on Power Apparatus and Systems,* Vol. PAS-84, No. 9, pp. 807–814.

Fink, L.H., and Smerke, J.J. (1958), "Control of the Thermal Environment of Buried Cables Systems, Part II," *AIEE Transactions on Power Apparatus and Systems,* Vol. 77, pp. 161–168.

Groeneveld, G.J., Snijders, A.L., Koopmans, G., and Vermeer, J. (1984), "Improved Method to Calculate the Critical Conditions for Drying Out Sandy Soils Around Power Cables," *IEE Proceedings,* Vol. 131, Part C, No. 2, pp. 42–53.

IEEE Task Force on Soil Thermal Resistivity: Compaction of Cable Trench Backfill (1966), IEEE/PES Winter Meeting, Conference paper 31 CP 66-75.

IEEE 442-1991, reaffirmed 2003, "Guide for Soil Thermal Resistivity Measurements."

Mason, V.V., and Kurtz, M. (1952), "Rapid Measurement of the Thermal Resistivity of Soil," in *AIEE Transactions, Summer Meeting,* pp. 570–577.

Neher, J.H., and McGrath, M.H. (1957), "The Calculation of the Temperature Rise and Load Capability of Cable Systems," *AIEE Transactions,* Vol. 76, Part 3, pp. 752–772.

Radakrishna, H.S. (1981). "Soil Thermal Resistivity and Thermal Stability Measuring Instrument," Report prepared by Ontario Hydro Research Division for Electric Power Research Institute, EPRI EL-2128.

Saleeby, K.E., Black, W.Z., and Hartley, J.G. (1979), "Effective Thermal Resistivity for Power Cables Buried in Thermal Backfill," *IEEE Transactions on Power Apparatus and Systems,* Vol. PAS-98, pp. 2201–2214.

Schmill, J.V. (1967), "Variable Soil Thermal Resistivity—Steady State Analysis," *IEEE Transactions on Power Apparatus and Systems,* Vol. PAS-86, pp. 215–223.

Shannon, W.L., and Wells, W.A. (1947), "Tests for Thermal Diffusivity of Granular Materials," in *American Society for Testing Materials, 15th Annual Meeting, June 1947.*

Stolpe, J. (1969), "Soil Thermal Resistivity Measured Simply and Accurately," in *IEEE Underground Conference,* Vol. 89, pp. 297–304.

Thornthwaite, C.W. (1948), "An Approach Toward a Rational Classification of Climate," *Geographical Review,* Vol. 38.

Wiedemann (1888), *Annalen Physik,* ed. by A.L.E.F. Schleirmacher, Vol. 34.

Vassallo, D.G. (1969), "The Influence of Climatic Conditions on the Moisture Content of Cable Backfills," Ontario Hydro Research Division Report, No. E69-9-K, February 1969.

CHAPTER 5

Special Considerations for Real-Time Rating Analysis and Deeply Buried Cables

5.1 INTRODUCTION

Because of the difficulties in siting new lines in cities, there is great pressure for increased power transfer on some older transmission lines. Recently, dynamic cable rating systems have gained a lot of attention since they offer an opportunity to overcome some of the design shortcomings.

Dynamic feeder rating (DFR) systems are an important alternative to analytical approaches for determining the real-time conductor operating temperature and corresponding real-time cable ratings. Operations of several such systems have been reported in the literature (U.S. Department of Energy, 1978; Nelson, 1988; Balog et al., 2000; Pragnell et al., 2000; Douglass et al., 2000; Walldorf & Engelhardt, 1998; Anders et al., 2003). All existing real-time rating systems utilize measured conductor and soil ambient temperatures and conductor current. Measurements of conductor current are normally available at terminating stations but the temperature measurements require special hardware installation. A preferable method for cable surface temperature measurements is the use the fiber-optic sensors. This is often done in new installations or in retrofits for cables installed in ducts. In existing cable circuits, the only reasonable option is to install thermocouples at the selected locations and record the data in the locally installed remote terminal units (RTUs). Because of this requirement, DFR systems are usually fairly expensive to install. In some cases, it might be simply impossible to install thermocouples and the RTUs because of the surrounding infrastructure, large distances, or lack of the local power supply or communication facilities. In cases like these, a reasonable alternative would be to estimate cable losses and the conductor temperature from the current measurements alone.

In this chapter, a method is discussed that allows estimation of conductor temperature and cable losses as a function of conductor current and some parameters corresponding to the nominal loading conditions (Brakelmann, 2003; Brakelmann and Anders, 2004). Such parameters include cable rating and loss factors computed

Rating of Electric Power Cables in Unfavorable Thermal Environment. By George J. Anders
ISBN 0-471-67909-7 © 2004 the Institute of Electrical and Electronics Engineers.

by cable manufacturers for the laying conditions specified by the user. There are two major applications of the proposed approach. One relates to rating of power cables with current measurements alone and the other to the computation of the external thermal resistance with a nonunity load factor used in the Neher/McGrath approach. Both applications are examined in this chapter. We will investigate in some detail calculation of the loss factor, particularly with the aim of exploring its application for rating of deeply buried cables.

As already mentioned, installation of new cable lines in large metropolitan areas in industrialized countries has become more and more difficult. Obtaining necessary permissions is only one of the problems. From a technical point of view, most of the available right-of-ways are already occupied, either by other power or communication circuits or by other infrastructures such as heating, sewage, and water pipes or underground transportation corridors.

Therefore, in spite of fairly high costs, laying of cables in deep, directly drilled tunnels becomes attractive. Such installations have several advantages in comparison with directly buried cables. The most important are:

- The civil engineering work does not have to disrupt pedestrian or vehicular traffic.
- The length of the cable route can be minimized.
- Cables can be laid independently of the structure of the buildings above the right-of-way.
- Access to the circuits is facilitated if laid in large tunnels
- The influence of electromagnetic fields at the earth surface is considerably diminished in comparison with the standard installations.
- In the case of large tunnels, it is possible to install additional circuits without loss of existing rating capabilities by introducing forced air circulation.
- Advantageous thermal environment permits improved heat dissipation.

The last bulleted advantage is of particular interest in the context of cable rating calculations and will be explored in this chapter.

An advantageous thermal environment for deeply buried cables is a result of two factors. On the one hand, the soil tends to have higher moisture content at large depth due to the penetration of ground waters. On the other hand, daily, weekly, and even yearly load variations have a profound effect on the ratings of deeply buried cables. This effect is not so pronounced for cables buried at the usual depths.

Thus, two applications will be discussed in this chapter. One is a prediction of the conductor temperature and losses from the current measurements alone, discussed in Sections 5.2 to 5.5, and the other is the modification of the Neher/McGrath procedure for the calculation of the load loss factor utilizing the results obtained in Section 5.5. This modification will be described in Sections 5.6 and 5.7, where an in-depth analysis of the loss factor applications will be offered.

5.2 PREDICTION OF CONDUCTOR TEMPERATURE FROM THE CONDUCTOR CURRENT

5.2.1 Introduction

The losses in a cable conductor are dependent on the operating temperature. Since in field installations this temperature is unknown, the estimation of the losses is a difficult task and, therefore, usually not performed at all or some approximations are used. The losses are also a function of daily load fluctuations. In the cyclic rating calculations discussed in Chapter 2 and in the Neher/McGrath method, these fluctuations are taken into account in the calculation of the external thermal resistance by considering a loss-load factor (referred to simply as a loss factor). In rating computations, the loss factor is calculated from the load shape alone, without consideration of the conductor temperature. In reality, the definition of the loss factor requires consideration of the conductor resistance, which is temperature dependent.

The aim of this section is to propose an improvement in the accuracy of the rating calculations by an introduction of the temperature dependence of the conductor losses in the rating formulae. Since, as mentioned above, the conductor temperature is usually unknown, we will need some other information to account for the absence of the real-time temperature values. The additional information comes from standard rating conditions as specified by the cable manufacturer and involves cable rating values at nominal conditions. The subject is explored in detail in the following subsection.

5.2.2 Mathematical Model

In this subsection, we will develop a relationship [Equation (5.18)] that allows calculation of the conductor temperature when the conductor current and certain values corresponding to the rated conditions are known. This relationship will take into account variation of the conductor resistance with temperature as well as a possibility of soil dry-out. Conductor temperature rise $\Delta\theta$ is related to the conductor losses W_c through the following equation [see Equations (1.42) and (1.49)]:

$$\Delta\theta = W_c T_t + \Delta\theta_d - \theta_x \qquad (5.1)$$

where
W_c = conductor losses (W/m)
T_t = total thermal resistance defined in Equation (1.41) (K · m/W)
$\Delta\theta_d$ = temperature rise caused by dielectric losses (°C)

A possible drying-out of the surrounding soil is taken into account in Equation (5.1), following the two-layer-model of Cox and Coates (1965) (see Section 1.4.1.2), by means of the variables v and θ_x with

$$\theta_x = (v_x - 1) \cdot \Delta\theta_{cr} \quad \text{and} \quad v_x = \frac{\rho_2}{\rho_1}(\theta_x = \theta_{xT}) \quad \text{for} \quad \Delta\theta_e > \Delta\theta_{cr}$$

$$v_x = 1 \ (\theta_x = 0) \quad \text{for} \quad \Delta\theta_e \leq \Delta\theta_{cr} \tag{5.2}$$

where
$\Delta\theta_e$ = temperature rise at the cable surface (°C)
$\Delta\theta_{cr}$ = critical temperature rise of the soil (°C)
θ_{xT} = correction temperature (°C)
ρ_1, ρ_2 = thermal resistivities of the moist and dry soils, respectively (K · m/W)

The thermal resistivity of the dry soil is normally larger than that of the moist soil by a factor of 2 to 6. An example of the standard values of these quantities is given in Table 5-1 (IEC 287-2-1, 1994).

We will rewrite Equation (5.1) to relate conductor losses to the current as a straightforward multiplication of the ac resistance and squared current I^2R. In order to achieve this, we will first introduce the following revised designation of the total thermal resistance of the cable:

$$T_{t,x} = \frac{T_1}{n \cdot (1 + \lambda_1 + \lambda_2)} + \frac{T_2 \cdot (1 + \lambda_1)}{1 + \lambda_1 + \lambda_2} + T_3 + v_x \cdot T_4 = T_I + v_x \cdot T_4 \tag{5.3}$$

and the electrical resistance representing all ohmic losses in the cable:

$$R_{t20} = R_{20} \cdot (1 + \lambda_1 + \lambda_2) \tag{5.4}$$

R is temperature dependent. Hence, the conductor temperature θ changes the total ohmic losses W_I of the cable according to

$$W_I = W_t - W_d = W_{20} \cdot [1 + \alpha_T \cdot (\theta - 20)] = n \cdot R_{t20} \cdot I^2 \cdot [1 + \alpha_T \cdot (\theta - 20)] \tag{5.5}$$

where W_{20} is the conductor loss at 20°C, (W/m) and α_T is the temperature coefficient of resistance (1/°C) of the conductor material.

With these notations, Equation (5.1) can be rewritten as

$$\Delta\theta = T_{t,x} \cdot n \cdot R_{t20} \cdot (\alpha_T \cdot \Delta\theta + c_\alpha) \cdot I^2 + \Delta\theta_d - \theta_x = T_{t,x} \cdot (W_t - W_d) + \Delta\theta_d - \theta_x \tag{5.6}$$

Table 5-1   Standard values for soil properties

Variable	Unit	Value
θ_{amb}	°C	20
$\Delta\theta_{cr}$	°C	15
ρ_1	(K · m)/W	1
ρ_2	(K · m)/W	2.5
θ_x	°C	22.5

with

$$c_\alpha = 1 + \alpha_T \cdot (\theta_{amb} - 20) \tag{5.7}$$

For the nominal load, characterized by the rated current I_R and the nominal total losses W_{tR}, the cable conductor reaches its maximum stationary temperature θ_{max}. In this case, Equation (5.6) takes the form

$$\Delta\theta_{max} = T_{t,R} \cdot (W_{tR} - W_d) + \Delta\theta_{d,R} - \theta_{x,R} \tag{5.8}$$

whereas

$$W_{tR} - W_d = n \cdot R_{t20} \cdot I_R^2 \cdot [1 + \alpha_T \cdot (\theta_{max} - 20)] = n \cdot R_{t20} \cdot c_m \cdot I_R^2 \tag{5.9}$$

and

$$c_m = [1 + \alpha_T \cdot (\theta_{max} - 20)] \tag{5.10}$$

The values $T_{t,R}$, $\Delta\theta_{d,R}$, and $\theta_{x,R}$ are defined above, but here they are related to the rated load. We will compute now the value of R_{t20} from Equations (5.9) and (5.8) and substitute it in Equation (5.6):

$$\begin{aligned}
\Delta\theta - \Delta\theta_d + \theta_x &= T_{t,x} \cdot n \cdot R_{t20} \cdot I^2 \cdot (\alpha_T \cdot \Delta\theta + c_\alpha) \\
&= (\Delta\theta_{max} - \Delta\theta_{d,R} + \theta_{x,R}) \cdot \frac{v_T}{c_m} \cdot \left(\frac{I}{I_R}\right)^2 \cdot (\alpha_T \cdot \Delta\theta + c_\alpha)
\end{aligned} \tag{5.11}$$

In Equation (5.11), the variable v_T is given by

$$\begin{aligned}
v_T &= \frac{T_{t,x}}{T_{t,R}} \quad \text{for} \quad \Delta\theta_e \leq \Delta\theta_{cr}, \quad \text{and} \\
v_T &= 1 \quad \text{for} \quad \Delta\theta_e > \Delta\theta_{cr}
\end{aligned} \tag{5.12}$$

A caution is required when interpreting Equation (5.12). Three different situations may arise. If there is already drying out for current I (we assume that $I \leq I_R$), then $v_T = 1$ for both currents. In another case, we may have no drying out for current I but the soil dries out for the rated current. In this case, the nominator in the first equation in (5.12) corresponds to the moist soil and the denominator to the dried-out soil. The temperature inequality associated with the first equation holds only for current I but not for the rated current. Finally, the third possibility exists when no drying out occurs for the rated current. In this case, $v_T = 1$. In the reasoning made above we implicitly assumed that the loss factors that appear in the definition of the total thermal resistance [Equation (5.3)] do not depend on the current values. In reality, this is not true but the error introduced by this assumption is negligible.

Solving now Equation (5.11) for the temperature rise, we obtain

$$\Delta\theta = \frac{v_T \cdot c_\alpha \cdot (\Delta\theta_{\max} - \Delta\theta_{d,R} + \theta_{x,R}) \cdot \left(\dfrac{I}{I_R}\right)^2 - c_m(\theta_x - \Delta\theta_d)}{c_m - v_T \cdot \alpha_T \cdot (\Delta\theta_{\max} - \Delta\theta_{d,R} + \theta_{x,R}) \cdot \left(\dfrac{I}{I_R}\right)^2} \tag{5.13}$$

Making substitutions,

$$\theta_1 = c_\alpha \cdot (\theta_{\max} - \theta_{amb} - \Delta\theta_{d,R} + \theta_{x,R}) \cdot v_T \tag{5.14}$$

$$\theta_2 = c_m \cdot (\Delta\theta_d - \theta_x) \tag{5.15}$$

$$c_1 = 1 + \alpha_T \cdot [\theta_{amb} - 20 + (\Delta\theta_{d,R} - \theta_{x,R})] \tag{5.16}$$

$$c_2 = \alpha_T \cdot (\theta_{\max} - \Delta\theta_{d,R} + \theta_{x,R}) \tag{5.17}$$

Equation (5.13) takes the form

$$\Delta\theta = \frac{\theta_1 \cdot \left(\dfrac{I}{I_R}\right)^2 + \theta_2}{c_1 + c_2 \cdot \left[1 - v_T \cdot \left(\dfrac{I}{I_R}\right)^2\right]} \tag{5.18}$$

Equation (5.18) represents a crucial relationship that allows calculation of the conductor temperature when the conductor current and certain values corresponding to the rated conditions are known. The issues related to the practical application of this formula are discussed in the next section.

From Equations (5.5), (5.9). and (5.18) we can compute the total cable losses corresponding to current I as

$$W_t = W_d + (W_{tR} - W_d) \cdot \left(\frac{I}{I_R}\right)^2 \cdot \left[\frac{c_\alpha}{c_m} + \frac{\alpha_T}{c_m} \cdot \frac{\theta_1 \cdot \left(\dfrac{I}{I_R}\right)^2 + \theta_2}{c_1 + c_2 \cdot \left[1 - v_T \cdot \left(\dfrac{I}{I_R}\right)^2\right]}\right] \tag{5.19}$$

$$= W_d + (W_{tR} - W_d) \cdot \left(\frac{I}{I_R}\right)^2 \cdot v_\theta$$

where the temperature coefficient v_θ is defined as

$$v_\theta = \frac{c_3}{c_1 + c_2 \cdot \left[1 - v_T\left(\dfrac{I}{I_R}\right)^2\right]} \tag{5.20}$$

with

$$c_3 = 1 + \alpha_T \cdot [\theta_{amb} - 20 + (\Delta\theta_d - \theta_x)] \tag{5.21}$$

An important characteristic of Equation (5.19) is that it takes into account the variation of conductor resistance with temperature as well as a possibility of the soil dry-out via the coefficient v_θ defined in Equation (5.20). Up to now, for long cable connections (e.g., submarine cables), the practice has been to compute cable losses by obtaining the voltage and current distributions by means of the line equations, which furnish the active power at the entry and at the end of the line and by this the total losses of the cable, without consideration of their temperature dependence. This means that $v_\theta = 1$ in Equation (5.19). Such an assumption may lead to an error of more than 20% in estimating cable losses, as illustrated in Example 5.1.

Equation (5.19) allows calculation of the total energy losses E of the cable of length L over a period of time in the case in which the current variations are known. For the duration of the time interval τ, these losses can be computed from

$$E = W_d \cdot L \cdot \tau + \frac{(W_{tR} - W_d) \cdot L}{I_R^2} \cdot \int_{t=0}^{\tau} I^2(t) \cdot v_\theta(t) \cdot dt \tag{5.22}$$

Factor v_T as well as the values of $\Delta\theta_d$ and θ_x in Equations (5.18) and (5.20) will change, depending whether drying out occurs or not; that is, whether the temperature rise at the cable surface exceeds the critical temperature rise for drying out of the soil, $\Delta\theta_{cr}$, or not. This can be correlated to a critical current I_{cr} and the corresponding critical total losses W_{tcr} by the relation

$$\Delta\theta_{cr} = W_{tcr} \cdot T_4 \tag{5.23}$$

with

$$W_{tcr} = W_d + n \cdot R_{t20} \cdot I_{cr}^2 \cdot (c_\alpha + \alpha_T \cdot \Delta\theta_{c,cr}) \tag{5.24}$$

where $\Delta\theta_{c,cr}$ is the conductor temperature rise corresponding to the critical current. Its value can be obtained from

$$\Delta\theta_{c,cr} = \Delta\theta_{cr} + T_1 \cdot (W_{tcr} - W_d) + T_d \cdot W_d \tag{5.25}$$

where

$$T_d = \frac{T_1}{2 \cdot n} + T_2 + T_3 \tag{5.26}$$

From Equations (5.23) to (5.26), we can compute the critical current as

$$I_{cr} = \frac{1}{\sqrt{n \cdot R_{t20}}} \cdot \sqrt{\frac{\Delta\theta_{cr}/T_4 - W_d}{1 + \alpha_T \cdot [\Delta\theta_{cr} + \theta_{amb} - 20 + T_1 \cdot \Delta\theta_{cr}/T_4 - (T_1 - T_d) \cdot W_d]}} \tag{5.27}$$

Example 5.1

We will consider the cable model No. 2. This is a 145 kV, three-conductor armored cable with XLPE insulation. We will compute the conductor temperature for various levels of loading of this cable. The parameters required for the analysis are extracted from Table A1 and shown in Table 5-2.

The temperature rise caused by the dielectric losses is equal to $\Delta\theta_{d,R} = 0.8°C$ for moist soil conditions and equal to $1.73°C$ when the soil dries out.

The analysis involves calculation of the conductor temperature and total losses for a range of currents from 0 to 966 A. We will show detailed calculations for the current of 600 A and display the remaining results in graphical form.

For the nominal loading conditions, the external cable temperature reaches 56.6°C. This value gives the cable surface temperature rise of 41.6°C, which is higher than the critical temperature rise limit of 35°C (the ampacity value in Table 5-2 reflects the drying out condition). Hence, in this example, $v_T \neq 1$ and we will start the analysis by computing the value of θ_x.

From Equation (5.2), we have

$$v_x = \frac{\rho_2}{\rho_1} = \frac{1.8}{0.6} = 3.0$$

$$\theta_{x,R} = (v_x - 1) \cdot \Delta\theta_{cr} = (3 - 1) \cdot 35 = 70°C$$

The equivalent thermal resistances corresponding to the moist and dry conditions, respectively, are obtained from Equation (5.3) as

$$T_{t,x} = \frac{T_1}{n \cdot (1 + \lambda_1 + \lambda_2)} + \frac{T_2 \cdot (1 + \lambda_1)}{1 + \lambda_1 + \lambda_2} + T_3 + v_x \cdot T_4$$

$$= \frac{0.754}{3 \cdot (1 + 0.123 + 0.210)} + \frac{0.04 \cdot (1 + 0.123)}{1 + 0.123 + 0.210} + 0.047 + 1 \cdot 0.299$$

$$= 0.269 + 0.299 = 0.568 \text{ K} \cdot \text{m/W}$$

$$T_{t,R} = \frac{T_1}{n \cdot (1 + \lambda_1 + \lambda_2)} + \frac{T_2 \cdot (1 + \lambda_1)}{1 + \lambda_1 + \lambda_2} + T_3 + v_x \cdot T_4$$

$$= 0.269 + 3 \cdot 0.299 = 1.165 \text{ K} \cdot \text{m/W}$$

The temperature coefficient v_T is calculated from Equation (5.12) as

$$v_T = \frac{T_{t,x}}{T_{t,R}} = \frac{0.568}{1.165} = 0.488$$

On the other hand, for the current $I = 600$ A, no drying out occurs and $\theta_x = 0$.

Next, we will calculate the values of the auxiliary variables. From Equations (5.7) and (5.10), we have

Table 5-2   Values for numerical example[a]

		Precomputed parameters
Cable ampacity	I (A)	966
Conductor resistance at θ_{max}	R (ohm/km)	0.033
Sheath loss factor	λ_1	0.123
Armor loss factor	λ_2	0.210
Thermal resistance of insulation	T_2 (K · m/W)	0.754
Thermal resistance of armor bedding	T_2 (K · m/W)	0.040
Thermal resistance of jacket	T_3 (K · m/W)	0.047
External thermal resistance, moist soil	T_4 (K · m/W)	0.299
External thermal resistance, dry soil	T_4 (K · m/W)	0.896
Losses in the conductor	W_c (W/m)	30.76
Dielectric losses per core	W_d (W/m)	0.52
Total joule losses per conductor	W_I (W/m)	41.01
Total losses per cable	W_t (W/m)	124.58

[a]Some of the parameters of the 145 kV cable are different from those in Appendix A because different laying conditions are assumed in this example.

$$c_\alpha = 1 + \alpha_T \cdot (\theta_{amb} - 20°C) = 1 + 0.00393 \cdot (15 - 20) = 0.98$$

$$c_m = [1 + \alpha_T \cdot (\theta_{max} - 20°C) = 1 + 0.00393 \cdot (90 - 20) = 1.275$$

From Equations (5.14) to (5.17) we obtain

$$\theta_1 = c_\alpha \cdot (\theta_{max} - \theta_{amb} - \Delta\theta_{d,R} + \theta_{x,R}) \cdot v_T = 0.98 \cdot (90 - 15 - 1.73 + 70) \cdot 0.488 = 68.49°C$$

$$\theta_2 = c_m \cdot (\Delta\theta_d - \theta_x) = 1.275 \cdot (0.8 - 0) = 1.02°C$$

$$c_1 = 1 + \alpha_T \cdot [\theta_{amb} - 20 + (\Delta\theta_{d,R} - \theta_{x,R})] = 1 + 0.00393 \cdot (15 - 20 + 1.73 - 70) = 0.712$$

$$c_2 = \alpha_T \cdot (\theta_{max} - \Delta\theta_{d,R} + \theta_{x,R}) = 0.00393 \cdot (90 - 15 - 1.73 + 70) = 0.563$$

We can now compute the conductor temperature rise caused by 600 A current. From Equation (5.18), we have

$$\Delta\theta = \frac{\theta_1 \cdot \left(\frac{I}{I_R}\right)^2 + \theta_2}{c_1 + c_2 \cdot \left[1 - v_T \cdot \left(\frac{I}{I_R}\right)^2\right]} = \frac{68.49 \cdot \left(\frac{600}{966}\right)^2 + 1.02}{0.712 + 0.563 \cdot \left[1 - 0.488 \cdot \left(\frac{600}{966}\right)^2\right]} = 23.5°C$$

Hence, conductor temperature for this current is equal to $\theta_c = \Delta\theta + \theta_{amb} = 23.5 + 15 = 38.5°C$.

Repeating the calculations for other values of the current, we obtain the results shown in Figure 5-1. Note that for $I = 0$ A, the conductor temperature reaches

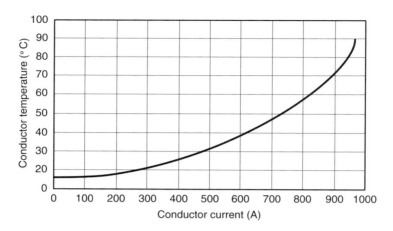

Figure 5-1 Conductor temperature as a function of conductor current for the three-conductor armored cable.

15.8°C because we assumed that the cable is energized all the time and dielectric losses are present. We can also observe a sharp temperature rise when the conductor current exceeds 900 A. As will be shown below, the drying out occurs at the current value of 952 A and this results in a steep increase in the cable surface temperature.

The cable losses are computed next. First, we obtain the temperature coefficient from Equations (5.21) and (5.20):

$$c_3 = 1 + \alpha_T \cdot (\theta_{amb} - 20 + (\Delta\theta_d - \theta_x)] = 1 + 0.00393 \cdot (15 - 20 + 0.8 - 0) = 0.983$$

$$v_\theta = \frac{c_3}{c_1 + c_2 \cdot \left[1 - v_T\left(\frac{I}{I_R}\right)^2\right]} = \frac{0.983}{0.712 + 0.563 \cdot \left[1 - 0.488 \cdot \left(\frac{600}{966}\right)^2\right]} = 0.841$$

From Equation (5.19), we now compute the total losses in the cable:

$$W_t = W_d + (W_{tR} - W_d) \cdot \left(\frac{I}{I_R}\right)^2 \cdot v_\theta = 3 \cdot \left[0.52 + (41.53 - 0.52) \cdot \left(\frac{600}{966}\right)^2 \cdot 0.841\right]$$

$$= 41.49 \text{ W/m}$$

The losses computed from Equation (5.19) as a function of conductor current are shown in Figure 5-2 (dashed line). The solid line represents the joule losses computed without consideration of temperature variation of resistance; that is, for $v_\theta = 1$.

The line at the top of the graph shows the ratio of the losses computed without and with the temperature correction. The maximum error in this case exceeds 20%

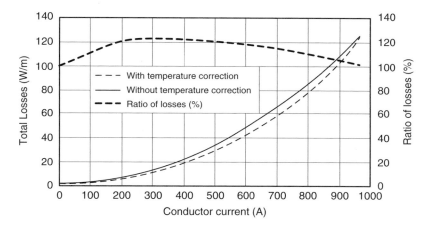

Figure 5-2 Total losses for the three-conductor armored cable as a function of conductor current.

for the currents in the usual loading range of this type of the cable; that is, between 300 and 500 A.

Assuming that the critical temperature rise at which drying out starts for this installation is 35°C, the critical loading current can be computed from Equation (5.27). In order to perform this calculation, we need to obtain first the internal resistance from Equation (5.26) and the conductor resistance at 20°C. The internal equivalent thermal resistance was computed above as $T_I = 0.263$ K · m/W:

$$T_d = \frac{T_1}{2 \cdot n} + T_2 + T_3 = \frac{0.754}{6} + 0.04 + 0.047 = 0.213 \text{ K · m/W}$$

$$R_{t20} = \frac{R \cdot (1 + \lambda_1 + \lambda_2)}{1 + \alpha_T \cdot (\theta_{max} - 20°C)} = \frac{0.000033 \cdot (1 + 0.123 + 0.210)}{1 + 0.00393 \cdot (90 - 20)} = 0.0000345 \text{ ohm/m}$$

The critical conductor current is equal to

$$I_{cr} = \frac{1}{\sqrt{n \cdot R_{t20}}} \cdot \sqrt{\frac{\Delta\theta_{cr}/T_4 - W_d}{1 + \alpha_T \cdot [\Delta\theta_{cr} + \theta_{amb} - 20 + T_I \cdot \Delta\theta_{cr}/T_4 - (T_I - T_d) \cdot W_d]}}$$

$$= \frac{1}{\sqrt{3 \cdot 0.0000345}}$$

$$\cdot \sqrt{\frac{35/0.299 - 0.52}{1 + 0.00393 \cdot [35 + 15 - 20 + 0.269 \cdot 35/0.299 - (0.269 - 0.213) \cdot 0.52]}}$$

$$= 952 \text{ A} \qquad \blacksquare$$

5.3 PRACTICAL APPLICATION OF THE TEMPERATURE PREDICTION EQUATION

A profitable use of Equations (5.18), (5.22), and (5.27) is possible if the corresponding cable data are known. In this case, for all operating conditions, the stationary conductor temperature as well as the cable losses can be determined by inserting the actual current values into these simple analytical formulae.

One possibility is to execute a complete current-rating analysis, thus determining all thermal resistances, loss factors, and other relevant parameters. This is normally done by the cable manufacturer, since they have at their disposal all construction data as well as suitable rating programs.

The extraction of the necessary coefficients θ_1, θ_2, c_1, c_2, and c_3 from the cable data without any rating calculations is possible if the following nominal values for full-load operation are given:

- rated current, I_R
- dielectric losses, W_d
- ambient soil temperature, θ_{amb}

These three parameters are easily obtained from the cable specifications. With these values, Equation (5.18) can be used for the prediction of the conductor temperature for a given current with an assumption that soil dry-out does not occur. Since this is the situation most often encountered in practice, the method presented in this chapter can be used as a simple alternative to the full real-time temperature-monitoring system. If, on the other hand, there is a need for calculation of cable losses, or if soil dry-out may occur during cable operation, we will require two additional quantities, namely,

1. The total rated losses, W_{tR} (W/m) of the cable system
2. The cable surface temperature, θ_e, °C

From these data, the external thermal resistance T_4 can be computed as

$$T_4 = \frac{\theta_e - \theta_{amb}}{W_{tR}} \tag{5.28}$$

The total thermal resistances of the cable can be approximated by

$$T_{t,x} \approx T_{d,x} \approx T'_{app} = (\theta_{max} - \theta_e)/W_{tR} + T_4 \tag{5.29}$$

The approximation given by Equation (5.29) results in a very small error for the cables with small dielectric losses.

As a substitute for the cable surface temperature and the total cable losses, we can assume soil thermal resistivities for moist and dry conditions, and proceed with the calculations accordingly.

5.4 FIELD VERIFICATION OF THE TEMPERATURE CALCULATIONS

To verify the accuracy of the temperature prediction obtained from Equation (5.18), data from an existing DFR system were extracted and used for comparative analysis (Anders et al., 2003). The circuit being monitored is a 138 kV pipe-type cable installed in New York City. The cable is very similar to the standard cable described in the Neher and McGrath (1957) paper, with only small modifications of some dimensions.

The design characteristics of this cable required for the temperature calculations are as follows:

- Rated current, $I_R = 1336$ A, computed for the soil thermal resistivity of 0.7 K · m/W
- Dielectric losses, $W_d = 2.73$ W/m
- Ambient soil temperature, $\theta_{amb} = 22°C$

Since, in this case, we are dealing with an existing DFR installation, the ambient temperature and soil thermal resistivity were taken from the field measurements for a selected time, and the steady-state rating was computed by the cable rating system for these quantities.

Figure 5-3 shows current values for one day (measurements taken every 20 minutes).

The results of the measurements and corresponding calculations are shown in Figure 5-4. The measured conductor temperature is actually computed by the DFR system from measured cable surface and ambient temperatures as well as cable current. The x-axis shows the time measurement points that span about 24 hours (measurements and calculations are reported every 20 minutes).

Several important conclusions can be drawn from the results shown in Figure 5-4. We can observe that the computed values change in a much greater range than the actual conductor temperatures. The reason for this is that the thermal capaci-

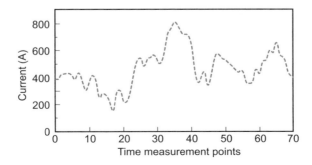

Figure 5-3 Load variations for a selected day in a field installation. Measurements taken every 20 minutes.

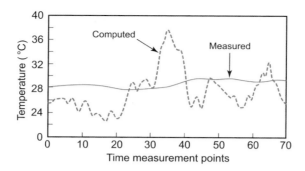

Figure 5-4 Comparison of measured and predicted conductor temperatures in a field installation. Measurements taken every 20 minutes.

tances are neglected in our analysis. They cause the temperature variations inside the cable to be much slower than the current changes alone would indicate.

It is also interesting to note that the peaks and valleys in the computed curve precede those obtained from measurements, again, due to lack of time considerations in our equations. Finally, we can observe that the differences are rather small, most of the time in the range of ±4°C. An exception is a measurement interval between the 30th and 40th time points. Here, our predicted temperature is about 9°C higher than the measured value. An analysis of the recorded data showed that there was a large surge of current at the two-hour interval at that time (see Figure 5-3). The current, which normally hovered in the range of 300 to 500 A, suddenly increased to over 800 A and stayed there for about one hour, dropping off as rapidly as it rose. Since our method does not have time memory, we had predicted a high conductor temperature but, in reality, this rise was tempered by high system capacitances and the short duration of the high-current period.

5.5 LOSS FACTOR CALCULATIONS IN RATING STANDARDS

In the first part of this section, we will use the results obtained above to investigate the effect of neglecting the variations of the loss factor with temperature in standard rating calculations. Next, we will examine the effect of daily and weekly load variations on cable ratings. It will be shown in Section 5.5.2 that the common practice of considering only the daily load cycle may lead to a substantial underestimation of the cable current carrying capabilities.

5.5.1 Daily Load Cycle

Computation of a cyclic rating requires an evaluation of the cable capacitances and conductor temperatures at several time points. Neher and McGrath (1957) proposed an alternative, simple method to account for the cyclic loading of a cable. Their approach requires a modification of the external thermal resistance of the cable. The underlying principles are discussed below.

In order to evaluate the effect of a cyclic load upon the maximum temperature rise of a cable system, Neher (1953) observed that one can look upon the heating effect of a cyclical load as a wave front that progresses alternately outwardly and inwardly in respect to the conductor during the cycle. He further assumed that with the total joule losses generated in the cable equal to W_I (W/m), the heat flow during the loss cycle is represented by a steady component of magnitude μW_I plus a transient component $(1 - \mu)W_I$ that operates for a period of time τ during each cycle. The transient component of the heat flow will penetrate the earth only to a limited distance from the cable, thus the corresponding thermal resistance T_{4et} will be smaller than its counterpart T_{4ss}, which pertains to steady-state conditions.

Assuming that the temperature rise over the internal thermal cable resistance is complete by the end of the transient period τ, the maximum temperature rise at the conductor may be expressed as

$$\Delta\theta = W_I v[T + \mu T_{4ss} + (1 - \mu)T_{4et}] + \Delta\theta_d - \theta_x \quad (5.30)$$

where T is the internal thermal resistance of the cable, T_{4ss} is the external thermal resistance with constant load, and T_{4et} is the effective transient thermal resistance in the earth. The effect of moisture migration is taken into account by two variables:

$$v = (\rho_c/\rho_s) \quad \text{for} \quad \theta_e \leq \theta_{cr}$$

$$v = 1 \quad \text{for} \quad \theta_e \geq \theta_{cr} \quad (5.31)$$

$$\theta_x = \theta_{cr} \cdot (v - 1)$$

where
ρ_c = thermal resistivity of the dry soil, K · m/W
ρ_s = thermal resistivity of the moist soil, K · m/W
θ_{cr} = critical temperature at which drying out starts, °C
θ_e = temperature of the cable surface, °C.

Further, Neher (1953) assumed that the last thermal resistance in Equation (5.30) may be represented with sufficient accuracy by an expression of the general form

$$T_{4et} = A\rho_s \ln \frac{B\sqrt{\delta \cdot \tau}}{D_e} \quad (5.32)$$

in which constants A and B were evaluated empirically to best fit the temperature rises calculated over a range of cable sizes. Using measured data, Neher (1953) obtained values for the constants $A = 1/2\pi$ and $B = 61{,}200$ when τ is expressed in hours and δ (soil diffusivity) is expressed in m²/s.

Introducing the notation

$$D_x = 61,200\sqrt{\delta(\text{length of cycle in hours})} \tag{5.33}$$

the external thermal resistance in Equation (5.30) can be written as,[1]

$$T_4 = \mu T_{4ss} + (1-\mu)T_{4et} = \frac{\rho_s}{2\pi}\left[\mu \ln \frac{4L}{D_e} + (1-\mu)\ln \frac{D_x}{D_e}\right] = \frac{\rho_s}{2\pi}\left[\ln \frac{D_x}{D_e} + \mu \ln \frac{4L}{D_x}\right] \tag{5.34}$$

where
ρ_s = soil thermal resistivity, K · m/W
L = depth of cable center, mm
D_e = external diameter of the cable or pipe, mm
μ = the loss factor

The right-hand side of Equation (5.34) can be interpreted as follows. Inside the circle of diameter D_x, the temperature changes according to the peak value of the losses. Outside this circle, it changes with the average losses.

From Equation (5.33), we observe that the fictitious diameter D_x at which the effect of loss factor commences is a function of the diffusivity of the medium δ and the length of the loss cycle. In the majority of cases, the soil diffusivity will not be known. In such cases, a value of $0.5 \cdot 10^{-6}$ m²/s can be used. This value is based on a soil thermal resistivity of 1.0 K · m/W and a moisture content of about 7% of dry weight [see Section 5.4 of Anders (1997) for more details on this subject]. The value of D_x for a load cycle lasting 24 hours and with a representative soil diffusivity of $0.5 \cdot 10^{-6}$ m²/s is 211 mm (or 8.3 in).

Figure 5-5 shows a characteristic diameter for a cable dissipating total losses of W_t (W/m).

The loss factor is defined by

$$\mu = \frac{1}{W_{\max} \cdot \tau} \cdot \int_{t=0}^{\tau} W(t) \cdot dt \tag{5.35}$$

This is usually approximated by

$$\mu_I = \frac{1}{I_{\max}^2 \cdot \tau} \cdot \int_{t=0}^{\tau} I^2(t) \cdot dt \tag{5.36}$$

In engineering practice, the loss factor is approximated from the known or assumed load factor LF as (Anders, 1997)

$$\mu_{LF} = 0.3 \cdot LF + 0.7 \cdot LF^2 \tag{5.37}$$

---

[1]Most European standards (excluding Germany), which in the most part are based on the IEC standards, do not introduce a notion of the loss factor. Instead, calculations involving varying loads are referred to a separate standard using full time-dependent analysis.

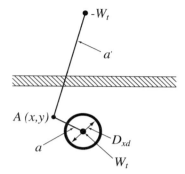

Figure 5-5 Characteristic isotherm of diameter D_x for the cable dissipating W_t in losses (from Brakelmann, 1995a, with permission from Elektrizitätswirtschaft).

The load factor, in turn, is defined by

$$LF = \frac{1}{I_{max} \cdot \tau} \cdot \int_{t=0}^{\tau} I(t) \cdot dt \qquad (5.38)$$

Depending on the load variations, the approximations given by Equations (5.36) and (5.37) can introduce a significant error, as shown later in this chapter.

Example 5.2 [extracted from Anders (1997)]
We will compute the rating of the cable model No.1 using the Neher/McGrath approach. The parameters of this circuit are $\lambda'_{11} = 0.206$, $\lambda'_{12} = 222$, $\lambda'_{1m} = 0.089$, $T_1 = 0.214$ K $\cdot$ m/W, and $T_3 = 0.104$ K $\cdot$ m/W. Cables are spaced two cable diameters apart.

The fictitious diameter D_x is first obtained from Equation (5.33) and is equal to 211 mm. The external thermal resistance of the middle cable is obtained from Equation (5.34) with the loss factor computed in Example 2.2 equal to 0.504. In this case, equation discussed in Section 1.6.6.5 is used for the external thermal resistance and the effect of the outer cables is also taken into account:

$$T_4 = \frac{\rho_s}{2\pi} \left\langle \ln \frac{D_x}{D_e} + \mu \left\{ \ln \frac{4L}{D_e} + \left[\frac{1 + 0.5(\lambda'_{11} + \lambda'_{12})}{1 + \lambda'_{1m}} \right] \cdot \ln \left[1 + \left(\frac{2L}{s_1} \right)^2 \right] \right\} \right\rangle$$

$$= \frac{1}{2\pi} \left\langle \ln \frac{211}{35.8} + 0.504 \cdot \left\{ \ln \frac{4 \cdot 1000}{211} + \frac{1 + 0.5(0.206 + 0.222)}{1 + 0.089} \right. \right.$$

$$\left. \left. \ln \cdot \left[1 + \left(\frac{2 \cdot 1000}{2 \cdot 35.8} \right)^2 \right] \right\} \right\rangle = 1.11 \text{ K} \cdot \text{m/W}$$

Assuming the sheath loss factor for the middle cable, the rating is equal to

$$I = \left[\frac{90 - 15}{0.781 \cdot 10^{-4}[0.214 + (1 + 0.09) \cdot (0.104 + 1.11)]} \right]^{0.5} = 790 \text{ A}$$

If we use Heinhold's Equation (5.51), the fictitious diameter D_x is equal to

$$D_x = \frac{103 + 246 \sqrt{0.504}}{\sqrt{1} \cdot 1^{0.4}} = 277.6 \text{ mm}$$

The external thermal resistance and the rating of the cable become

$$T_4 = \frac{1}{2\pi} \left\langle \ln \frac{277.6}{35.8} + 0.504 \cdot \left\{ \ln \frac{4 \cdot 1000}{277.6} + \frac{1 + 0.5(0.206 + 0.222)}{1 + 0.089} \right. \right.$$
$$\left. \left. \ln \cdot \left[1 + \left(\frac{2 \cdot 1000}{2 \cdot 35.8} \right)^2 \right] \right\} \right\rangle = 1.14 \text{ K} \cdot \text{m/W}$$

$$I = \left[\frac{90 - 15}{0.781 \cdot 10^{-4}[0.214 + (1 + 0.09) \cdot (0.104 + 1.14)]} \right]^{0.5} = 782 \text{ A}$$

We will continue this example later with the new developments presented in this chapter. ∎

In order to include the effect of the resistance variations with temperature in loss factor calculations, Equation (5.36) is modified as follows:

$$\mu_I = \frac{1}{I^2_{\max} \cdot \tau} \cdot \int_{t=0}^{\tau} I^2(t) \cdot v_\theta dt \tag{5.39}$$

where v_θ is obtained from Equation (5.20):

Example 5.3
We will consider the three-conductor armored cable model No. 2 analyzed in Example 5.1. This time, we will consider a two-step load applied to this cable. The shape of the load is shown in Figure 5-6.

The current in Figure 5-6 is characterized by 8 hours of full load ($I = I_1$), followed by 16 hours of a lower load ($I = I_2$; possibly an example for an industrial load). It is assumed, that the peak value $I_{\max}$ of the load cycle corresponds to 60% of the rated current I_R of the cable. Under those conditions, the soil will not dry out (see Example 5.1).

The load and loss factors calculated from the traditional Equations (5.38), (5.37), and (5.36) are computed next:

$$LF = \frac{1}{I_{\max} \cdot \tau} \cdot \int_{t=0}^{\tau} I(t) \cdot dt = \frac{1}{1 \cdot 24} \sum_{t=1}^{24} I(t) = 0.73$$

$$\mu_{LF} = 0.3 \cdot LF + 0.7 \cdot LF^2 = 0.3 \cdot 0.73 + 0.7 \cdot 0.73^2 = 0.596$$

$$\mu_I = \frac{1}{I^2_{\max} \cdot \tau} \cdot \int_{t=0}^{\tau} I^2(t) \cdot dt = \frac{1}{1 \cdot 24} \sum_{t=1}^{24} I^2(t) = 0.573$$

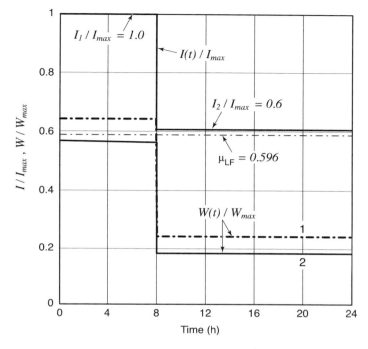

Figure 5-6 Two-step load function applied to the cable model No. 2.

In order to compute the temperature-dependent loss factor, we need to evaluate the temperature factor from Equation (5.20). We will show the calculations for the maximum current (1 p.u.) with the necessary constants computed in Example 5.1:

$$v_\theta = \frac{c_3}{c_1 + c_2 \cdot \left[1 - v_T \left(\dfrac{I}{I_R}\right)^2\right]} = \frac{0.983}{0.712 + 0.563 \cdot \left[1 - 0.488 \cdot \left(\dfrac{1}{1/0.6}\right)^2\right]} = 0.836$$

From Equation (5.39), we have

$$\mu_\theta = \frac{1}{I_{max}^2 \cdot \tau} \cdot \int_{t=0}^{\tau} I^2(t) \cdot v_\theta dt = \frac{1}{1 \cdot 24} \sum_{t=1}^{24} I^2(t) \cdot v_\theta = 0.469$$

The corresponding values of the external thermal resistance and the ratings are shown in Table 5-3.

When the temperature dependence of the loss factor is introduced, the ampacity of the cable increases by 47 A or about 4.2%.

We will conclude this example with the analysis of the possible error that can occur when traditional loss factor computations are performed. Figure 5-7 shows the

Table 5-3   Rating as a function of the applied loss factor

Loss factor		T_4	Ampacity
μ_{LF}	0.596	0.185	1112
μ_I	0.573	0.179	1120
μ_θ	0.469	0.150	1159

ratio of the loss factors obtained from Equations (5.36) and (5.37) to the one obtained from Equation (5.39) as a function of I_2/I_{max} (Brakelmann and Anders, 2004).

We can observe that substantial errors can occur in the evaluation of the loss factor when the temperature dependence is neglected. For this particular load curve, the traditional loss factor as computed from Equation (5.37) can be up to 27% larger in comparison with the one obtained, which takes into account variation of the conductor resistance with temperature. The maximum error occurs when loss factor computed from Equation (5.37) is used with small values of the current I_2.   ■

It should be stressed that the above discussion pertains to a uniform load cycle, for example, daily load variations during weekdays. In the following section, we will explore more complex load cycle variations.

5.5.2   Consideration of Weekly Load Variations

In the majority of practical cases, the load variations will exhibit a more complex pattern than the one described by a daily load cycle. For example, loading of cables

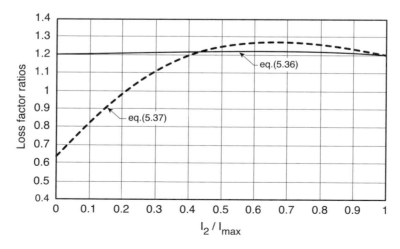

Figure 5-7   Ratio of traditional loss factors to the temperature-dependent one.

is usually much lighter during the weekend than during the weekdays. In this section, we will take advantage of this fact and show how the cable rating can be improved by considering both daily and weekly load variations [Equations (5.46) to (5.48)]. Most of the material in this section is based on the work of Brakelmann (1995a,b).

Figure 5-8 shows an approximation of a typical weekly load cycle. In this figure, curve 1 represents a load cycle during which the peak conditions prevail for 10 hr each day and for the remaining 14 hr the load is 70% of the peak value. During the sixth and seventh days of the week, the loads are decreased by 80% in comparison with the weekday values. Curve 2 represents corresponding loss values. This means that (Brakelmann, 1995a)

$$W_I(t) = W_I \cdot F_1(t) \cdot F_2(t) = W_I[\mu_d(1 - \mu_d) \cdot f_1(t)] \cdot [\mu_w + (1 - \mu_w) \cdot f_2(t)] \quad (5.40)$$

where W_I now represents the peak value of the losses and subscripts d and w correspond to the daily and weekly load variations, respectively.

The weekly load cycle is described in Equation (5.40) by two components corresponding to weekdays and weekend days. The modulating functions $F_1(t)$ and $F_2(t)$, representing load variations during the week days and weekend days, respectively, are calculated by taking the ratios of the current squared at a given hour to the squared value of the maximum current. By definition, the function $F_1(t)$ has value of 1 during the weekends and function $F_2(t)$ is equal to 1 during the weekdays. These functions are then represented by the average loss factors and the modulating functions $f_1(t)$ and $f_2(t)$, with both functions taking always the values less or equal to one.

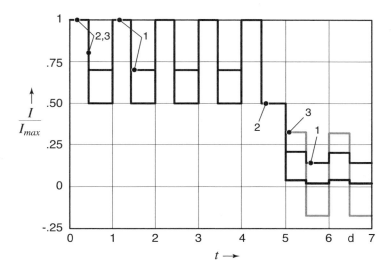

Figure 5-8   An approximation of a typical weekly load cycle (from Brakelmann, 1995a, with permission form Elektrizitätswirtschaft).

From Equation (5.40), we obtain

$$W_I(t) = W_I[\mu_d\mu_w + \mu_w(1 - \mu_d) \cdot f_1(t) + \mu_d(1 - \mu_w) \cdot f_2(t) + (1 - \mu_d)(1 - \mu_w) \cdot f_1(t) \cdot f_2(t)]$$

(5.41)

To facilitate calculations, we will assume $f_2(t) = 1$ in the last term of Equation (5.41), which leads to

$$W_I(t) = W_I[\mu_d\mu_w + (1 - \mu_d) \cdot f_1(t) + \mu_d \cdot (1 - \mu_w) \cdot f_2(t)] = W_I[\mu_r + w_1(t) + \mu_d w_2(t)]$$

(5.42)

Equation (5.42) represents loss variations during the week with a resultant loss factor

$$\mu_r = \mu_d\mu_w$$

(5.43)

The first component

$$w_1(t) = (1 - \mu_d)f_1(t)$$

(5.44)

of the loss variations represents the daily cycle and we can associate with it a daily characteristic diameter D_{xd}.

The second component $\mu_d w_2(t)$ represents weekly load variations reduced by the daily load factor μ_d and we can associate with it a weekly characteristic diameter D_{xw}. The effect of these approximations is shown in Figure 5-8, where curve 3 represents the losses computed with Equation (5.42). We can observe that during the weekend, the losses can be higher during the peak hours and lower (even negative) during the low-load period. However, the average losses are the same as for curve 2. A Fourier analysis performed by Brakelmann (1989) has shown that the earth thermal resistance computed with this approximation has an error not exceeding 2%, hence it is negligible in rating calculations.

Figure 5-9 presents the characteristic diameters for the daily and weekly load cycles. When calculating the temperature rise at a reference point A, there are three possible situations.

a. Point A is located outside the outer characteristic diameter. The temperature rise at this point depends on the average loss factor μ_r; that is,

$$\Delta\theta = W_I v \mu_d\mu_w \frac{\rho_s}{2\pi} \ln \frac{a'}{a} + \Delta\theta_d - \theta_x$$

(5.45)

b. Point A is located between the two isotherms D_{xd} and D_{xw}. The temperature at this point is calculated with the Neher's assumption that the average daily losses $\mu_d W_I$ heat this point and the maximum weekly losses will be applied, so that

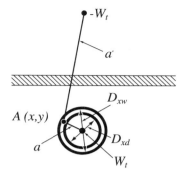

Figure 5-9 Characteristic isotherms for daily and weekly load cycles (from Brakelmann, 1995a, with permission from Elektrizitätswirtschaft).

$$\Delta\theta_d = W_I v \mu_d \left[\left(\frac{\rho_s}{2\pi} \ln \frac{a'}{a} - T_{4w} \right) - \mu_w T_{4w} \right] + \Delta\theta_d - \theta_x$$

$$= W_I v \mu_d \left[\frac{\rho_s}{2\pi} \ln \frac{a'}{a} - (1 - \mu_w) T_{4w} \right] + \Delta\theta_d - \theta_x \tag{5.46}$$

where T_{4w} is the thermal resistance between the weekly isotherm and the earth surface.

c. Point A is located inside the daily isotherm. The temperature at this point is computed as if the peak losses W_I were applicable in the entire region, but the losses between the two isotherms are governed by the weekly load variations and the average daily variations, and outside both isotherms by the average mixed variations. Thus, we have

$$\Delta\theta = W_I v \left[\frac{\rho_s}{2\pi} \ln \frac{a'}{a} - (1 - \mu_d) T_{4d} - \mu_d (1 - \mu_w) T_{4w} \right] + \Delta\theta_d - \theta_x \tag{5.47}$$

where T_{4d} denotes the thermal resistance between the daily isotherm and the earth surface.

For rating calculations, we will need the external thermal resistance of the cable surface. In this case, Equation (5.47) takes the form

$$\Delta\theta = W_I v \left[\frac{\rho_s}{2\pi} \ln \frac{2L}{D_e} - (1 - \mu_d) T_{xd} - \mu_d (1 - \mu_w) T_{xw} \right] + \Delta\theta_d - \theta_x \tag{5.48}$$

5.5.2.1 Characteristic Diameter. Alternative expressions for D_x (mm) are given by Heinhold (1990):

$$D_x = \frac{205}{\sqrt{w} \rho_s^{0.4}} \qquad \text{for sinusoidal load variation} \tag{5.49}$$

$$D_x = \frac{493\sqrt{\mu}}{\sqrt{w}\rho_s^{0.4}} \qquad \text{for rectilinear load variation} \qquad (5.50)$$

$$D_x = \frac{103 + 246\sqrt{\mu}}{\sqrt{w}\rho_s^{0.4}} \qquad \text{for other load variations} \qquad (5.51)$$

where $w = \tau_0/\tau$, τ is the length of the period, and $\tau_0 = 24$ hr $= 1$ day.

The weekly characteristic diameters for two values of soil resistivity utilizing Equations (5.49) to (5.51) are shown in Figure 5-10.

Equations (5.49) to (5.51) are independent of the cable diameter. Investigations by Brakelmann (1986, 1989) have shown that these equations are valid for cables with diameters between 5 and 150 mm.

From Figure 5-10, we can observe that the characteristic diameter for weekly load variations with unknown shape and moist soil varies between 0.7 and 0.9 m. For the daily load cycle, this diameter would vary between 0.25 and 0.3 m. For dry soil conditions, the characteristic diameter with mixed load shape varies between 0.5 and 0.65 m, whereas for the daily load cycle, the corresponding values are 0.17 to 0.23 m.

Example 5.4

Let us consider again the cable model No. 1, with the daily load curve as described in Example 5.2. In addition, let us assume a weekly load factor of 0.55, which yields a loss factor of 0.377 and a yearly load factor of 0.4, giving a loss factor of 0.232 [see Equation (5.37)]. Since the load shape is neither sinusoidal

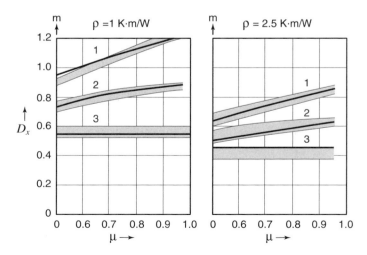

Figure 5-10   Weekly characteristic diameters as a function of the loss factors for (1) rectangular, (3) sinusoidal, and (2) other load variations (from Brakelmann, 1995a, with permission from Elektrizitätswirtschaft).

nor rectilinear, we will use Equation (5.51) to calculate daily and weekly characteristic diameters:

$$D_{xd} = \frac{103 + 246\sqrt{\mu}}{\sqrt{w}\rho_s^{0.4}} = \frac{103 + 246\sqrt{0.504}}{\sqrt{1/1} \cdot 1^{0.4}} = 277.6 \text{ mm}$$

$$D_{xw} = \frac{103 + 246\sqrt{\mu}}{\sqrt{w}\rho_s^{0.4}} = \frac{103 + 246\sqrt{0.377}}{\sqrt{1/7} \cdot 1^{0.4}} = 672.0 \text{ mm}$$

Thus, the middle cable, for which we will compute the rating, is inside the weekly characteristic diameter for both the outer cables.

The external resistances between the characteristic diameters and the earth are, respectively, equal to

$$T_{4d} = \frac{\rho_s}{2\pi} \ln \frac{4L}{D_{xd}} = \frac{1}{2\pi} \ln \frac{4 \cdot 1000}{278} = 0.425 \frac{\text{K} \cdot \text{m}}{\text{W}}$$

$$T_{4w} = \frac{\rho_s}{2\pi} \ln \frac{4L}{D_{xw}} = \frac{1}{2\pi} \ln \frac{4 \cdot 1000}{672} = 0.284 \frac{\text{K} \cdot \text{m}}{\text{W}}$$

The external thermal self-resistance of the middle cable and the mutual thermal resistances are equal to

$$T_{22} = T_{4\_} = \frac{\rho_s}{2\pi} \ln \frac{4L}{D_e} = \frac{1}{2\pi} \ln \frac{4 \cdot 1000}{38.5} = 0.750 \frac{\text{K} \cdot \text{m}}{\text{W}}$$

$$T_{12} = \frac{\rho_s}{2\pi} \frac{1 + 0.5(\lambda'_{11} + \lambda'_{12})}{1 + \lambda'_m} \ln \left[\sqrt{\left(\frac{2L}{s_1}\right)^2 + 1} \right]$$

$$= \frac{1}{2\pi} \frac{1 + 0.5(0.206 + 0.222)}{1 + 0.0890} \ln \left[\sqrt{\left(\frac{2 \cdot 1000}{2 \cdot 38.5}\right)^2 + 1} \right] = 0.590 \frac{\text{K} \cdot \text{m}}{\text{W}}$$

The external thermal resistance of the middle cable with a unity load factor is, thus, equal to

$$T_4 = T_{22} + 2T_{12} = 0.750 + 2 \cdot 0.59 = 1.932 \text{ K} \cdot \text{m/W}$$

Taking into consideration the daily characteristic diameter and remembering that the middle cable is inside the characteristic diameter of the outer cables, we have from Equation (5.47),

$$T_4 = T_{22} + 2T_{12} - (1 - \mu_d)T_{4d} = 1.932 - (1 - 0.504) \cdot 0.425 = 1.722 \frac{\text{K} \cdot \text{m}}{\text{W}}$$

and the cable rating neglecting the moisture migration in the soil is equal to

$$I = \left(\frac{\theta_c - \theta_{amb}}{R[T_1 + (1 + \lambda_m)(T_3 + T_4)]} \right)^{0.5}$$

$$= \left[\frac{90 - 15}{0.781 \cdot 10^{-4}[0.214 + (1 + 0.09) \cdot (0.104 + 1.722)]} \right]^{0.5} = 660 \text{ A}$$

When a 100% load factor is assumed, the ampacity of this cable is 629 A. We can observe that the ampacity computed in Example 5.2 is much higher than the one obtained above. This is a result of an incorrect assumption often made when applying the Neher/McGrath (1957) method that the load shape is sinusoidal and that the middle cable lies outside the characteristic isotherms of the outer cables. The second assumption is the main reason why the ampacities are so different. Even with the Neher/McGrath characteristic diameter of 211 mm, the middle cable is inside the daily characteristic isotherm of the outer cables. This underscores the importance of the accurate analysis of the influence of the loss factor in cable rating calculations

We will now consider the weekly load variations. Applying Equation (5.48), the external thermal resistance of the middle cable is equal to

$$T_4 = T_{22} + 2T_{12} - (1 - \mu_d)T_{4d} - \mu_d(1 - \mu_w)T_{4w}$$

$$= 1.932 + (1 - 0.504) \cdot 0.425 - 0.504 \cdot (1 - 0.377) \cdot 0.284 = 1.633 \text{ K} \cdot \text{m/W}$$

and the cable rating is equal to

$$I = \left(\frac{\theta_c - \theta_{amb}}{R[T_1 + (1 + \lambda_m)(T_3 + T_4)]} \right)^{0.5}$$

$$= \left[\frac{90 - 15}{0.781 \cdot 10^{-4}[0.214 + (1 + 0.09) \cdot (0.104 + 1.633)]} \right]^{0.5} = 675 \text{ A}$$

Inclusion of the weekly load variations increases cable ampacity by about 2.2% in comparison with the case when only daily loss factor is considered.

We will now consider a larger cable spacing of 0.75 m. In this case, the rating with a unity load factor is equal to 635 A. The increase in circulating current losses is compensated for by a decrease in the mutual heating. The leading phase is the hottest, and for simplicity of calculation, we will consider the rating of this cable.

The external thermal self-resistance of this cable is equal to 0.750 K · m/W. We will compute the mutual thermal resistances neglecting the effect of sheath loss factors. We have

$$T_{12} = \frac{\rho_s}{2\pi} \ln\left[\sqrt{\left(\frac{2L}{s_1}\right)^2 + 1} \right] = \frac{1}{2\pi} \ln\left[\sqrt{\left(\frac{2 \cdot 1000}{750}\right)^2 + 1} \right] = 0.167 \frac{\text{K} \cdot \text{m}}{\text{W}}$$

$$T_{13} = \frac{\rho_s}{2\pi} \ln\left[\sqrt{\left(\frac{2L}{s_1}\right)^2 + 1}\right] = \frac{1}{2\pi} \ln\left[\sqrt{\left(\frac{2 \cdot 1000}{1500}\right)^2 + 1}\right] = 0.0813 \frac{K \cdot m}{W}$$

When considering the daily and weekly load cycles, the leading cable is outside both characteristic diameters of cables 2 and 3. Therefore, its external thermal resistance is equal to

$$T_4 = T_{4_-} - (1 - \mu_d)T_{4d} - \mu_d(1 - \mu_w)T_{4w} + \mu_d\mu_w(T_{12} + T_{13})$$

$$= 0.750 - (1 - 0.504) \cdot 0.425 - 0.504 \cdot (1 - 0.377) \cdot 0.284 + 0.504 \cdot 0.377$$

$$\cdot (0.167 + 0.0813) = 0.498 \frac{K \cdot m}{W}$$

The rating of the cable is now equal to

$$I = \left(\frac{\theta_c - \theta_{amb}}{R[T_1 + (1 + \lambda_1)(T_3 + T_4)]}\right)^{0.5}$$

$$= \left[\frac{90 - 15}{0.781 \cdot 10^{-4}[0.214 + (1 + 0.971) \cdot (0.104 + 1.494)]}\right]^{0.5} = 828 \text{ A}$$

If we include now only the daily load variations, the external thermal resistance of cable 1 becomes

$$T_4 = T_{4ss} - (1 - \mu_d)T_{4d} + \mu_d(T_{12} + T_{13})$$

$$= 0.750 - (1 - 0.504) \cdot 0.425 + 0.504 \cdot (0.167 + 0.0813) = 0.665 \frac{K \cdot m}{W}$$

with the resulting current equal to 745 A.

Thus, inclusion of the weekly loss factor in the calculations increases ampacity of this cable by 11%. ∎

5.6 DEEPLY BURIED CABLES

When the cable circuit is located at large depths, for example, in a deeply buried tunnel, the heating time constant is very large and yearly load cycle variations should also be taken into account. Applying similar reasoning as for the weekly load variations, the variable losses can be described by an equation similar to (5.40):

$$W_I(t) = W_I \cdot F_1(t) \cdot F_2(t) \cdot F_3(t)$$

$$= W_I[\mu_d + (1 - \mu_d) \cdot f_1(t)] \cdot [\mu_w + (1 - \mu_w) \cdot f_2(t)] \cdot [\mu_y + (1 - \mu_y) \cdot f_3(t)] \quad (5.52)$$

where μ_y is the average loss factor for yearly load variations.

Making similar simplifying assumptions to those that led to Equation (5.42), the last equation can be written as

$$W_I(t) = W_I[\mu_d \cdot \mu_w \cdot \mu_y + (1 - \mu_d) \cdot f_1(t) + \mu_d\mu_y \cdot (1 - \mu_w) \cdot f_2(t)$$

$$+ \mu_d\mu_w \cdot (1 - \mu_y) \cdot f_3(t)] = W_I[\mu_d + w_1(t) + \mu_d \cdot \mu_y \cdot w_2(t) + \mu_d \cdot \mu_w \cdot w_3(t)] \quad (5.53)$$

Since now we have three characteristic diameters, the temperature rise at the cable (or the tunnel) surface is obtained applying reasoning similar to that which led to Equation (5.48):

$$\Delta\theta = W_I \cdot v\{T_{4\_} \cdot [1 - \mu_d \cdot (1 + \mu_w\mu_y - \mu_w - \mu_y)] - (1 - \mu_d) \cdot T_{4d}$$

$$-\mu_d\mu_y \cdot (1 - \mu_w) \cdot T_{4w} - \mu_d\mu_w \cdot (1 - \mu_y) \cdot T_{4y}\} + \Delta\theta_d - \theta_x \quad (5.54)$$

5.6.1 Characteristic Diameter

Brakelmann (1995b) performed an analysis of the changes in characteristic diameter for cables in a deeply buried tunnel. The characteristic diameters are shown in Figure 5-11.

As mentioned above, the validity of Equations (5.49) to (5.51) has been verified for normal cable diameters. However, for large tunnels additional studies were required. Brakelmann (1995b) used Fourier analysis to derive the curves shown in Figure 5-11 for tunnel diameters ranging from 2 m (the upper boundary of the region) to 5 m (lower boundary) for two shapes of the load cycle: sinusoidal (region bounded by straight lines) and rectilinear (region bounded by curved lines).

Analysis of the results shown in Figure 5-11 reveals that, for soil resistivity of 1 K · m/W, the effect of load variations can be represented by the average values beyond the following distances from the tunnel surface:

- 0.15 m for the daily cycle
- 0.5 m for the weekly cycle
- 3.5 m for the yearly cycle

Example 5.5

We will compute the ratings of six circuits, each in trefoil formation and located in a horizontal tunnel as shown in. Figure 5-12. Cable model No. 1 will be used with the following parameters (we will use the concentric neutral loss factor as in Table A1, even though the cables are now in a triangular formation): $D_e = 0.0358$ m, $R = 0.0781 \cdot 10^{-3}$, Ω/m, $\lambda_1 = 0.09$, $T_1 = 0.214$ K · m/W, and $T_3 = 0.104$ K · m/W.

The tunnel has a square cross section with a height of 2.0 m, with 0.5 m concrete walls and 18 m of soil above the roof. The thermal resistivities of concrete and soil are 0.6 and 1.2 K · m/W, respectively. The soil ambient temperature is 15°C.

The external thermal resistances of the tunnel wall and the soil are obtained from Equations (3.4) and (10.41) in Anders (1997), respectively, and are equal to

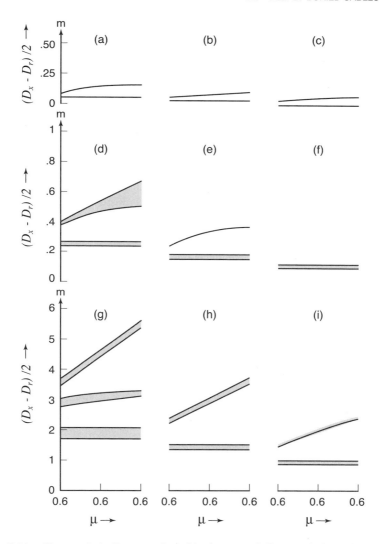

Figure 5-11 Characteristic diameters for cables in a tunnel (from Brakelmann, 1995b, with permission form Elektrizitätswirtschaft).

a, b, c	daily load cycle
d, e, f	weekly load cycle
g, h, I	yearly load cycle
a, d, g	$\rho_s = 0.4$ K · m/W
b, e, h	$\rho_s = 1.0$ K · m/W
c, f, i	$\rho_s = 2.5$ K · m/W

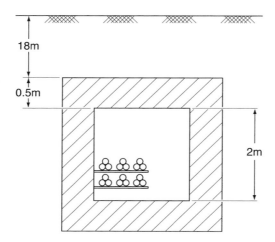

Figure 5-12 Installation of cables in tunnel (see Example 5.5). (From Anders 1997, with permission from IEEE Press.)

$$T_4'' = \frac{\rho_{th} \cdot l}{S} = \frac{0.6 \cdot 0.5}{2 \cdot 1} = 0.15 \text{ K} \cdot \text{m/W}$$

$$T_{4_-} = T_4''' = \frac{\rho_s}{2\pi} \ln\left(3.388 \frac{L_T}{a}\right) = \frac{1.2}{2\pi} \ln\left(3.338 \frac{19.5}{2}\right) = 0.665 \text{ K} \cdot \text{m/W}$$

We will assume the following average load factors for daily, weekly, and yearly load variations with the loss factors computed from Equation (5.37):

$$LF_d = 0.85 \qquad \mu_d = 0.761$$

$$LF_w = 0.82 \qquad \mu_w = 0.717$$

$$LF_y = 0.78 \qquad \mu_y = 0.660$$

Assuming undefined load shapes, curves in Figure 5-11 give the following characteristic diameters:

$$D_{xd} = 2.72 \text{ m}, \qquad D_{xw} = 3.09 \text{ m}, \qquad D_{xy} = 7.20 \text{ m}$$

The external thermal resistances between characteristic diameters and the earth surface are

$$T_{4d} = 0.626 \text{ K} \cdot \text{m/W}, \qquad T_{4w} = 0.601 \text{ K} \cdot \text{m/W}, \qquad T_{4y} = 0.440 \text{ K} \cdot \text{m/W}$$

Considering now daily, weekly, and yearly load cycles, from Equation (5.54), we have

$$T_4 = T_{4\_} \cdot [1 - \mu_d \cdot (1 + \mu_w \mu_y - \mu_w - \mu_y)] - (1 - \mu_d) \cdot T_{4d} - \mu_d \mu_y \cdot (1 - \mu_w)$$
$$\cdot T_{4w} - \mu_d \mu_w \cdot (1 - \mu_y) \cdot T_{4y}$$
$$= 0.665 \cdot [1 - 0.761 \cdot (1 + 0.717 \cdot 0.660 - 0.717 - 0.660)] - (1 - 0.761) \cdot 0.626$$
$$- 0.761 \cdot 0.660 \cdot (1 - 0.717) \cdot 0.601 - 0.761 \cdot 0.717 \cdot (1 - 0.660) \cdot 0.440$$
$$= 0.300 \text{ K} \cdot \text{m/W}$$

The rating calculations for cables in a tunnel are performed in an iterative manner. First, the temperature of the air inside the tunnel is assumed, from which the thermophysical parameters of the air are computed. The next step is to formulate the heat balance equations for the system composed of the cables with surface temperature θ_s, internal wall of the tunnel temperature θ_w, and external wall temperature θ_o.

The heat balance equations for this system with the unity load factor were derived in Example 10.2 in Anders (1997), with an assumed air temperature of 30°C in the first iteration. We will use the same set of equations, but this time we will utilize the value of the external thermal resistance computed above. We have

$$5.03 \cdot 10^{-4} \cdot I^2 = 0.735 \left(\theta_s - \frac{\theta_s + \theta_w}{2} \right)^{1.2} + 6.91 \cdot 10^{-8} (\theta_s^{*4} - \theta_w^{*4})$$

$$\frac{\theta_w - \theta_o}{0.15} = 12.26 \cdot \left(\frac{\theta_s + \theta_w}{2} - \theta_w \right)^{1.25} + 6.91 \cdot 10^{-8} (\theta_s^{*4} - \theta_w^{*4})$$

$$\frac{\theta_w - \theta_o}{0.15} = \frac{\theta_o - 15}{0.300}$$

$$2.8 \cdot 10^{-5} \cdot I^2 = 90 - \theta_s$$

where the quantities with an asterisk denote the absolute temperatures.

The solution of these equations yields

$$I = 462 \text{ A}, \qquad \theta_s = 84°C, \qquad \theta_w = 75°C, \qquad \theta_o = 55°C$$

This can be compared with the ampacity of 365 A obtained for the case with the unity load factor. When only the daily loss factor is considered, the external thermal resistance is

$$T_4''' = T_{4\_}''' - (1 - \mu_d)T_{4d} = 0.665 - (1 - 0.761) \cdot 0.626 = 0.516 \text{ K} \cdot \text{m/W}$$

and the resulting ampacity is equal to 399 A.

Since the film temperature is higher than assumed in the computations, the iterations should be repeated with new air properties until convergence is achieved.

■

5.6.2   Temperature Changes for Deeply Buried Cables

The time constants for deeply buried cables are very large. Figure 5-13 show the temperature rise with time when a current jump at $t = 0$ is applied (Brakelmann, 1989).

We can observe in Figure 5-13 the following temperature rises as a function of time:

After one month	rise about 22%
After one year	rise about 50%
After ten years	rise about 80%
After 80 years	rise about 90% of the final value

The line representing temperature rise computed with a use of an exponential integral is also shown in this figure. Both lines merge after about one year of heating. Since, in general, the differences between the two lines are not significant, a simple method utilizing an exponential integral formula can be used for evaluation of the temperature rise of air in a deep tunnel.

Another conclusion that can be drawn from the shapes of the curves in Figure 5-13 is that in the initial period of tunnel operation, the cables can be heavily loaded. At a later time, fans can be installed to increase cable ratings.

5.7   CONCLUDING REMARKS

In order to obtain accurate conductor operating temperature and corresponding losses, temperature sensing devices are sometimes installed on a cable or pipe sur-

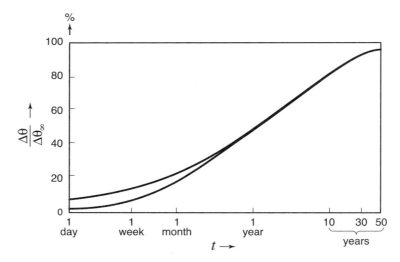

Figure 5-13   Air temperature change in a deep tunnel when a current jump at $t = 0$ is applied (from Brakelmann, 1995b, with permission from Elektrizitätswirtschaft).

face. Installation of such devices is often very costly, or outright impossible. In such cases, an approximate method of real-time rating analysis based on the current measurements alone may prove to be very beneficial. Such a method has been presented in this chapter.

The calculation of the loss factor as a function of conductor temperature has also been introduced. The error that is obtained in estimating the conductor temperature will depend on the loading of the cable with reference to the nominal load. At low load conditions, the error is the largest. It can also be quite large for the usual cable loading conditions in the range of 40 to 80% of the rating. The effect on ampacity will depend on the cable construction. In the case presented in this chapter, the error was in the range of 4%. Fortunately, traditional calculation methods give conservative ratings.

The last part of this chapter examined the effect of daily, weekly, and yearly load variations on cable rating. The last two factors are particularly important for deeply buried tunnels for which the time constants are very large.

5.8 CHAPTER SUMMARY

In this section, several developments presented in this chapter will be summarized for easy reference.

Prediction of Conductor Temperature from Conductor Current

Conductor temperature rise $\Delta\theta$ is related to the conductor current I through the following equation:

$$\Delta\theta = \frac{\theta_1 \cdot \left(\dfrac{I}{I_R}\right)^2 + \theta_2}{c_1 + c_2 \cdot \left[1 - v_T \cdot \left(\dfrac{I}{I_R}\right)^2\right]}$$

where

$$\theta_1 = c_\alpha \cdot (\theta_{\max} - \theta_{amb} - \Delta\theta_{d,R} + \theta_{x,R}) \cdot v_T$$

$$\theta_2 = c_m \cdot (\Delta\theta_d - \theta_x)$$

$$c_1 = 1 + \alpha_T \cdot [\theta_{amb} - 20 + (\Delta\theta_{d,R} - \theta_{x,R})]$$

$$c_2 = \alpha_T \cdot (\Delta\theta_{\max} - \Delta\theta_{d,R} + \theta_{x,R})$$

$$v_T = \frac{T_{t,x}}{T_{t,R}} \qquad \text{for } \Delta\theta_e \leq \Delta\theta_{cr}$$

$$v_T = 1 \qquad \text{for } \Delta\theta_e > \Delta\theta_{cr}$$

$$c_m = [1 + \alpha_T \cdot (\theta_{max} - 20)]$$

$$c_\alpha = 1 + \alpha_T \cdot (\theta_{amb} - 20)$$

In order to relate conductor losses to the current as a straightforward multiplication of the ac resistance and squared current $I^2 R$, we introduced the following revised designation of the total thermal resistance of the cable:

$$T_{t,x} = \frac{T_1}{n \cdot (1 + \lambda_1 + \lambda_2)} + \frac{T_2 \cdot (1 + \lambda_1)}{1 + \lambda_1 + \lambda_2} + T_3 + v_x \cdot T_4 = T_I + v_x \cdot T_4$$

and the electrical resistance representing all ohmic losses in the cable:

$$R_{t20} = R_{20} \cdot (1 + \lambda_1 + \lambda_2)$$

For the nominal load, characterized by the rated current I_R and the nominal total losses W_{tR}, the cable conductor reaches its maximum stationary temperature θ_{max}. Subscript d relates to dielectric losses.

A possible drying out of the surrounding soil is taken into account by means of the variables and θ_x, with

$$\theta_x = (v_x - 1) \cdot \Delta\theta_{cr} \qquad \text{and} \qquad v_x = \frac{\rho_2}{\rho_1} \ (\theta_x = \theta_{xT}) \qquad \text{for } \Delta\theta_e > \Delta\theta_{cr}$$

$$v_x = 1 \ (\theta_x = 0) \qquad \text{for } \Delta\theta_e \le \Delta\theta_{cr}$$

where
$\Delta\theta_e$ = temperature rise at the cable surface (°C)
$\Delta\theta_{cr}$ = critical temperature rise of the soil (°C)
θ_{xT} = correction temperature (°C)
ρ_1, ρ_2 = thermal resistivities of the moist and dry soils, respectively (K $\cdot$ m/W)

The total cable losses corresponding to current I are computed from

$$W_t = W_d + (W_{tR} - W_d) \cdot \left(\frac{I}{I_R}\right)^2 \cdot v_\theta$$

where

$$v_\theta = \frac{c_3}{c_1 + c_2 \cdot \left[1 - v_T \left(\frac{I}{I_R}\right)^2\right]}$$

$$c_3 = 1 + \alpha_T \cdot [\theta_{amb} - 20 + (\Delta\theta_d - \theta_x)]$$

Loss Factor in Rating Standards

The dependence of the conductor load-loss factor on its temperature can be expressed as

$$\mu_\theta = \frac{1}{I_{max}^2 \cdot \tau} \cdot \int_{t=0}^{\tau} I^2(t) \cdot v_\theta \cdot dt$$

Loss Factors for Deeply Buried Cables

For deeply buried cables, the weekly and yearly load variations could be taken into account to significantly increase cable rating. The conductor temperature rise is then computed from

$$\Delta\theta = W_I \cdot v\{T_{4_-} \cdot [1 - \mu_d \cdot (1 + \mu_w\mu_y - \mu_w - \mu_y)] - (1 - \mu_d) \cdot T_{4d} - \mu_d\mu_y \cdot (1 - \mu_w)$$
$$\cdot T_{4w} - \mu_d\mu_w \cdot (1 - \mu_y) \cdot T_{4y}\} + \Delta\theta_d - \theta_x$$

where T_{4_-} denotes the external thermal resistance of the cable subjected to a load with a unity load factor. Subscripts d, w, and y represent daily, weekly, and yearly quantities, respectively. When we want to consider only the daily and weekly load variations, all quantities with a subscript y are set to zero.

When we consider load cycles, we need to evaluate characteristic diameters for each cable beyond which the loss factors are considered. The following formulae give various definitions of this diameter.

Neher/McGrath expression:

$$D_x = 61,200\sqrt{\delta(\text{length of cycle in hours})}$$

where δ is the thermal diffusivity of the soil.

Heinhold expressions:

$$D_x = \frac{205}{\sqrt{w}\rho_s^{0.4}} \qquad \text{for sinusoidal load variation}$$

$$D_x = \frac{493\sqrt{\mu}}{\sqrt{w}\rho_s^{0.4}} \qquad \text{for rectilinear load variation}$$

$$D_x = \frac{103 + 246\sqrt{\mu}}{\sqrt{w}\rho_s^{0.4}} \qquad \text{for other load variations}$$

where $w = \tau_0/\tau$, τ is the length of the period and $\tau_0 = 24$ hr $= 1$ day.

REFERENCES

Anders G.J. (1997), *Rating of Electric Power Cables. Ampacity Computations for Transmission, Distributions, and Industrial Applications.* IEEE Press Power Engineering Series. IEEE Press, New York; McGraw-Hill, New York, 1998.

Anders, G.J., Napieralski, A., Orlikowski M., and Zubert M. (2003) "Advanced Modeling Techniques for Dynamic Feeder Rating Systems," *IEEE Transactions on Industry Applications,* Vol. 39, No. 3, pp. 619–626.

Balog, G.E., Kaldhussaeter, E., and Nerby, T.I. (2000), "Cable Temperature Monitoring," in *Proceedings of IEEE Insulated Conductors Meeting,* St. Petersburg, Florida.

Brakelmann, H. (1986), "Kabelbelastbarkeitals Funktion der Tagelastgangkennlinie," *Bull. SEV,* pp. 767–771.

Brakelmann, H. (1989), *Balastbarkeiten der Energiekabel—Berechnungsmetoden und Parameteranalysen,* VDE-Verlag, Berlin/Offenburg.

Brakelmann, H. (2003), "Loss Determination for Long Three-Phase High-Voltage Submarine Cables," *ETEP,* pp. 193–198.

Brakelmann, H. (1995a), "Kabelbelastbarkeit bei Berücksichtigung von Tages und Wochenlastzyklen," *Elektrizitästwirtschaft,* Jg. 94, Heft 7, pp. 368–372.

Brakelmann, H. (1995b), "Kabelbelastbarkeit im Unbelüfteten Tunnel," *Elektrizitästwirtschaft,* Jg. 94, Heft 26, pp. 368–372.

Brakelmann, H., and Anders, G.J. (2004), "Improvement in Cable Rating Calculations by Consideration of Dependence of Losses on Temperature," *IEEE Transactions on Power Delivery,* Vol. 19, No. 3, pp. 919–925.

Brent, R.P. (1973), *Algorithms for Mineralization without Derivatives,* Prentice-Hall, Englewood Cliffs, N.J.

Cox, H.N., and Coates, R. (1965), "Thermal Analysis of Power Cables in Soils of Temperature-Responsive Thermal Resistivity," *Proc. IEE S,* pp. 2275–2283.

Douglass, D.A., Lawry, D.C., Edris, A., and Bascom., E.C. (2000), "Dynamic Thermal Ratings Realize Circuit Load Limits," *IEEE Computer Applications in Power,* Vol. 13, No. 1, pp. 38–44.

Gear, C.W. (1971), *Numerical Initial Value Problems in Ordinary Differential Equations,* Prentice-Hall, Englewood Cliffs, N.J.

IEC 287-1-1 (1994), "Electric Cables—Calculation of the Current Rating—part I: Current Rating Equations (100% Load Factor) and Calculations of Losses—Section 1: General," IEC Publication 60287.

IEC 287-2-1 (1994), "Electric cables—Calculation of the Current Rating—Part II: Current Rating Equations (100% Load Factor) and Calculations of Thermal Resistance," IEC Publication 60287.

Neher J.H. (1953), "Procedures for Calculating the Temperature Rise of Pipe Cable and Buried Cables for Sinusoidal and Rectangular Loss Cycles," *AIEE Transactions,* Vol. 72, Part III, pp. 541–545.

Neher, J.H., and McGrath, M.H. (1957), "The Calculation of the Temperature Rise and Load Capability of Cable Systems," *AIEE, Trans.* Vol. 76, Pt. III, pp. 752–772.

Nelson, R.J. (1988), "Computer Techniques Enhance Cable Ratings," *Transmission and Distribution,* July, pp. 52–60.

Pragnell, R.G., Gaspari, R., and Larsen, S.T. (1999), "Real Time Thermal Rating of HV Cables," in *Proceedings of IEEE Insulated Conductors Meeting,* St. Petersburg, Florida, 2000.

U.S. Department of Energy (1978), "The Design and Testing for a System to Monitor Temperature in Underground Electric Power Transmission Cables," Report HCP/T-2079, prepared by System Control Inc., Palo Alto, California.

Walldorf, S.P., and Engelhardt, J.S. (1998), "The First Ten Years of Real-Time Ratings on Underground Transmission Circuits, Overhead Lines, Switchgear and Power Transformers," *Electricity Today,* Feb. 1998, pp. 86–88.

CHAPTER 6

Installations Involving Multiple Cables in Air

6.1 INTRODUCTION

Rating methods for common cable installations are well established. In addition to the vast literature on the subject, there are international and national standards containing computational methods and tabulated ratings for a wide range of cable sizes and laying conditions. However, in practice we often encounter situations that cannot be treated by the traditional approaches. In particular, tabulated ratings pertain to either a single three-core cable or a single circuit composed of three single-core cables. Occasionally, two- or even three-circuit installations are considered. In all the cases, the cables are assumed to be identical and equally loaded.

The cable rating standards, in particular IEC Publication 60287 (IEC Standard 60287, 1994), provide computational algorithms for unequally loaded buried cable circuits. However, the laying conditions are restricted to cables either buried in a uniform soil or in a thermal backfill. For cables installed in free air, either on ladders, on cleats, or attached to a wall, identical loading conditions are assumed. A limited number of cables are usually considered in this case. The basic formulae for rating of cables laid underground and in air are reviewed in Chapter 1.

In order to deal with unusual cable laying conditions, derating factors are sometimes proposed. Such factors are most often developed under many simplifying assumptions (Jong, 2003), or, in many instances, are arbitrarily selected by cable engineers. In this chapter, we will consider two approaches to deal with multiple cable installations. First, we will develop rating equations for multicore cables installed in air, and then we will present tables with derating factors that were developed through many years of experience by cable engineers.

6.2 CURRENT RATING OF MULTICORE CABLES

In this section, we present an algorithm to calculate the current-carrying capability of multicore cables (Brakelmann et al., 2004). The method takes into account the presence of air gaps and metallic conductors in the cable bundle. Under an assumption of a uniform loss density, analytical expressions that take into account conduction inside the bundle and convection and radiation outside it are presented. The uniform loss density assumption is then relaxed, and two-zone heating models are considered. The method is illustrated with a numerical example of a telecommunication cable with 96 stranded cores. Experimental results are also reported for the same cable.

6.2.1 Introduction

The need for bundle rating arises from the fact that modern telecommunication cables or cables installed in aircraft are grouped together for a portion of the cable run, and even with a small current flowing in each cable, the heat generated in the entire bundle can be significant. An example of such a cable is shown in Figure 6-1. In this figure, a bundle of N separate cores (here, $N = 96$) is held together by means of an aluminum band (static shield) and a steel-wire conductor, and enclosed by a PVC sheath. Each two cores are twisted into a couple and, additionally, four couples can be stranded into a bundle. These bundles of eight cores are again stranded layer by layer. Such cables are used in large-scale power plants or chemical plants for the supply of solenoid valves or in an aircraft from a distribution bus to individual load devices.

The thermal analysis of such cable constructions is difficult, since the separate cores contact each other only in small, not-well-defined areas. The gaps between the

Figure 6-1   Telecommunication cable. Top: cross-section. Bottom: stranding of a bundle with eight cores (four twisted pairs).

cores are air filled, so that the thermal fluxes in the cores are not radially symmetrical. Thus, the complex bundle geometry gives rise to elaborate boundary conditions for conduction, convection, and radiation. Therefore, as was the case with cable bundles in the trays discussed in the literature, we will need several simplifying assumptions in order to derive an analytical expression for cable rating calculations. The simplifying assumptions and the resulting analytical model are presented below following a brief review of earlier work and some background information on this subject in the next section. In Example 6.1, we will show a comparison of the calculated ratings with the experimental results for the communication cable bundle in Figure 6-1.

6.2.2 Background

The first attempt to address the ampacity calculations of bundled cables was reported by Schach and Kidwell (1952). Their study was conducted under the following simplifying assumptions:

1. Cables were bound at intervals not exceeding 24 inches so as to form a cylindrical bundle.
2. The relative positions of the component cables were maintained throughout the bundle length.
3. Bundles were suspended horizontally in still air.
4. Each loaded cable carried the same current.

With these restrictions, current rating becomes a function of the dimensions, resistance, and thermal properties of the component cables, the total number of cables in the bundle, the number of cables loaded simultaneously, the position of the loaded cables in the bundle, and the spacing between adjacent cables.

The rating of the cables in the bundle is obtained by considering a standard relation between the maximum allowable conductor temperature rise $\Delta\theta_{max} = \theta_c - \theta_{amb}$ and the losses W_t (W/m) generated in the cable bundle [see Equation (1.42)]:

$$\Delta\theta_{max} = W_t \cdot (T_1 + T_3 + T_4) \tag{6.1}$$

where T_1, T_3, and T_4 (K · m/W) are insulation, jacket and external thermal resistances, respectively. Calculation of T_3 and T_4 follows a standard procedure, briefly reviewed below. Evaluation of T_1 is discussed in Section 1.6.6.1.

6.2.2.1 *Evaluation of the Jacket Thermal Resistance.* The jacket thermal resistance is calculated from a standard formula:

$$T_3 = \frac{\rho_J \cdot \ln \left[D_e/(D_e - 2t_J) \right]}{2\pi} \tag{6.2}$$

where t_J (mm) is the jacket thickness, ρ_J (K · m/W) is its thermal resistivity, and D_e (mm) is the cable outside diameter.

6.2.2.2 *Rating Equations.* The permissible current in each core I_c is obtained from

$$I_c = \sqrt{\frac{W_t}{N \cdot R}} \tag{6.3}$$

and the total current rating of the cable is

$$I = N \cdot I_c \tag{6.4}$$

where N is the number of loaded conductors and R (ohm/m) is the ac resistance of a single conductor at the operating temperature.

In the following, the factors in Equation (6.3) will be evaluated.

When computing the value of R, the increase of the conductor resistance because of the stranding of the cores should be taken into account. In our example, the stranding factor is not included in the resistance calculations because the resistance at 20°C is specified by the manufacturer and the area of the conductor is selected such that the desired resistance is obtained. Therefore, for the conductor temperature θ_{max}, the conductor resistance becomes

$$R = \frac{[1 + \alpha_T \cdot (\theta_{max} - 20°C)] \cdot \rho_{20}}{A_{core}} \tag{6.5}$$

where

α_T = temperature coefficient of the conductor resistance, 1/K
A_{core} = a single conductor area, m^2

To evaluate W_t, we first observe that since cables put in bundles have negligible dielectric losses, the energy balance equation at the surface of the cable bundle can be written as

$$W_t = W_{conv} + W_{rad} \tag{6.6}$$

where W_{conv} and W_{rad} are the heat losses due to convection and radiation, respectively. Since the cables are located indoors, the effect of solar radiation is neglected. Computation of the losses generated inside the cable bundle is discussed in the next section. Substituting appropriate formulas for the convective and radiative heat losses, and neglecting the effect of solar radiation, the following form of the heat balance equation is obtained from Equation (1.13):

$$W_t = \pi D_e^* h_s(\theta_e^* - \theta_{amb}^*) + \pi D_e^* \varepsilon \sigma_B(\theta_e^{*4} - \theta_{amb}^{*4}) \tag{6.7}$$

where

h_s = the convection coefficient at the surface of the cable, W/K $\cdot$ m^2
θ_e^* = cable surface temperature, K
σ_B = Stefan–Boltzmann constant, equal to $5.67 \cdot 10^{-8}$ W/m$^2 \cdot$ K^4

ε = emissivity of the cable outer covering
D_e^* = cable bundle external diameter, m
θ_{amb}^* = ambient temperature, K

This equation is usually solved iteratively. In steady-state rating computations, the effect of heat loss caused by convection is taken into account by suitably modifying the value of the external thermal resistance of the cable.

Evaluation of the convection coefficient can become quite involved. Over the years, cable engineers have developed several simple formulae to account for the convective and radiative heat transfer outside the cable surface. Some of these methods will be reviewed below.

6.2.2.3 Evaluation of the External Thermal Resistance. To obtain an expression for the external thermal resistance of a cable in air, we first rewrite Equation (6.7) as

$$W_t = \pi D_e^* h_t \Delta \theta_s \qquad (6.8)$$

where $\Delta \theta_s = \theta_e^* - \theta_{amb}^*$ is the cable surface temperature rise above ambient and h_t the total heat transfer coefficient. From Equation (6.8), the external thermal resistance, T_4 (K · m/W) of the cable is given by

$$T_4 = \frac{\Delta \theta_s}{W_t} = \frac{1}{\pi D_e^* h_t} \qquad (6.9)$$

Before the external thermal resistance is computed from Equation (6.9), the value of the heat transfer coefficient h_t has to be determined. From a large number of tests on various cables in various configurations, carried out in the United Kingdom in the 1930s, Whitehead and Hutchings (1939) deduced that the total thermal dissipation from the surface of a cable in air might conveniently be expressed as (see Section 1.6.6.5 for a discussion of this topic)

$$W_t = \pi D_e^* h (\Delta \theta_s)^{5/4} \qquad (6.10)$$

where h (W/m^2 $K^{5/4}$) is the heat transfer coefficient embodying convection, radiation, conduction, and mutual heating [see Anders (1997) for an explanation how the mutual heating is taken into account]. From Equation (6.10), the following expression for the external thermal resistance of cables in free air is obtained:

$$T_4 = \frac{1}{\pi D_e^* h (\Delta \theta_s)^{1/4}} \qquad (6.11)$$

The values of the heat transfer coefficient h were obtained experimentally and plotted as a function of the cable diameter for various cable arrangements (IEC 60287, 1994; Whitehead & Hutchings, 1939). The curves of Whitehead & Hutchings were later fitted with the following analytical expression:

$$h = \frac{Z}{(D_e^*)^g} + E \tag{6.12}$$

where for a cable installed in free air, $Z = 0.21$, $E = 3.94$, and $g = 0.6$ can be selected (see Table 1-7).

Equation (6.11) was developed under the assumption that the cable is subjected to natural cooling only and that h is constant for a fixed value of D_e^*. Cables installed outdoors or in ventilated rooms may be subjected to wind, resulting in forced convective cooling. With low wind speed, natural convective cooling and forced convective cooling may occur at the same time. We will briefly review the issues involved in computation of the heat transfer coefficient when both modes of cooling are present.

Neglecting conductive heat transfer, which is very small in free air, the heat transfer coefficient is given by

$$h_t = h_{conv} + h_{rad} \tag{6.13}$$

The radiative heat transfer coefficient h_{rad} can be computed from Equation (6.7), and can then be approximated by (Black and Rehberg, 1985)

$$h_{rad} = \frac{\sigma \varepsilon [(\theta_e + 273)^4 - (\theta_{amb}\, 273)^4]}{\Delta \theta_s} \approx \sigma_B \varepsilon (1.38 \cdot 10^8 + 1.39 \cdot 10^6 \theta_{amb}) \tag{6.14}$$

Another approximation was given by Neher and McGrath (1957)

$$h_{rad} = 4.2 \cdot \varepsilon \cdot (1 + 0.0167 \cdot \theta_s) \tag{6.15}$$

The convective heat transfer h_{conv} will be composed of two parts, $h_{o,n}$ and $h_{o,f}$, corresponding to the natural and forced convective heat transfer, respectively. The value of $h_{o,n}$ can be computed from Equation (6.10) as

$$h_{o,n} = h\Delta \theta_s^{1/4} \tag{6.16}$$

This coefficient is defined by the following relationships (Incropera & de Witt, 1990):

$$h_{o,n} = \mathrm{Nu} \cdot k_{air}/D_e^*$$
$$\mathrm{Nu} = 0.48 \cdot (\mathrm{Gr} \cdot \mathrm{Pr})^{0.25} \tag{6.17}$$
$$\mathrm{GR} = g \cdot \beta \cdot \Delta \theta_s \cdot D_e^{*3}/v^2$$

where
k_{air} = thermal conductivity of air, which can be computed from the relations given in Appendix D of (Anders, 1997), W/K · m
β = volumetric thermal expansion coefficient, K^{-1}

g = acceleration due to gravity m/s^2
v = air viscosity, m^2/s
Pr = Prandtl number (approximately equal to 0.71)

Morgan (1993) proposed another approximation for the natural convection coefficient. In the temperature ranges $-10°C \leq \theta_{amb} \leq 40°C$ and $10°C \leq \theta_e \leq 100°C$, and for $10^4 < (Gr \cdot Pr) \leq 10^7$, the following approximation may be made (Morgan, 1993):

$$0.48 k_{air} \left(\frac{\beta g \, Pr}{v^2} \right)^{0.25} \approx 1.23 (\pm 4\%) \qquad (6.18)$$

Thus,

$$h_{o,n} \approx 1.23 \cdot \left(\frac{\Delta \theta_s}{D_e^*} \right)^{0.25} \qquad (6.19)$$

Other possible approximations are summarized in Table 6-1 (Dorison et al., 2003).

In Winkler's formulas, k' and k'' are corrective factors to deal with the variable characteristics of air with respect to the temperature:

$$k' = 0.919 + \frac{\theta_s + \theta_{amb} + 546}{738} \qquad k'' = 1.033 - \frac{\theta_s + \theta_{amb} + 546}{1818} \qquad (6.20)$$

Table 6-1 Convection models used for cables installed in air

Model	Convection coefficient $h_{o,n}$
McAdams (Neher & McGrath, 1957)	$1.32 \left(\dfrac{\Delta \theta_s}{D_e^*} \right)^{0.25}$
Morgan (1993)	$1.23 \left(\dfrac{\Delta \theta_s}{D_e^*} \right)^{0.25}$
Lauriat (1996)	$\dfrac{10^{-2}}{D_e^*} + 1.2 \Delta \theta_s^{1/3} + 0.217 \dfrac{\Delta \theta_s^{1/6}}{D_e^{*1/2}}$
Neher and McGrath (1957)	$1.05 \left(\dfrac{\Delta \theta_s}{D_e^*} \right)^{0.25}$
Winkler 1 (1963)	$k' \dfrac{0.185}{D_e^*} + k'' \cdot 1.342 \left(\dfrac{\Delta \theta_s}{D_e^*} \right)^{0.25}$
Winkler 2 (1963)	$k' \dfrac{0.0185}{D_e^*} + k'' \cdot 1.08 \left(\dfrac{\Delta \theta_s}{D_e^*} \right)^{0.25}$
Slaninka (1969)	$\dfrac{0.02}{D_e^*} + 1.08 \left(\dfrac{\Delta \theta_s}{D_e^*} \right)^{0.25}$

The forced convection coefficient is given by

$$h_{o,f} = \frac{k_{air}b(\text{Re})^p}{D_e^*}$$ (6.21)

where Re is the Reynolds number, which is defined by

$$\text{Re} = \frac{w \cdot D_e^*}{v}$$ (6.22)

with
w = wind velocity, m/s
b, p = constants depending on the value of the Reynolds number and defined, for example in Anders (1997)

An approximate value for the forced convection coefficient can be obtained from Neher and McGrath (1957):

$$h_{o,f} = 2.87 \sqrt{\frac{w}{D_e^*}}$$ (6.23)

Morgan (1992) has shown that for cables with an outside surface temperature rise above ambient not exceeding 20 K, as is often the case in outdoor cable installations, the natural convective cooling can be ignored when the wind velocity exceeds 1 m/s.

6.2.3   Heat Conduction Inside the Cable Bundle

Inside the cable bundle, the heat is transferred by conduction. In the following investigations, we will consider two situations of cable loading. In the first case, we will assume a homogeneous loss density and a homogeneous insulation anywhere in the bundle. In order to take into account the presence of the air gaps, the effective thermal resistivity, ρ_{eff}, of the bundle will be defined by means of special filling factors. In the second case, we will consider a partial loading of the cable with either an inner or the outer part loaded and the other part unloaded.

6.2.3.1 Uniform Loss Density

Thermal Resistance of the Bundle of Cores. Schach and Kidwell (1952) found the effective thermal conductivity k_{eff} by treating the bundle as a cylinder and measuring the quantities required by the heat conduction relation

$$k_{eff} = \frac{I^2 R \cdot \ln(D_1/D_0)}{2\pi \cdot (\theta_c - \theta_1)}$$ (6.24)

when I current was flowing through the center cable only, and where
D_1, D_0 = diameters of the bundle and the center conductor, respectively, mm

θ_1, θ_c = temperatures of bundle surface and the center conductor, respectively, °C
R = conductor resistance at operating temperature, ohm/m

They performed temperature measurements for a single cable and bundles of 18 and 8 cables. The calculated thermal conductivity values were 0.7, 0.4 and 0.5 W/K · m, respectively.

Equation (6.24) does not take into account the presence of the air gaps and the metallic conductors. In order to consider these, we will assume that the conductor areas in the cross section can be considered as thermal short circuits and the air gaps, which have negligible thermal conductivities, are considered to be homogeneously distributed over the total cross section. Then, for the bundle composed of N cables, the filling factor for the air gaps can be obtained from

$$f_1 = \frac{N \cdot D_{core}^2}{D_1^2} \cdot f_{strand} \tag{6.25}$$

where D_{core} (mm) is the diameter over one cable and the stranding factor is defined by

$$f_{strand} = \sqrt{1 + \left(\frac{\pi \cdot D_{strand}}{L_{strand}}\right)^2} \tag{6.26}$$

D_{strand} = mean diameter of the stranding, mm
L_{strand} = stranding length, mm

The filling factor for the conductor areas is equal to

$$f_C = \frac{4 \cdot N \cdot A_{core}}{\pi \cdot D_1^2 \cdot f_{cond}} \cdot f_{strand} \tag{6.27}$$

with A_{core} (mm^2) being the conductor cross section of each core and f_{cond} takes into account the compaction of a single conductor.

The effective thermal resistivity is equal to

$$\rho_{eff} = \rho \frac{(1 - f_C)}{f_1} \tag{6.28}$$

where ρ is the thermal resistivity of the insulating material.

Temperature of the Hottest Conductor. We will now assume that the same current I_c flows in every cable. Since the separate cores are approximately regularly distributed over the cross section and, on the other hand, they change their positions in the cross section along the cable axis, following the stranding, a homogeneously distributed loss density q is assumed:

$$q = \frac{W_t}{\pi \cdot D_1^2/4} = \frac{4 \cdot N \cdot R \cdot I_c^2}{\pi \cdot D_1^2} \tag{6.29}$$

In order to calculate the temperature distribution in the cable bundle, we will modify the procedure proposed by Schach and Kidwell (1952). In their approach, they looked at three layers of cables surrounding a centrally located core. In our approach, the insulation is divided in coaxial layers of thickness Δr. Considering a point in the distance r_x from the axis, we have to multiply the losses in each layer between r_x and the bundle surface [$r_x < r < R_1$, ($R_1 = D_1/2$)] by the thermal resistance between this layer and the bundle surface. Superimposing all these temperature rises leads to the total temperature rise of the point at the distance r_x with respect to the temperature of the bundle surface:

$$\Delta\theta(r_x) = \lim_{\Delta r \to 0}\left\{\sum q \cdot 2\pi \cdot r_x \cdot \Delta r \cdot \frac{\rho_{eff}}{2\pi} \cdot \ln\left(\frac{R_1}{r_x}\right)\right\} \tag{6.30}$$

This sum converges to the integral

$$\Delta\theta(r) = \frac{q \cdot \rho_{eff}}{2\pi} \cdot \int_r^{R_1} \ln\left(\frac{R_1}{r}\right) \cdot 2\pi \cdot r \cdot dr = q \cdot \rho_{eff} \cdot \int_r^{R_1} r \cdot \ln\left(\frac{R_1}{r}\right) dr \tag{6.31}$$

By inserting the power losses from Equation (6.29), Equation (6.31) leads to

$$\Delta\theta(r) = \frac{\rho_{eff} \cdot W_t}{4\pi} \cdot \left\{\left(\frac{r}{R_1}\right)^2 \cdot \left[2 \cdot \ln\left(\frac{r}{R_1}\right) - 1\right] + 1\right\} \tag{6.32}$$

The center conductor is the hottest and its temperature is obtained by letting $r \to 0$ in Equation (6.32). Thus, we have

$$\Delta\theta(0) = \Delta\theta_{max} = W_t \cdot \frac{\rho_{eff}}{4\pi} = W_t \cdot T_{eff} \tag{6.33}$$

Therefore, Equation (6.32) can also be rewritten as

$$\Delta\theta(r) = \Delta\theta(0) \cdot \left\{\left(\frac{r}{R_1}\right)^2 \cdot \left[2 \cdot \ln\left(\frac{r}{R_1}\right) - 1\right] + 1\right\} \tag{6.34}$$

Figure 6-2 shows the temperature distribution resulting from Equation (6.34).

Temperature Rise for Cables in Free Air. From Equation (6.11), for the cable is installed in free air, the temperature rise of the cable surface is given by

$$\Delta\theta_s = W_t \cdot T_4 = \frac{W_t}{\pi \cdot D_e^* \cdot h \cdot \Delta\theta_s^{1/4}} \tag{6.35}$$

From Equation (6.35), we obtain

$$\Delta\theta_s = \frac{W_t^{4/5}}{(\pi \cdot D_e \cdot h)^{4/5}} = W_t \cdot \frac{1}{(\pi \cdot D_e \cdot h \cdot W_t^{1/4})^{4/5}} \tag{6.36}$$

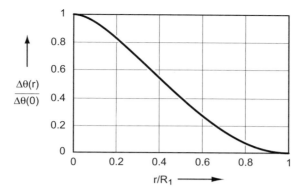

Figure 6-2 Radial temperature distribution in the multicore bundle (from Brakelmann et al., 2004, with permission from IEEE Press).

The highest temperature rise in the axis of the cable can be derived from Equations (6.33), (6.2), and (6.36) as

$$\Delta\theta_{\max} = W_t \cdot \left\{ \frac{\rho_{eff}}{4\pi} + \frac{\rho_J \cdot \ln[D_e/(D_e - 2t_J)]}{2\pi} + \frac{1}{[\pi \cdot D_e^* \cdot h \cdot W_t^{1/4}]^{4/5}} \right\} \qquad (6.37)$$

Given $\Delta\theta_{\max}$, the power losses W_t are calculated from Equation (6.37) by iterations. The procedure converges quite quickly, and after less than five iteration steps the error is smaller than 0.1%.

An alternative approach is to replace the last term in Equation (6.37) by Equation (6.9) with h_t defined in Equation (6.13), as illustrated in Example 6.1.

Temperature Rise for Cables in Moving Air. Taking into account Equations (6.33), (6.2), and (6.15), the temperature rise in the axis of the cable can here be written as

$$\Delta\theta_{\max} = W_t \cdot \left\{ \frac{\rho_{eff}}{4\pi} + \frac{\rho_J \cdot \ln[D_e/(D_e - 2t_J)]}{2\pi} \right.$$
$$\left. + \frac{1}{\pi \cdot D_e^* \cdot [h_{conv} + 4.2 \cdot \varepsilon_S \cdot (1 + 0.0167 \cdot \theta_s)]} \right\} \qquad (6.38)$$

$$\theta_s = W_t \cdot T_4 + \theta_{amb} \qquad (6.39)$$

Again, given $\Delta\theta_{\max}$, the power losses W_t are calculated from Equations (6.38) and (6.39) by iterations. The convergence of the iterations is comparably quick, as discussed above.

On the other hand, if we use Black's approximation (Black and Rehberg, 1985) Equation (6.38) takes the form

$$\Delta\theta_{max} = W_t \cdot \left\{ \frac{\rho_{eff}}{4\pi} + \frac{\ln[D_e/(D_e - 2t_J)]}{2\pi \cdot k_J} \right.$$

$$\left. + \frac{1}{\pi \cdot D_e^* \cdot [h_{conv} + \sigma_B \varepsilon_s (1.38 \cdot 10^8 + 1.39 \cdot 10^6 \theta_{amb})]} \right\} \tag{6.40}$$

The last equation can be solved directly for W_t.

Example 6.1

We will illustrate the algorithm presented above with a numerical example involving a telecommunication cable with 96 cores. Each core is composed of a copper conductor enclosed in PVC insulation. The outer jacket is also made of PVC. The dimensions required for the calculations are as follows (the external diameter includes the thickness of the static shield equal to 0.75 mm): $A_{core} = 0.5$ mm^2, $D_1 = 19$ mm, $t_J = 1.8$ mm, $D_e = 24.1$ mm, $D_{strand} = 10$ mm, $L_{strand} = 130$ mm.

We will show the calculations for $\theta_{max} = 73°C$, the measured conductor temperature at $w = 0$ m/s, and $\theta_{amb} = 20°C$. The stranding factor is obtained from Equation (6.26) as

$$f_{strand} = \sqrt{1 + \left(\frac{\pi \cdot D_{strand}}{L_{strand}} \right)^2} = \sqrt{1 + \left(\frac{\pi \cdot 10}{130} \right)^2} = 1.03$$

and the resistance of a single conductor is equal to

$$R = \frac{[1 + \alpha_T \cdot (\theta_{max} - 20°C)] \cdot \rho_{20}}{A_{core}} = \frac{[1 + 0.00393 \cdot (73 - 20)] \cdot 1.72E\text{-}8}{0.5E\text{-}6}$$

$$= 0.0417 \text{ ohm/m}$$

We will start by evaluating the effective thermal conductivity of the insulation inside the bundle from Equations (6.25) to (6.28). The air gaps and conductor-filling factors are obtained first. The air-filling factor is equal to [Equation (6.25)]

$$f_1 = \frac{N \cdot f_{strand} \cdot D_{core}^2}{D_1^2} = \frac{96 \cdot 1.3 \cdot 1.6^2}{19^2} = 0.70$$

For the conductor-filling factor, we will assume $f_{cond} = 0.55$ in Equation (6.27):

$$f_C = \frac{4 \cdot N \cdot f_{strand} \cdot A_{core}}{\pi \cdot D_1^2 \cdot f_{cond}} = \frac{4 \cdot 96 \cdot 1.03 \cdot 0.5}{\pi \cdot 19^2 \cdot 0.55} = 0.317$$

With the resistivity of the PVC equal to 5.0 K $\cdot$ m/W, the effective thermal resistivity of the insulation is equal to

$$\rho_{eff} = \rho \frac{(1 - f_C)}{f_1} = 5 \cdot \frac{(1 - 0.317)}{0.70} = 4.9 \text{ K} \cdot \text{m/W}$$

Natural Convection. The convection coefficient is obtained from Equation (6.19). We will start the iterations by assuming that the cable surface temperature is 10 degrees higher than the ambient:

$$h_{o,n} = 1.23 \cdot \left(\frac{\Delta\theta_s}{D_e^*} \right)^{0.25} = 1.23 \cdot \left(\frac{10}{0.0241} \right)^{0.25} = 5.55 \text{ W/m}^2 \cdot \text{K}$$

The radiation heat transfer coefficient will be computed from Equation (6.14):

$$h_{rad} = \sigma_B \varepsilon_s (1.38 \cdot 10^8 + 1.39 \cdot 10^6 \theta_{amb}) = 5.67 \cdot 10^{-8} \cdot 0.9 \cdot (1.38 \cdot 10^8 + 1.39 \cdot 10^6 \cdot 20)$$
$$= 8.46 \text{ W/m}^2 \cdot \text{K}$$

The total losses generated in this cable are computed from Equation (6.40):

$$53 = W_t \cdot \left\{ \frac{\rho_{eff}}{4\pi} + \frac{\rho_J \ln[D_e/(D_e - 2t_J)]}{2\pi} \right.$$

$$+ \left. \frac{1}{\pi \cdot D_e^* \cdot [h_{conv} + \sigma_B \varepsilon_s (1.38 \cdot 10^8 + 1.39 \cdot 10^6 \theta_{amb})]} \right\}$$

$$= W_t \cdot \left\{ \frac{4.9}{4\pi} + \frac{5 \cdot \ln[24.1/(24.1 - 2 \cdot 1.8)]}{2\pi} + \frac{1}{\pi \cdot 0.0241 \cdot (5.55 + 8.46)} \right\}$$

resulting in

$$W_t = 36.3 \text{ W/m}$$

This value is used to recompute the cable surface temperature and the convection coefficient. The process converges after four iterations with losses of 39.4 W/m. The permissible current in each conductor is now computed from Equation (6.3):

$$I_c = \sqrt{\frac{W_t}{N \cdot R}} = \sqrt{\frac{39.4}{96 \cdot 0.0413}} = 3.14 \text{ A}$$

and the total current in the cable is equal to

$$I = N \cdot I_c = 96 \cdot 3.14 = 302 \text{ A}$$

Forced Convection. In this case, we need to compute the heat transfer coefficient for forced convection in addition to the natural convection and radiation heat transfer coefficients obtained earlier. Again, we will start the iterations by assuming that the cable surface temperature is 10 degrees higher than the ambient.

For wind velocity of 0.2 m/s, the Neher/McGrath approximation [Equation (6.23)] of the forced convection coefficient becomes

$$h_{o,f} = 2.87 \sqrt{\frac{w}{D_e^*}} = 2.87 \cdot \sqrt{\frac{0.2}{0.0241}} = 8.27 \text{ W/m}^2 \cdot \text{K}$$

The convective heat transfer coefficient is obtained by combining the natural and forced convection coefficients. The external thermal resistance is now obtained from Equation (6.9):

$$T_4 = \frac{1}{\pi D_e^* h_t} = \frac{1}{\pi \cdot 0.0241 \cdot (8.46 + \sqrt{5.55^2 + 8.27^2})} = 0.717 \frac{K \cdot m}{W}$$

The total losses are equal to

$$W_t = \frac{\Delta \theta_{max}}{T_1 + T_3 + T_4} = \frac{73 - 20}{0.416 + 0.312 + 0.717} = 42.9 \text{ W/m}$$

and the corresponding cable surface temperature is

$$\theta_s = \theta_{amb} + W_t \cdot T_4 = 20 + 42.9 \cdot 0.717 = 50.8°C$$

This temperature is now used to recompute the heat transfer coefficients. After four iterations, we obtain $W_t = 44.4$ W/m, which results in the conductor current of 3.33 A and the total cable current of 320 A. The total cable current is given for comparison with the test results that follow, rather than being of practical use in normal installations where the current per conductor is usually of interest.

Experimental Results. The same cable was installed in free air in a high-current laboratory (Brakelmann et al., 2004). The cable was heated by current transformers. The uniform distribution of the total current onto the separate cores was screened by means of clamp-on current probes.

Following the procedure for cables in still air, the permissible current of the cable was calculated for a range of ambient air temperatures. The results are shown in Figure 6-3. Sample calculations for ambient temperature of 20°C are shown above. Additionally, the experimental results are shown on the line with wind velocity of 0 m/s for the same ambient temperature (311 A at 20°C).

Calculations were also performed to examine the influence of air flows with different flow velocities ($w = 0.20$ m/s and $w = 0.50$ m/s). The results are also shown in Figure 6-3. We can observe that even small air velocities may have a remarkable, beneficial influence on the current rating.

There is a possibility that a very small movement of air existed in the laboratory during the measurements. The measured current of about 311 A could be achieved with air velocity of about 0.09 m/s, which could not be measured with the instruments used in the experiment.

Figure 6-4 shows the calculated and the measured conductor temperatures of the hottest cable as well as bundled cables outside surface temperature as functions of the total current I in still air. An excellent correspondence between calculations and measurements can be observed for all three operating states.

In this example, the conductor resistance is taken as the resistance at maximum operating temperature. This is a slight overestimate, as only the conductors near the center of the cable will be at maximum temperature. It might be worth calculating

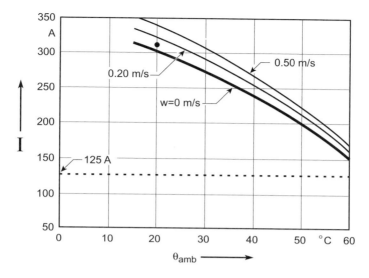

Figure 6-3 Permissible total current as a function of ambient air temperature. The dot represents the measured result (from Brakelmann et al., 2004, with permission from IEEE Press).

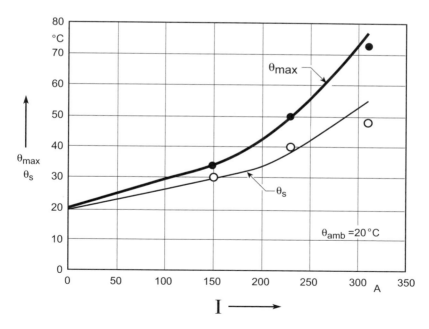

Figure 6-4 Maximum cable temperature θ_{max} and bundle surface temperature θ_s ($w = 0$ m/s). Dots represent the measured results (from Brakelmann et al., 2004, with permission from IEEE Press).

the temperature of the outer layer of conductors using a conductor resistance based on the average of the outer conductor temperature and the maximum temperature. This approach will bring the calculated values slightly closer to the test results shown above.   ■

6.2.3.2 Unequally Loaded Bundle. We will consider two situations with a non-uniform loss density in the bundle. In one case, the cables in the center of the bundle will be loaded, and in the other, the current will flow in the outer cables. As a special case of the second scenario, we will consider a cable constructed with a center duct.

In both cases, we will consider the heat dissipated in the inner and the outer parts of the cable bundle. Figure 6-5 will serve as an illustration of the developments presented below.

The temperature rise in the cable bundle is now composed of two components: one corresponding to the inner part of the bundle and the other to the outer part. Calculation of these temperature rises is examined in the following sections and is based on the ideas of Mark Coates, Convener of Working Group 19 of the IEC.

Cables Loaded in the Central Part. We will consider a situation illustrated in Figure 6-6. If the center part of the cable bundle with effective diameter R_2 ($R_2 < R_1$) is loaded, the loss density in Equation (6.29) becomes

$$q = \frac{W_t}{\pi \cdot R_2^2} \tag{6.41}$$

Changing the integration limits in Equation (6.31), to the new limits from 0 to R_2 and substituting the loss density from Equation (6.41), we obtain the temperature rise $\Delta \theta_2$ at the distance R_2 from the cable center as

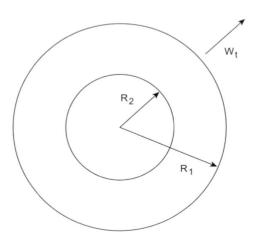

Figure 6-5   Heat dissipation in the inner and outer parts of the cable bundle. R_1 and R_2 are the effective radii of the composite mass of the cables.

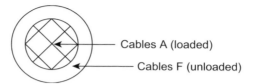

Figure 6-6   Center part of the cable bundle loaded.

$$\Delta\theta_2 = \frac{\rho_{eff} \cdot W_t}{4\pi} \tag{6.42}$$

To obtain the temperature difference $\Delta\theta_1$ between R_2 and R_1, we will use an expression for the thermal resistance of a concentric layer with the inner and outer radii R_2 and R_1, respectively. We have

$$\Delta\theta_1 = \frac{\rho_{eff} \cdot W_t}{2\pi} \ln \frac{R_1}{R_2} \tag{6.43}$$

The highest temperature rise in the axis of the cable is now obtained by substituting Equations (6.42) and (6.43) for the first term in the brackets in Equation (6.37):

$$\Delta\theta_{max} = W_t \cdot \left\{ \frac{\rho_{eff}}{4\pi} \left(1 + 2 \ln \frac{R_1}{R_2} \right) + \frac{\rho_J \cdot \ln[D_e/(D_e - 2t_J)]}{2\pi} + \frac{1}{[\pi \cdot D_e^* \cdot h \cdot W_t^{1/4}]^{4/5}} \right\} \tag{6.44}$$

Alternatively, we can use Equation (6.38) to obtain

$$\Delta\theta_{max} = W_t \cdot \left\{ \frac{\rho_{eff}}{4\pi} \left(1 + 2 \ln \frac{R_1}{R_2} \right) + \frac{\rho_J \cdot \ln[D_e/(D_e - 2t_J)]}{2\pi} \right.$$

$$\left. + \frac{1}{\pi \cdot D_e^* \cdot [h_{conv} + 4.2 \cdot \varepsilon \cdot (1 + 0.0167 \cdot \theta_s)]} \right\} \tag{6.45}$$

With given $\Delta\theta_{max}$, the power losses W_t are calculated from Equation (6.44) or (6.45) by iterations. The procedure converges quite quickly, and after less than five iteration steps the error is smaller than 0.1%.

An alternative approach is the replace the last term in Equation (6.44) by Equation (6.9), with h_t defined in Equation (6.13).

Cables Loaded in the Outer Part.   The situation considered in this section is illustrated in Figure 6-7.

If the cables located between R_2 and R_1 are loaded, the loss density in Equation (6.29) becomes

$$q = \frac{W_t}{\pi \cdot (R_1^2 - R_2^2)} \tag{6.46}$$

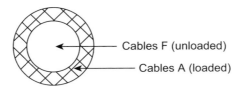

Figure 6-7   Outer part of the cable bundle loaded.

This yields the temperature distribution

$$\Delta\theta(r) = \frac{\rho_{eff} \cdot W_t}{4\pi \cdot (R_1^2 - R_2^2)} \cdot \left\{ r^2 \cdot \left[2 \cdot \ln\left(\frac{r}{R_1}\right) - 1 \right] + R_1^2 \right\} \tag{6.47}$$

with the highest temperature rise in the axis ($r = R_2$):

$$\Delta\theta(R_2) = \rho_{eff} \cdot W_t \cdot \frac{R_2^2 \cdot \left[2 \cdot \ln\left(\frac{R_2}{R_1}\right) - 1 \right] + R_1^2}{4\pi \cdot (R_1^2 - R_2^2)}$$

$$= \rho_{eff} \cdot W_t \cdot \frac{\left(\frac{R_2}{R_1}\right)^2 \cdot \left[2 \cdot \ln\left(\frac{R_2}{R_1}\right) - 1 \right] + 1}{4\pi \cdot \left[1 - \left(\frac{R_2}{R_1}\right)^2 \right]} \tag{6.48}$$

If the cable construction has a central duct with the radius R_2, the highest temperature rise is also obtained from Equation (6.48).

From Equations (6.48) and (6.33), the relation between the thermal resistance of a multicore bundle with and without a central duct becomes

$$\frac{T_{eff}}{T_{eff,2}} = \frac{\left(\frac{R_2}{R_1}\right)^2 \cdot \left[2 \cdot \ln\left(\frac{R_2}{R_1}\right) - 1 \right] + 1}{1 - \left(\frac{R_2}{R_1}\right)^2} \tag{6.49}$$

Example 6.2
We will consider the communication cable examined in Example 6.1. Let us assume that half, that is, 48 cables are loaded. Similarly to Example 6.1, we will show the calculations for $\theta_{max} = 73°C$, the measured conductor temperature at $w = 0$ m/s, and $\theta_{amb} = 20°C$. We will start the iterations by assuming that the cable surface temperature is 10 degrees higher than the ambient.

With half of the cables loaded, we first need to determine the radius R_2. Denoting by N_L the number of loaded cables and including the air-filling factor f_1 computed in Example 6.1, we have

$$R_2 = \tfrac{1}{2}(D_{core}^2 \cdot N_L/f_1)^{1/2} = \tfrac{1}{2}(1.6^2 \cdot 48/0.7)^{1/2} = 6.62 \text{ mm}$$

Center of the Bundle Loaded. With the partial results taken from Example 6.1, Equation (6.45) becomes

$$
53 = W_t \cdot \left\{ \frac{\rho_{eff}}{4\pi}\left(1 + 2\ln\frac{R_1}{R_2}\right) + \frac{\rho_J \ln[D_e/(D_e - 2t_J)]}{2\pi} \right.
$$

$$
\left. + \frac{1}{\pi \cdot D_e^* \cdot [h_{conv} + \sigma_B \varepsilon_s(1.38 \cdot 10^8 + 1.39 \cdot 10^6 \theta_{amb})]} \right\}
$$

$$
= W_t \cdot \left\{ \frac{4.9}{4\pi}\left(1 + 2\ln\frac{9.5}{6.62}\right) + \frac{5 \cdot \ln[24.1/(24.1 - 2 \cdot 1.8)]}{2\pi} \right.
$$

$$
\left. + \frac{1}{\pi \cdot 0.0241 \cdot (5.55 + 8.46)} \right\}
$$

resulting in

$$
W_t = 33.1 \ \text{W/m}
$$

This value is used to recompute the cable surface temperature and the convection coefficients. The process converges after four iterations with losses of 35.7 W/m. The permissible current in each conductor is now computed from Equation (6.3):

$$
I_c = \sqrt{\frac{W_t}{N_L \cdot R}} = \sqrt{\frac{35.7}{48 \cdot 0.0143}} = 4.23 \ \text{A}
$$

and the total current in the cable is equal to

$$
I = N_L \cdot I_c = 48 \cdot 4.23 = 203 \ \text{A}
$$

Outer Part of the Bundle Loaded. In this case, the first tem in the right-hand side of Equation (6.44) or (6.45) is replaced by the right-hand side of Equation (6.48). The ratio $R_2/R_1 = 0.70$. Thus, we have

$$
53 = W_t \cdot \left\{ \frac{\rho_{eff} \cdot \left(\frac{R_2}{R_1}\right)^2 \cdot \left[2 \cdot \ln\left(\frac{R_2}{R_1}\right) - 1\right] + 1}{4\pi \cdot \left[1 - \left(\frac{R_2}{R_1}\right)^2\right]} + \frac{\rho_J \ln[D_e/(D_e - 2t_J)]}{2\pi} \right.
$$

$$
\left. + \frac{1}{\pi \cdot D_e^* \cdot [h_{conv} + \sigma_B \varepsilon(1.38 \cdot 10^8 + 1.39 \cdot 10^6 \theta_{amb})]} \right\}
$$

$$
= W_t \cdot \left\{ 4.9 \cdot \frac{0.7^2 \cdot [2\ln(0.7) - 1] + 1}{4\pi \cdot (1 - 0.7^2)} + \frac{5 \cdot \ln[24.1/(24.1 - 2 \cdot 1.8)]}{2\pi} \right.
$$

$$
\left. + \frac{1}{\pi \cdot 0.0241 \cdot (5.55 + 8.46)} \right\}
$$

resulting in

$$W_t = 2.22 \text{ W/m}$$

The process converges after four iterations with losses of 23.3 W/m. The permissible current in each conductor is now computed from Equation (6.3):

$$I_c = \sqrt{\frac{W_t}{N_L \cdot R}} = \sqrt{\frac{23.3}{48 \cdot 0.0413}} = 3.41 \text{ A}$$

and the total current in the cable is equal to

$$I = N_L \cdot I_c = 48 \cdot 3.41 = 164 \text{ A} \qquad \blacksquare$$

6.3   EXAMPLES OF DERATING FACTORS

In this section, we will consider groups of identical low- or medium-voltage cables installed in air. In many practical installations, such cables are either touching or placed in a very close proximity to each other. The circuits are often attached to walls or ceilings or laid on floors or ladders. The traditional rating methods cannot be applied in these situations directly. In practice, ampacity of a single circuit is calculated first, using standard procedures, and a suitable derating factor is then applied. The magnitude of this derating depends on the cable construction and on the installation conditions. In this section, we will present current rating factors for cables having extruded insulation and a rated voltage from 3.6/6 kV up to 18/30 kV, installed under a range of conditions.[1] In the majority of cases, the deratig factors are based on engineering experience or on laboratory experiments. In such cases, we will simply state their values. On some occasions, the derating factors can be computed. In these cases, we will point out the methods used.

The guidance provided in this section is intended to complement the current ratings given in several IEC and IEEE standards and other sources on cable ratings. The factors given in this chapter are intended to be indicative only, they are averages over a range of conductor sizes and cable types. For particular cases, the correction factor should be calculated using the methods in IEC 60287-2-2 (1995)[2] or obtained from the cable supplier. The method given in IEC 60287-2-2 for groups in air was derived from test work carried out by ERA in the 1970s (Parr, 1974). This report describes tests carried out to determine the h values in Equation (6.10) for cables under a range of installation conditions. The results from this work were also used to slightly modify the h curves given in IEC 60287-2-1. The h values for cables fixed to a wall came from this work.

---

[1]This section is based on the document being prepared for the IEC by Working Group 19 of Technical Committee 20.
[2]All diagrams in this section come from IEC Standard 60364-5-52 (2001) with permission from IEC.

Tests were carried out using PVC-covered aluminium tubes of four different diameters to represent cables. The diameters were 76.5 mm, 38.4 mm, 19.4 mm, and 13 mm. Tests were carried out with the following arrangements of tubes:

1. A single tube spaced at various distances from a vertical wall
2. A single trefoil group at various distances from a vertical wall
3. Up to 6 cables and trefoil groups, side by side with various spacings
4. Up to 6 cables and trefoil groups, above each other with various spacings

The results for the horizontal and vertical single-layer groups were combined to derive factors for the grid arrangements covered by the IEC Standard.

6.3.1 Rating Factors

The rating factors given in the following sections are based on the assumption that the nominal current rating of the cable is given for the following conditions:

- Maximum conductor temperature 90°C
- Ambient air temperature 30°C

The rating factors given in this chapter should be applied as multipliers to the nominal current rating of the cable to obtain the current rating of the cable under the particular installation conditions that are applicable. Rating factors are given to take account of the following conditions:

1. Laying in a single layer on walls, floors, or in cable trays
2. Laying in a single layer under ceilings
3. Laying in a single layer in ventilated cable trays
4. Laying in a single layer on cable ladders, brackets, or wire mesh
5. Laying in several layers
6. Multiple cables per phase

Where one circuit consists of a number of single-core cables in parallel, the number of circuits should be taken as the number of cables per phase.

6.3.1.1 Laying in a Single Layer on Walls, Floors, or in Cable Trays.
The factors given in Table 6-2 apply to touching cables installed in a single layer in

Table 6-2 Single layer of touching cables installed on a floor

Number of circuits or 3-core cables								
1	2	3	4	5	6	7	8	9
1.00	0.85	0.79	0.75	0.73	0.72	0.72	0.71	0.70

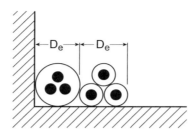

Figure 6-8   Cables laid on a floor.

ventilated cable channels on floors or in unperforated cable trays (see Figure 6-8). Where the clearance between adjacent cables exceeds twice their overall diameter D_e, no reduction factor need be applied.

The factors given above have been taken from IEC 60364-5-52 (2001). These factors were generated by ERA in the early 1970s from test work carried out to derive the h factors for cables in trays.

6.3.1.2   Laying in a Single Layer Under Ceilings. The factors given in Table 6-3 apply to touching cables installed in a single layer on the underside of a ceiling (see Figure 6-9). Where the clearance between adjacent cables exceeds twice their overall diameter D_e, no reduction factor need be applied.

6.3.1.3   Laying in a Single Layer on Ventilated Cable Trays. The factors given in Table 6-4 apply to touching cables installed in a single layer on a perforat-

Table 6-3   Single layer of touching cables under a ceiling*

Number of circuits or three-core cables								
1	2	3	4	5	6	7	8	9
0.95	0.81	0.72	0.68	0.66	0.64	0.63	0.62	0.61

*Factors taken from IEC 60364-5-52 (2001).

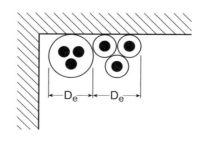

Figure 6-9   Cables installed on brackets under a ceiling.

Table 6-4   Single layer of touching cables under a ceiling*

Number of circuits or three-core cables								
1	2	3	4	5	6	7	8	9
1.00	0.88	0.82	0.77	0.75	0.73	0.73	0.72	0.72

*Taken from IEC 60364-5-52 (2001).

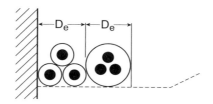

Figure 6-10   Single layer on perforated trays.

ed cable tray, as illustrated in Figure 6-10. Where the clearance between adjacent cables exceeds twice their overall diameter, D_e, no reduction factor need be applied.

6.3.1.4 *Laying in a Single Layer on Cable Ladders, Brackets, or Wire Mesh.* The factors given in Table 6-5 apply to touching cables installed in a manner that allows free air circulation around the group. Where the clearance between adjacent cables exceeds twice their overall diameter (see Figure 6-11), no reduction factor need be applied.

Table 6-5   Single layer of touching cables in free air*

Number of circuits or three-core cables								
1	2	3	4	5	6	7	8	9
1.00	0.87	0.82	0.80	0.80	0.79	0.79	0.78	0.78

*Taken from IEC 60364-5-52 (2001).

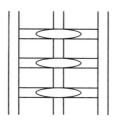

Figure 6-11   Cables in single layer installed on ladders.

6.3.1.5 Laying in Several Layers. When cables are laid in several layers, the ampacity of cables is usually lower than computed with the formulas given in the IEC Standard 60287, because of the very strong mutual heating effects. Such arrangements occur for cables laid in trays. Since 1975, the accepted technique in North America for calculating the ampacity of power cables in trays has been the use of tables provided by ICEA/NEMA Standard P54-440 (ICEA/NEMA, 1986). These tables are based upon a thermal model originally proposed by Stolpe (1970) that assumes that every cable in the cable tray carries the maximum current producing the maximum cable bundle temperature. The ICEA standard also assumes that the heat generated in the cables is uniformly distributed throughout the cross section of the cable mass.

In order to remove the conservatism in the thermal model based on the above assumptions, Harshe and Black (1994) proposed a simple thermal model that can be used to determine the maximum temperature of a mass of cables in a single, open-top horizontal cable tray. The model accounts for load diversity within the tray by providing two different loading options. The thermal model has been verified by comparing predicted cable bundle temperatures with temperatures measured at an actual installation. The derivation of the model is described in Anders (1997).

6.3.1.6 Several Cables Connected in Parallel. Where cables are connected in parallel, they should be made to the same specification, have the same cross section, the same earthing system, the same length, and should not have branches along their route.

In general, it is recommended that the number of cables in parallel should be limited to a minimum. If more than four feeders are needed to carry the assigned power, the design of the network should be reconsidered. When many feeders are connected in parallel, the current sharing between them is poor, and there is a risk of abnormal heating.

A method for calculating the current sharing between parallel cables in any arrangement is given in IEC 60287-1-3 (1994) and in Anders (1997).

It should be noted that if there are n cables in parallel per phase, the grouping factor appropriate to n circuits should be applied to the nominal single circuit rating of the cables. An additionnal derating factor, f_s, is recommended to account for the unequal current sharing that is expected for certain cable arrangements. The value of this factor was computed using the methods described in IEC Standard 60287-1-3 (1994) and Anders (1997).

Figure 6-12 (a) and (b) and Figure 6-13 (a) and (b) show the symmetrical arrangements of cables recommended by the IEC. The reduction factor is $f_s = 1$ in these cases.

Figure 6-14 (a) to (d) shows nonsymmetrical arrangements of cables. The reduction factor is $f_s = 0.8$ in these cases. Parallel cables should not be arranged so that all the cables of one phase are grouped together.

6.4 CONCLUDING REMARKS

Multicore cables are used in large-scale power plants or chemical plants, primarily for the supply of magnetic valves with a nominal voltage of 24 V (dc), the nominal

Triangular formation

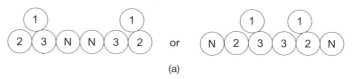

(a)

Flat formation

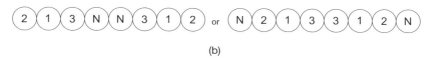

(b)

Figure 6-12 Two cores per phase with or without a neutral conductor. (a) Triangular and (b) flat formation.

power of which is presently limited to 30 W. If all circuits of the cable are used in this way, a total current of 125 A will result. Figure 6-3 shows that this continuous current can be transmitted by this cable, installed in still air, even for the highest possible ambient temperatures of 50°C.

From Figure 6-3, for standard conditions (cables in still air; ambient temperature 30°C) the current rating of the multicore cable is approximately equal to. 270 A. This multicore cable has 96 cores spaced at 0.5 mm^2, giving a total cross-sectional conductor area of 48 mm^2. The VDE standard gives a rating of 214 A for a single-core, 1 kV cable with a 50 mm^2 copper conductor (VDE 0298 T2; Heinhold, 1999). IEC 60364-5-52 gives a current rating of 275 A for a single-core, unarmored, 1 kV, XLPE-insulated 50 mm^2 cable installed spaced from other cables in free air at 30°C. This is a similar installation condition to that for which the rating for the mul-

Triangular formation

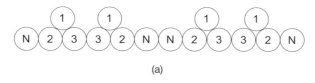

(a)

Flat formation

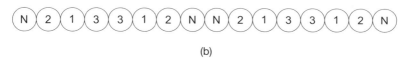

(b)

Figure 6-13 Four cores per phase and possible neutral conductor. (a) Triangular and (b) flat formation.

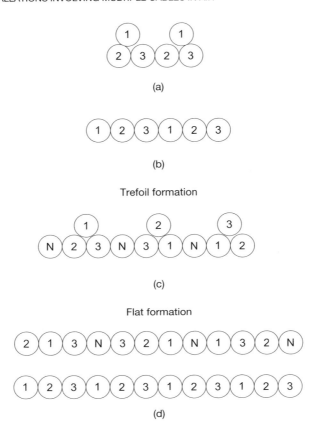

Figure 6-14 (a) Nonsymmetrical arrangements for two cores per phase in trefoil formation. Also applicable for two multi-core cables laid in parallel. (b) Nonsymmetrical arrangements for two cores per phase in flat formation. (c) Three cores per phase and possible neutral conductor. (d) Nonsymmetrical arrangements for four cores per phase in flat formation.

ticore cable was calculated. For three cables flat touching, the rating in IEC 6034-5-52 is given as 216 A, and for three cables in a trefoil pattern the rating is 207 A. In spite of the slightly larger cross section, these are a much smaller currents than the one computed above. Of course, the cable construction and dimensions are different, but these results give a good indication of the rating capabilities of multicore cables.

The derating factors are intended to take account of the fact that the current may not be shared equally between parallel cables, irrespective of the earthing and bonding arrangements. Occasionally, there is confusion between the designers and installers over the difference between the terms "earthing" and "bonding." Earthing is connecting to earth and bonding is joining together. If the armor is bonded at both ends but only earthed at one end, there will be circulating currents. Screens and armor are almost always earthed at one end but may or may not be bonded or earthed at the other end. In LV installations, it is common to bond and earth at both ends. In higher-voltage installations, single-point bonding and cross bonding are more likely to be used. Calculation methods for parallel cables outlined in Anders (1997) could be applied for different bonding arrangements. When a circuit is single-point bonded, the length of the cable should be limited so that the standing voltages that will develop at the ungrounded end do not exceed the prescribed standard value. This length should usually not exceed about 200 m.

6.5 CHAPTER SUMMARY

The most important topic discussed in this chapter is the calculation of current ratings of multicore communication cables installed in air. The permissible current in each core I_c is obtained from

$$I_c = \sqrt{\frac{W_t}{N \cdot R}}$$

and the total current rating of the cable is

$$I = N \cdot I_c$$

where N is the number of loaded conductors and R (ohm/m) is the ac resistance of a single conductor at the operating temperature. The total losses W_t are evaluated as shown below for two distinct external conditions.

6.5.1 Cables in Free Air

The total losses are computed from the following equation by iterations:

$$\Delta\theta_{max} = W_t \cdot \left\{ \frac{\rho_{eff}}{4\pi} + \frac{\rho_J \cdot \ln[D_e/(D_e - 2t_J)]}{2\pi} + \frac{1}{[\pi \cdot D_e^* \cdot h \cdot W_t^{1/4}]^{4/5}} \right\}$$

where
$\Delta\theta_{max}$ is the maximum permissible conductor temperature rise (°C)
D_e is the cable external diameter (mm)
t_J is the thickness of the jacket (mm)
ρ_J is its thermal resistivity (K · m/W)

The effective thermal resistivity of the insulating material inside the bundle is obtained from

$$\rho_{eff} = \rho \frac{(1 - f_C)}{f_1}$$

where ρ is the thermal resistivity of the insulating material and

$$f_1 = \frac{N \cdot D_{core}^2}{D_1^2} \cdot f_{strand}$$

D_{core} (mm) is the diameter over one cable, and the stranding factor is defined by

$$f_{strand} = \sqrt{1 + \left(\frac{\pi \cdot D_{strand}}{L_{strand}}\right)^2}$$

where
D_{strand} is a mean diameter of the stranding, (mm)
L_{strand} is the stranding length (mm)
D_1 is the diameter of the bundle (just below the external jacket) (mm)

The filling factor for the conductor areas is equal to

$$f_C = \frac{4 \cdot N \cdot A_{core}}{\pi \cdot D_1^2 \cdot f_{cond}} \cdot f_{strand}$$

where A_{core} (mm^2) is the conductor cross section of each core and f_{cond} takes into account the compaction of a single conductor.

The heat transfer coefficient, h, can be computed using one of several published approaches. For example, using an IEC approach, its value is given by

$$h = \frac{0.21}{(D_e^*)^{0.6}} + 3.94$$

where the asterisk denotes dimensions in meters.

6.5.2   Cables in Moving Air

In this case, forced convection needs to be considered and the total losses are evaluated iteratively from the following equation:

$$\Delta\theta_{max} = W_t \cdot \left\{ \frac{\rho_{eff}}{4\pi} + \frac{\rho_J \cdot \ln[D_e/(D_e - 2t_J)]}{2\pi} \right.$$

$$\left. + \frac{1}{\pi \cdot D_e^* \cdot [h_{conv} + 4.2 \cdot \varepsilon_S \cdot (1 + 0.0167 \cdot \theta_s)]} \right\}$$

$$\theta_s = W_t \cdot T_4 + \theta_{amb}$$

Alternatively, we can use Black's approximation:

$$\Delta\theta_{max} = W_t \cdot \left\{ \frac{\rho_{eff}}{4\pi} + \frac{\ln[D_e/(D_e - 2t_J)]}{2\pi \cdot k_J} \right.$$

$$\left. + \frac{1}{\pi \cdot D_e^* \cdot [h_{conv} + \sigma_B \varepsilon_s (1.38 \cdot 10^8 + 1.39 \cdot 10^6 \theta_{amb})]} \right\}$$

from which we can obtain W_t directly.

An approximate value for the forced convection coefficient can be obtained from (Neher & McGrath, 1957)

$$h_{conv} = 2.87 \sqrt{\frac{w}{D_e^*}}$$

where w (m/s) is the wind velocity and the natural convection is ignored.

6.5.3 Unequally Loaded Bundle

We will consider two situations with a nonuniform loss density in the bundle. In one case, the cables in the center of the bundle will be loaded, and in the other, the current will flow in the outer cables. The above equations are slightly modified as shown below.

6.5.3.1 *Central Part Loaded.* If the center part of the cable bundle with effective diameter R_2 ($R_2 < R_1$) is loaded, the temperature rise equation for free convection becomes

$$\Delta\theta_{max} = W_t \cdot \left\{ \frac{\rho_{eff}}{4\pi}\left(1 + 2\ln\frac{R_1}{R_2}\right) + \frac{\rho_J \cdot \ln[D_e/(D_e - 2t_J)]}{2\pi} + \frac{1}{[\pi \cdot D_e^* \cdot h \cdot W_t^{1/4}]^{4/5}} \right\}$$

and for forced convection,

$$\Delta\theta_{max} = W_t \cdot \left\{ \frac{\rho_{eff}}{4\pi}\left(1 + 2\ln\frac{R_1}{R_2}\right) + \frac{\rho_J \cdot \ln[D_e/(D_e - 2t_J)]}{2\pi} \right.$$

$$\left. + \frac{1}{\pi \cdot D_e^* \cdot [h_{conv} + 4.2 \cdot \varepsilon_s \cdot (1 + 0.0167 \cdot \sigma_s)]} \right\}$$

6.5.3.2 *Outer Part Loaded.* The highest temperature rise is obtained, in this case, from

$$\Delta\theta_{max} = \rho_{eff} \cdot W_t \cdot \frac{\left(\dfrac{R_2}{R_1}\right)^2 \cdot \left[2 \cdot \ln\left(\dfrac{R_2}{R_1}\right) - 1\right] + 1}{4\pi \cdot \left[1 - \left(\dfrac{R_2}{R_1}\right)^2\right]}$$

REFERENCES

Anders G.J. (1997), *Rating of Electric Power Cables: Ampacity Computations for Transmission, Distribution and Industrial Applications,* IEEE Press, New York, McGraw Hill, New York, 1998.

Black, Z.W., and Rehberg, R.L. (1985), "Simplified Model for Steady State and Real-Time Capacity of Overhead Conductors," *IEEE Trans. on Power Systems,* Vol. PAS-104, pp. 2942–2963.

Brakelmann, H. (1985), *Belastbarkeiten der Energiekabel,* VDE-Verlag, Berlin/Offenburg.

Brakelmann, H, Lauter, P., and Anders, G.J. (2004), "Current Rating of Multicore Cables," submitted for publication in *IEEE Transactions on Industry Applications.*

Dorison, E., De Kepper, B., Protat, F., Gurkahraman A., and Masure L. (2003), "Current Rating of Cables Installed in Tunnels," in *Proceedings of Jicable '03,* Versailles, France, June 2003, pp. 600–605.

Goldenberg, H. (1957), "The Calculation of Cyclic Rating Factors for Cables Laid Direct or in Ducts," *Proceedings of IEE,* Vol. 104, Pt. C, pp. 154–166.

Goldenberg, H. (1958), "The Calculation of Cyclic Rating Factors and Emergency Loading for One or More Cables Laid Direct or in Ducts," *Proceedings of IEE,* Vol. 105, Pt. C, pp. 46–54.

Harshe, B.L., and Black, W.Z. (1994), "Ampacity of Cables in Single Open-Top Cable Trays," *IEEE Transactions on Power Delivery,* Vol. PWRD-9, No.4, pp. 1733–1740.

Heinhold, L. (1999), *Kabel und Leitungen für Starkstrom,* Publicis MCD Verlag, Erlangen.

ICEA/NEMA (1986), "Ampacities of Cables in Open-Top Cable Trays," *ICEA Publication No. P54-440, NEMA Publication No. WC51,* Washington, DC.

IEC Standard 60853-1 (1985), "Calculation of the Cyclic and Emergency Current Ratings of Cables. Part 1: Cyclic Rating Factor for Cables up to and Including 18/30 (36) kV," Publication 853-1.

IEC Standard 60853-2 (1989), "Calculation of the Cyclic and Emergency Current Ratings of Cables. Part 2: Cyclic Rating Factor of Cables Greater than 18/30 (36) kV and Emergency Ratings for Cables of All Voltages," Publication 853-2.

IEC Standard 60287 (1994), "Calculation of the Continuous Current Rating of Cables (100% load factor)."

IEC Standard 60287-2-2 (1995), "Calculation of the Continuous Current Rating of Cables (100% load factor)—Groups of Cables in Air."

IEC Standard 60364-5-52 (2001), "Electrical Installations of Buildings—Part 5-52: Selection and Erection of Electrical Equipment—Wiring Systems."

Incropera, F.P., and De Witt, D.P. (1990) *Introduction to Heat Transfer,* Wiley, New York.

Jong, Jan de (2003), "The Reduction Factors for a Cable in a Group of Power Cables," private communication.

Lauriat G. (1969), *Thermique—Cours A 1ère Anneee: Itroduction aux Transferts Thermiques,* CNAM/MEDIAS, Paris.

Leake, H.C. (1997), "Sizing of Cables in Randomly Filled Trays with Consideration for Load Diversity," *IEEE Transaitons on Power Delivery,* Vol. 12, No. 1, pp. 39–44.

Morgan, V.T. (1992), "Effect of Mixed Convection on the External Resistance of Single-core and Multicore Bundled Cables in Air," *IEE Proceedings-C,* Vol. 139, No. 2, pp. 109–116.

Morgan, V.T. (1993), "External Thermal Resistance of Aerial Bundled Cables," *IEE Proceedings-C,* Vol. 140, No. 2, pp. 65–62.

Neher, J.H., and McGrath, M.H. (1957), "The Calculation of the Temperature Rise and Load Capability of Cable Systems," *AIEE Transactions,* Vol. 76, Part 3, pp. 752–772.

Parr, R.G. (1974), "Heat Emissions from Cable in Air," ERA report 74-27.

Schach, M., and Kidwell Jr, R. E. (1952), "Continuous Current and Temperature Rise in Bundle Cables for Aircraft," *AIEE Transactions,* Vol. 71, Pt. II, pp. 376–384.

Slaninka, P. (1969), "External Thermal Resistance of Air-Installed Power Cables," *Proceedings of IEE,* Vol. 116, No. 9, pp. 1547–1552.

Song Ho-Liu, Xu Han, "Analytical Method of Calculating the transient and Steady State Temperature Rises for Cable Bundle in Tray and Ladder," *IEEE Transactions on Power Delivery,* Vol. 13, pp. 691–698.

Stolpe, J. (1970), "Ampacities for Cables in Randomly Filled Cable Trays," *IEEE Transactions on Power Apparatus and Systems,* Vol. PAS-90, pp. 962–973.

Whitehead, S., and Hutchings, E. E. (1939), "Current Ratings of Cables for Transmission and Distribution," *Journal of IEE,* Vol. 38, pp. 517–557.

Winkler F. (1963), "Current-Carrying Capacity of Cables Installed Horizontally in Free Air," *Siemens Review,* Vol. XXX, No. 11, pp. 413–418.

VDE 0298 T2, "VDE-Bestimmung für die Verwendung von Kabeln und isolierten Leitungen für Starkstromanlagen."

CHAPTER 7

Rating of Pipe-Type Cables with Slow Oil Circulation of Dielectric Fluid

This chapter presents a rigorous approach to calculation of current ratings in pipe-type cables with slow circulation of fluid. Detailed mathematical analysis is presented for the calculation of the hot-spot oil temperature as a function of the cable geometry and the oil-inlet temperature. The emphasis is placed on the real-time applications in which the pipe surface temperature is measured. A comparison with the commonly applied method based on Burrell's model is offered.

7.1 NOMENCLATURE

The notation used in this chapter is listed below.

D	diameter (m)
P	oil pressure (Pa)
M	rate of oil flow (m^3/s)
$S, \|S\|$	oil filled cross section and its area, respectively (m^2)
$T_{f\text{–}amb}$	external thermal resistance between oil and ambient earth (K · m/W)
T_{grav}	thermal resistance corresponding to the free convection/radiation/conduction heat flow (K · m/W)
T_{fs}, T_{fp}	oil thermal resistance at the surface of the cables and the pipe walls, respectively (K · m · W^{-1})
Z	length of the hot-spot section (m)
θ	temperature (°C)
$\Delta\theta_u$	ultimate oil temperature rise over ambient with no circulation (°C)
W_c	conductor losses (W/m)
W_d	dielectric losses (W/m)
W_t	total losses seen by the fluid $= n \cdot [W_c \cdot (1 + \lambda_1) + W_d]$
$c_{p.f}$	oil specific heat at constant pressure (J · kg^{-1} · K^{-1})
d_h	hydraulic diameter (m)

Rating of Electric Power Cables in Unfavorable Thermal Environment. By George J. Anders
ISBN 0-471-67909-7 © 2004 the Institute of Electrical and Electronics Engineers.

n	number of conductors in the pipe
q_v	internal heat source (W/m)
r	radial distance from the pipe center (m)
w	oil velocity along pipe axis (m/s)
$a_f(\theta)$	oil thermal diffusivity for specific temperature (m²/s)
β	ratio of the equivalent cable diameter to the inner diameter of the pipe
$\beta_f(\theta)$	oil expansion coefficient for specific temperature (K⁻¹)
λ_1	skid wire/reinforcing tape loss factor
$\lambda_f(\theta)$	oil thermal conductivity for given temperature (W · K⁻¹ · m⁻¹)
μ_f	dynamic viscosity (N · s/m²)
ρ	thermal resistivity, (K · m/W)
$\rho_f(\theta)$	oil density for specific temperature (kg/m³)

Dimensionless numbers

$(Gr)_{f,d_h}$	Grashoff number for hydraulic diameter d_h
$(\mathrm{Pe})_{f,d_h} = (\mathrm{Pr})_f \cdot (\mathrm{Re})_{f,d_h}$	the Peclet number
$(\mathrm{Pr})_f$	Prandtl number;
$(\mathrm{Ra})_{f,d_h}$	Rayleigh number for hydraulic diameter d_h
$(\mathrm{Re})_{f,d_h}$	Reynolds number for hydraulic diameter d_h

Subscripts and superscripts

c	conductor
s	cable surface or cable-oil film
f	fluid
p	internal pipe surface or oil-pipe film
co	coating
o	outside pipe coating
amb	ambient
in	fluid at inlet ($x = 0$)
out	fluid at outlet ($x = Z$)
∞	fluid parameter in infinity ($x \rightarrow +\infty$)
$-$	bar above a symbol represents an average value

7.2   INTRODUCTION

Artificial cooling of pipe-type cable circuits has been used since early 1950s. Two distinctly different approaches have found application in practice. On the one hand, oil is slowly circulated along the cable route and, on the other, forced cooling plants are employed. The cooling of long cable lines by artificial means can be difficult and expensive. However, it is often possible to improve a difficult hot-spot situation by artificial cooling without undue expense, especially in or near a generating station or substation.

In the case of an underground cable, it is possible to mitigate hot-spot conditions by saturating the soil with water, thereby improving its conductivity. While this practice has been considered, it turned out that with time the water lines laid along the cable route can become irreparably damaged. In the case of a pipe-type cable, the pressure medium (oil or gas) can be circulated and cooled. In forced-cooled installations, the fluid is cooled by means of a heat exchanger and returned to the cable pipe for further circulation. A diagram of such an installation is shown in Figure 7-1.

Slow movement of the oil has been used as a means of averaging out temperature variations along the cable route. There are three methods by which this movement is implemented in practice. One method is to circulate the oil via a return pipe installed parallel to the pipe-type cable circuit. Another method is to circulate the oil around a loop consisting of two parallel circuits that carry approximately the same load. This approach is illustrated schematically in Figure 7-2. The third approach is to oscillate the oil between two tanks at the ends of the circuit. The theoretical calculations (Buller, 1952) supported by some experimental evidence, reported by Burrell (1965), indicate that with a slow oil circulation of up to 10 gallons per minute (gpm), a substantial reduction in the conductor and pipe temperatures can be achieved, as illustrated in Figure 7-3.

The mathematical foundations for the analysis of the thermal considerations involved in the cooling of cable circuits have been extensively treated in the literature (Buller et al., 1951; Buller, 1952; Giaro, 1956; Ralstone and West, 1960; Oudin et al., 1960; Kitogawa, 1964; CIGRE, 1979; Chato, 1977; Glicksman et al., 1978; Chato and Crowley, 1980; Purnhagen, 1984; CIGRE, 1985). The last reference gives an extensive overview of the calculation of pipe-type cable ratings when forced cooling is involved. Therefore, in this chapter, we will focus our attention on pipe-type cables with slow fluid circulation.

The calculations routinely performed in North America for cables with slow oil circulation are based on the equations developed by Buller (1952). These equations, developed over 50 years ago, give reasonably good approximations of the heat transfer phenomena involved in slow circulation cooling of pipe-type cables. However, some simplifications embedded in these equations, justified at the time when

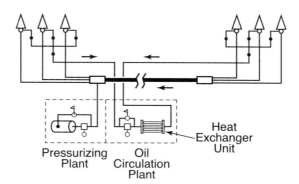

Figure 7-1 Diagram of artificial cooling connections (from Burrell, 1965, with permission from IEEE).

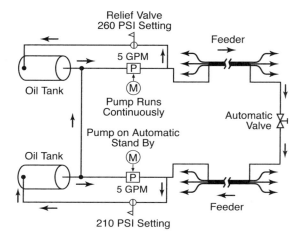

Figure 7-2   Piping schematic for slow oil-circulation connections (from Burrell, 1965, with permission from IEEE).

there were no computers, are no longer necessary today. The advent of fast computational techniques as well as more recent developments in heat transfer theory permit application of more accurate models.

Availability of more accurate methods does not justify a revision of the computational techniques that served the industry well for many years. However, a detailed analysis of the Buller's model revealed that the simplifying assumptions that were made to make the approach amenable to hand calculations limit the applicability of the models. This prompted the author, assisted by Dr. M. Zubert from the Technical University of Lodz, to perform detailed analysis of the heat transfer phe-

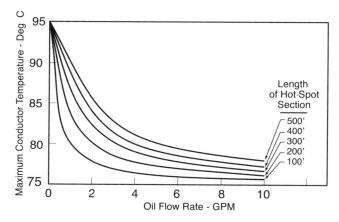

Figure 7-3   Effect of oil-flow rate on hot-spot temperature reduction (from Burrell 1965, with permission from IEEE).

nomenon in the case of slow oil circulation in a pipe-type cable. An additional incentive for a development of new rating equations was an increase in real-time cable rating applications. In such installations, the temperatures of the pipe covering and the ambient earth, as well as conductor current, are measured. Therefore, there is an opportunity to develop calculation techniques with these quantities as the input parameters. The results of these developments are reported in this chapter. A comparison with the Buller's results is also offered.

7.3 THERMAL EFFECTS OF DIELECTRIC FLUID CIRCULATION

The mathematical model for pipe and conductor temperature calculations in the presence of oil circulation will be developed in this chapter. When dielectric fluid circulates in a pipe, the premise is that its temperature, and hence the thermal resistance, will be smaller than in the static case. This, in turn, will lead to an increased current rating. We recall that in the static case, the thermal resistance of oil is given by the following equation, as discussed in Sections 1.6.6.4 and 1.6.6.5:

$$T_4' = \frac{0.26}{1 + 0.00026\,\theta_f D_s^*} \tag{7.1}$$

where θ_f (°C) is the average oil temperature and $D_s^* = 2.15 D_s$ (m) is a diameter of a circle circumscribing the three cores. This resistance includes the effects of conduction, radiation, and free convection.

When oil circulates in the pipe, we will consider two separate thermal resistances, denoted by T_{fs} and T_{fp}. They correspond to the oil film between the cables and the bulk oil and between the pipe surface and the oil, respectively. Occasionally, we may want to consider a third thermal resistance, in parallel with the film resistances that represents, for example, additional terms for free convection, radiation, and conduction. In that case, we could use Equation (7.1); however, for clarity of presentation, in the developments presented below we will concentrate on the film resistances only. The parallel resistance could be easily included in a computer implementation of the procedure presented below.

The thermal resistance between the oil and ambient earth will, therefore, be computed from

$$T_{f\text{-}amb} = T_{fp} + T_4'' + T_4''' \tag{7.2}$$

The last two terms in Equation (7.2) represent the thermal resistance of the pipe covering and the external thermal resistance of the pipe, respectively. Calculation of these quantities is discussed in Section 1.6.6.5 (Anders, 1997) and, assuming a unity load factor, is given by the following equation:

$$T = T_4'' + T_4''' = \frac{\rho_{co}}{\pi} \cdot \ln\frac{D_o}{D_{pe}} + \frac{\rho_e}{\pi} \cdot \ln\frac{4L}{D_o} \tag{7.3}$$

With a nonunity load factor, the last term in Equation (7.3) is replaced by $T_{4\mu}$, defined in Section 1.6.6.5. In the following sections, we will develop equations for the calculation of the oil temperature and the oil film resistances in the case of fluid circulation. We will start with some preliminary information about fluid mechanics, followed by a review of the Buller's equations, and conclude with the new model development.

7.3.1 Background Information

Rating calculations for high-pressure fluid-filled, pipe-type cables with slow fluid circulation are divided into two basic parts.

1. Calculation of coolant parameters based on the fluid velocity, the physical characteristics of the cable, and the geometry of the cable system
2. The solution of the temperature distribution in the system based on the selection of certain boundary conditions

These are briefly discussed below.

7.3.1.1 *Calculation of Coolant Parameters.* Calculation of film resistances requires the determination of a dimensionless group of variables: Reynolds number (Re), Prandtl number (Pr), Grashoff number (Gr), Nusselt number (Nu) and Peclet number (Pe).

The Reynolds number characterizes the flow conditions in the pipe. For the case of slow fluid circulation, laminar flow will exist when the Reynolds number is smaller than 800. The Prandtl number is a ratio of the kinematic viscosity (which affects velocity distribution) and the thermal diffusivity (which affects the temperature profile). The Grashoff number is used for natural convection heat transfer correlations and is the ratio of buoyant forces to friction forces. The Nusselt number is used to calculate the heat transfer coefficient of the fluid due to convection. The Peclet number is the ratio of the product of the Prandtl and Renolds numbers. Calculation of these numbers is discussed in more detail in the text.

Another significant geometric parameter for thermal hydraulic correlations that will often appear in the discussion below is a hydraulic diameter. The classical definition is

$$d_h = 4\,\frac{\text{free hydraulic cross-sectional area}}{\text{wetted perimiter of the surface within pipe}} \tag{7.4}$$

For a clear, round pipe, the hydraulic diameter is equal to the pipe inside diameter. This definition has proven helpful is establishing geometric similarities between

---

[1]In cable design practice, the computation of the pressure drop in the cable system based on oil flow and temperature distribution may be performed (CIGRE, 1985). However, for rating calculations, we do not require this step.

systems with simple cross sections such as square ducts or annuli. All the dimensionless numbers listed above will be defined for a specific value of d_h.

In the case of HPFF cables, the classical definition of hydraulic diameter does not provide an adequate translation between one cable and pipe geometry and another, especially when snaking occurs.[2] The definition recommended by CIGRE (1985) is given by the following formula:

$$d_h = \frac{D_p^2 - 3D_s^2}{D_p - 3D_s} \tag{7.5}$$

where subscripts p and s refer to pipe and cable shield diameters. Another approximation that is sufficient in engineering practice is given by

$$d_h = D_p - 2.15D_s = D_p - D_s^* \tag{7.6}$$

7.3.1.2 Thermal Calculations. The thermal analysis is performed in radial and axial directions. The following heat sources can be identified in a typical pipe type cable:

- Cable conductors
- Cable shields
- Steel pipe
- Dielectric losses
- Pipe oil coolant

The first four sources are mentioned in many parts of this book. The last source of losses is due to friction, lack of uniformity of loss distribution between phases, possible lack of isothermality of the cable shields and a potential growth of a boundary layer. These may play sometimes a significant role in forced-cooled cable installations (CIGRE, 1985) but will be ignored for the slow flow of coolant.

For the axial analysis, all three cable conductors in a given pipe are assumed to be at the same temperature at a given cross section along the pipe. Similarly, all three cable shields will be at the same temperature at a given cross section.

The temperature calculations are performed in two steps. First, the average oil temperature at a given location is computed, and then the temperature rise between the oil and the conductor is obtained.

7.3.2 Buller's Model

This model is characterized by its stark simplicity and minimal use of the fluid mechanics theory. These undoubtedly determines its popularity with cable designers and, as will be shown later, the model gives reasonably accurate results.

---

[2]Cable snaking inside the pipe can result in additional heat generation in the coolant due to increased friction. However, for slow oil circulation, we can ignore this phenomenon.

Slow oil circulation is used to smooth out oil temperature in a hot-spot section of the cable route. Assuming steady-state conditions of loading and oil flow rate, oil flowing in a pipe with a normal environment moves into a hot-spot section of length Z at an initial temperature designated by θ_f^{in}. When passing through, the oil temperature rises at an exponential rate to a maximum value [designated by $\theta_f(Z)$] at the outlet end of the hot-spot section. As oil moves again into a long section with a normal environment, the temperature decreases at an exponential rate toward an asymptotic value of θ_f^{in}. Figure 7-4 illustrates the meaning of the symbols used below in the development of the rating equations.

Heat generated by the cable in length dx is $W_o dx$. The heat rate W_o includes the losses in the conductor, insulation, and metallic tapes/skid wires. Heat dissipated to earth in length dx is $[(\theta_x - \theta_{amb})/T_{f-amb}]dx$. Heat removed by moving fluid in length dx is $Mc_{p,f}\rho_f d\theta_x$. The heat balance equation will then take the form

$$W_o dx = \frac{\theta_x - \theta_{amb}}{T_{f-amb}}\, dx + Mc_{p,f}\,\rho_f d\theta_x \tag{7.7}$$

which leads to the following differential equation:

$$\frac{d\theta_x}{dx} + \frac{\theta_x}{Mc_{p,f}\rho_f T_{f-amb}} = \frac{W_o}{Mc_{p,f}\rho_f} + \frac{\theta_{amb}}{Mc_{p,f}\rho_f T_{f-amb}} \tag{7.8}$$

The solution of this equation is

$$\theta_x = (W_o T_{f-amb} + \theta_{amb}) + Ce^{-x/Mc_{p,f}\rho_f T_{f-amb}} \tag{7.9}$$

At $x = 0$, $\theta = \theta_f^{in}$, so that

$$C = \theta_f^{in} - (W_o T_{f-amb} + \theta_{amb}) \tag{7.10}$$

Substituting Equation (7.10) into Equation (7.9), we obtain for $x = Z$

$$\theta_f(Z) = \theta_f^{in} + (\theta_f^\infty - \theta_f^{in}) \cdot \{1 - \exp[-3Z/(Mc_{p,f}\rho_f T_{f-amb})]\} \tag{7.11}$$

where $\theta_f^\infty = W_o T_{f-amb} + \theta_{amb}$ is oil temperature rise with the assumption of no circulation.

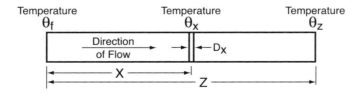

Figure 7-4   Illustration of oil flow through the cable pipe.

Buller (1952) proposed two different approaches to the calculation of the oil film resistances. In the case of a turbulent flow, following the technique described by McAdams (1942) for annular spaces, and substituting the constants for petroleum oils at 50°C (where these vary slightly with temperature) and using the SI system of units, he obtained

$$T_{fp} = \frac{9.512 \cdot 10^{-3} \cdot (\mu_f)^{0.4}}{(D_s^* \cdot w)^{0.8}} \cdot \left[1 - \left(\frac{D_s^*}{D_p} \right)^2 \right]^{0.2} \tag{7.12}$$

$$T_{fs} = \frac{9.512 \cdot 10^{-3} \cdot (\mu_f)^{0.4}}{(D_s^* \cdot w)^{0.8}} \cdot \left[\left(\frac{D_p}{D_s^*} \right)^2 - 1 \right]^{0.2} \tag{7.13}$$

For a laminar flow, Buller argued that there is little to be gained by using the relatively complex expressions available and it is better to neglect flow altogether and use the technique based on natural convection and conduction developed by Buller and Neher (1950). For approximate work, the following simple equations can be used:

$$T_{fs} = \frac{0.0057}{D_s^*} \qquad T_{ps} = \frac{0.0057}{D_p} \tag{7.14}$$

Denoting by $T_{4\mu}^{eff} = (1 + \lambda_1 + \lambda_2)T_{4\mu}$, where the subscript μ signifies the possibility of a nonunity load factor, the current rating is computed from

$$I = \left[\frac{\theta_c^{max} - \theta_{amb} - (\theta_f^{in} - \theta_{amb})e^{\alpha Z} - \theta_d}{R_{ac}[T_i + 3(1 + \lambda_1)T_{fs} + 3[(1 + \lambda_1)T_{fp} + T_{4\mu}^{eff}](1 - e^{\alpha Z})]} \right]^{0.5} \tag{7.15}$$

where

$$\alpha = \frac{-3}{Mc_{p,f}\rho_f T_{f-amb}} \tag{7.16}$$

The temperature rise caused by dielectric losses is obtained as

$$\theta_d = W_d[\tfrac{1}{2}T_1 + 3T_{fs} + 3(T_{fp} + T)(1 - e^{\alpha Z})] \tag{7.17}$$

We can observe that when $M = 0$, $\exp(\alpha Z) = 0$ and Equation (7.15) reduces to Equation (1.40).

Temperature rise of the conductors at any point is obtained in the standard manner by adding to the oil stream temperature at that point the temperature rise product of the heat losses, including dielectric losses, and the appropriate thermal resistances from conductor to oil stream.

7.3.2.1 Effective Cooling Distance.

In order to determine the applicability of the model, we need to find out the distance from the oil inlet at which the initial colder fluid ceases to have any cooling effect. The effective cooling distance can be obtained by considering, for example, a distance from the inlet to a position where $[\theta_f(Z_{cooling}) - \theta_f^{\infty}] = 95\% \cdot (\theta_f^{\infty} - \theta_f^{in})$. Solving this equation for Z, we obtain

$$Z_{cooling} = \frac{-3}{\alpha} = Mc_{p,f}\rho_f T_{f-amb} \qquad (7.18)$$

Example 7.1

We will use model cable model No. 3 to analyze the effect of slow oil circulation on cable rating. This is a 2000 kcmil pipe-type cable with the following parameters required in the analysis: $D_s = 67.59$ mm, $D_p = 206.38$ mm, $D_{pe} = 219.08$ mm, and $D_o = 244.48$ mm. The computed parameters of this cable are given in Appendix A and summarized in Table 7-1. The length of the hot-spot section is $Z = 10$ m.

When oil is not circulated, the rating of this cable is 902 A, with the thermal resistance of the oil equal to 0.082 Km/W. The rating calculations can proceed in several directions. In the simplest, but approximate, approach, we could use Equations (7.14) to substitute for $T_2 = T'_4$ in rating Equation (1.40). Applicability of this approach requires checking whether the flow is laminar or turbulent. We will consider two cases.

a) *Slow oil circulation.* We will assume oil velocity of 10 gallons per minute (gpm), which in metric units gives $M = 0.00063$ m³/s and $w = 0.037$ m/s. Before calculating the Reynolds number (see Appendix D), we will need the density and dynamic viscosity of the oil. Let us assume that at the hot-spot section, oil temperature is equal to 60°C. From Appendix D, we have

$$\mu_f(\theta_f) = 164.25 \cdot 60^{-2.0072} = 0.0443 \text{ N} \cdot \text{s/m}^2$$

$$\rho_f(\theta_f) = 903.6 - 0.566 \cdot 60 = 869.6 \text{ kg/m}^3$$

Table 7-1 Pipe-type cable parameters

		Computed parameters
Cable ampacity	I (A)	902
Conductor resistance at θ_{max}	R (ohm/km)	0.0246
Concentric/skid wires loss factor	λ_1	0.010
Pipe loss factor	λ_2	0.311
Thermal resistance of insulation	T_1 (K · m/W)	0.422
Thermal resistance of oil	T_2 (K · m/W)	0.082
Thermal resistance of jacket/pipe covering	T_3 (K · m/W)	0.017
External thermal resistance, 100% LF	T_4 (K · m/W)	0.343
External thermal resistance nonunity LF	T_4 (K · m/W)	0.289
Losses in the conductor	W_c (W/m)	19.93
Dielectric losses	W_d (W/m)	4.83

The Reynolds number is equal to

$$(Re)_{f,D_p} = w \cdot D_p \cdot \rho_f(\theta_f)/\mu_f(\theta_f) = 0.037 \cdot 0.206 \cdot 0.0443 \cdot 869.6 = 151$$

For a flow inside a circular tube, the turbulent flow is usually observed for Re > 2300. However, this critical value is strongly dependent on the surface roughness, the inlet conditions, and the fluctuations in the flow. In general, the transition may occur in the range of 2000 < Re < 4000. In our case, the flow is laminar and we will use Equation (7.14) to compute the oil film resistances:

$$T_{fs} = \frac{0.0057}{D_s^*} = \frac{0.0057}{0.145} = 0.0393 \text{ K} \cdot \text{m/W}$$

$$T_{fp} = \frac{0.0057}{D_p} = \frac{0.0057}{0.206} = 0.0276 \text{ K} \cdot \text{m/W}$$

Substituting for $T_2 = T_4' = T_{fs} + T_{fp} = 0.067$ in Equation (1.40), we obtain $I = 915$ A.

We will now use the same film resistances but apply Equation (7.15) to compute the current rating. The thermal resistance between oil and ambient earth for dielectric losses is obtained from Equation (7.2) as

$$T_{f-amb}^d = T_{fp} + T_4'' + T_4''' = 0.0276 + 0.017 + 0.343 = 0.388 \text{ K} \cdot \text{m/W}$$

Similarly, for the joule losses, we obtain $T_{f-amb} = 0.334$ K $\cdot$ m/W, and from Equation (7.16),

$$\alpha = \frac{-3}{Mc_{p,f}\rho_f T_{f-amb}} = \frac{-3}{0.00063 \cdot 1931 \cdot 869.6 \cdot 0.388} = -0.00730$$

Similarly, for the joule losses, we obtain $\alpha = -0.00848$. The temperature rise caused by dielectric losses is computed next from Equation (7.17):

$$\theta_d = W_d[\tfrac{1}{2}T_1 + 3T_{fs} + 3T_{f-amb}^d(1 - e^{\alpha Z})] = 4.83$$
$$\cdot [0.5 \cdot 0.422 + 3 \cdot 0.039 + 3 \cdot 0.388 \cdot (1 - e^{-0.0073})] = 1.99°C$$

From Equation (7.15), we have

$$I = \left[\frac{\theta_c^{max} - \theta_{amb} - (\theta_f^{in} - \theta_{amb})e^{\alpha Z} - \theta_d}{R_{ac}[T_1 + 3(1 + \lambda_1)T_{fs} + 3[(1 + \lambda_1)T_{fp} + T_4^{eff}](1 - e^{\alpha Z})]} \right]^{0.5}$$

$$= \left[\frac{70 - 25 - (50 - 25)e^{-0.073} - 1.99}{0.0000246[0.422 + 3 \cdot 1.01 \cdot 0.0393 + 3 \cdot [1.01 \cdot 0.0278 + 1.312 \cdot (0.017 + 0.289)](1 - e^{-0.0848})]} \right]$$

$$= 956 \text{ A}$$

With this current, the total losses dissipated by the cable are equal to

$$W_t = 3[I^2 R_{ac}(1 + \lambda_1 + \lambda_2) + W_d] = 3[956^2 \cdot 0.0000246 \cdot (1 + 0.01 + 0.311) + 4.83]$$
$$= 103.4 \text{ W/m}$$

out of which 20.9 W/m is generated in the pipe.

The oil temperature rise in infinity is equal to

$$\theta_f^\infty = 3I^2 R_{ac}(1 + \lambda_1)T_{f-amb} + 3W_d T_{f-amb}^d + \theta_{amb}$$
$$= 3 \cdot (955^2 \cdot 2.46 \cdot 10^{-5} \cdot 2.46 \cdot 10^{-5} \cdot 1.01 \cdot 0.334 + 4.83 \cdot 0.388) + 25 = 53.3°C$$

This temperature is lower than the initially assumed value of 60°C, hence, the parameters of the oil should be recalculated and the analysis repeated. At the end of the hot-spot section, the temperature is computed from Equation (7.11):

$$\theta_f(Z) = \theta_f^{in} + (\theta_f^\infty - \theta_f^{in}) \cdot [1 - \exp(\alpha Z)] = 50 + 53.3 - 50)$$
$$\cdot [1 - \exp(-0.00848 \cdot 10)] \cong 50.3°C$$

We can observe that, in this case, the temperature is about three degrees cooler compared to the static case, and at $Z = 10$ m the oil is warmed up only slightly. Since the main purpose of slow oil circulation is to smooth out the oil temperature differences along the cable route, one may argue that the steady-state Equation (1.40) could be used for rating calculations with the value of $T_2 = T_{fs} + T_{fp} = 0.0672$ K · m/W. As already indicated above, the current rating is equal to 915 A in this case. We can observe a substantial difference in the computed rating between these two approaches. The approach based on Equation (7.15) is most commonly used and it results in about a 6% improvement in cable rating in this case.

Finally, we can compute the effective cooling distance for this example. From Equation (7.18), we obtain

$$Z_{cooling} = M \cdot c_{p,f} \cdot \rho_f \cdot T_{f-amb} = 0.00063 \cdot 1931 \cdot 869.6 \cdot 0.334 = 354 \text{ m}$$

b) *Forced Cooling.* Even though the theory presented above pertains to slow oil circulation, for illustrative purposes we will consider a forced cooling application with oil inlet temperature equal to 10°C and oil velocity of about 1 fps (foot per second). In metric units, this gives $w = 0.3048$ m/s and $M = 0.00517$ m^3/s.

The bulk oil temperature will now be reduced, but for the first iteration we will assume the same values of viscosity and density as in the case of slow circulation. The Reynolds number is equal to

$$(Re)_{f,D_p} = w \cdot D_p \cdot \rho_f(\theta_f)/\mu_f(\theta_f) = 0.3048 \cdot 0.206 \cdot 869.6/0.0443 = 1235$$

The flow is still laminar, but this time we will use Equations (7.12) and (7.13) for illustration purposes. Note that since the oil temperature is now expected to be lower, the viscosity of the oil will increase, resulting in a decrease of the Reynolds number. The oil density will increase slightly. The film resistances are equal to

$$
\begin{aligned}
T_{fp} &= \frac{9.512 \cdot 10^{-3} \cdot (\mu_f)^{0.4}}{(D_s^* \cdot w)^{0.8}} \cdot \left[1 - \left(\frac{D_s^*}{D_p}\right)^2\right]^{0.2} \\
&= \frac{9.512 \cdot 10^{-3} \cdot (0.0443)^{0.4}}{(0.145 \cdot 0.3048)^{0.8}} \cdot \left[1 - \left(\frac{0.145}{0.206}\right)^2\right]^{0.2} = 0.0331 \text{ K} \cdot \text{m/W}
\end{aligned}
$$

Similarly, $T_{fs} = 0.0288$ K $\cdot$ m/W is obtained from Equation (7.13). We can observe that these values are very similar to those obtained with the approximate equations. The oil ambient thermal resistance for dielectric losses becomes

$$
T_{f\text{-}amb} = T_{fp} + T_4'' + T_4''' = 0.0331 + 0.017 + 0.343 = 0.393 \text{ K} \cdot \text{m/W}
$$

and similarly, for joule losses $T_{f\text{-}amb} = 0.339$ K $\cdot$ m/W. The value of exponent α is computed for dielectric and joule losses from Equation (7.16). The corresponding values are equal to $-9.0 \cdot 10^{-4}$ and $-1.0 \cdot 10^{-3}$, respectively. The temperature rise caused by dielectric losses is computed next from Equation (7.17):

$$
\begin{aligned}
\theta_d &= W_d[\tfrac{1}{2}T_1 + 3T_{fs} + 3T_{f\text{-}amb}(1 - e^{\alpha Z})] \\
&= 4.83 \cdot [0.5 \cdot 0.422 + 3 \cdot 0.0288 + 3 \cdot 0.393 \cdot (1 - e^{-0.0009 \cdot 10})] = 1.55°C
\end{aligned}
$$

From Equation (7.15), we have

$$
\begin{aligned}
I &= \left[\frac{\theta_c^{\max} - \theta_{amb} - (\theta_f^{in} - \theta_{amb})e^{\alpha Z} - \theta_d}{R_{ac}[T_1 + 3(1+\lambda_1)T_{fs} + 3[(1+\lambda_1)T_{fp} + T_4^{eff}](1 - e^{\alpha Z})]}\right]^{0.5} \\
&= \left[\frac{70 - 25 - (10 - 25)e^{-0.001} - 1.55}{0.0000246[0.422 + 3 \cdot 1.01 \cdot 0.0288 + 3 \cdot [1.01 \cdot 0.0331 + 1.312 \cdot (0.017 + 0.289)](1 - e^{-0.01})]}\right] \\
&= 2131 \text{ A}
\end{aligned}
$$

Thus, forced circulation doubles the rating of the circuit in comparison with the case in which the oil is slowly circulated.

If we used Equations (7.14) for film resistance, the ampacity of the cable would have been 2010 A, hence, not much different from the one obtained above.

The effective cooling distance is, in this case, equal to

$$
Z_{cooling} = M \cdot c_{p,f} \cdot \rho_f \cdot T_{f\text{-}amb} = 0.0051 \cdot 1931 \cdot 869.6 \cdot 0.339 = 2919 \text{ m} \quad \blacksquare
$$

7.3.3 Model for Real-Time Rating Computations

The model presented here was developed for applications in real-time cable monitoring installations. In such systems, the temperatures of the pipe surface θ_p and ambient soil temperatures are measured and constitute the input parameters in the computations. Therefore, the oil outlet temperature and oil film thermal resistances are expressed as a function of these parameters.[3] The developments will lead to Equation (7.26), representing the temperature distribution along the cable route, and Equation (7.32) to compute the oil film thermal resistances.

The oil flow and temperature changes inside the pipe can be described with the Fourier–Kirchoff, Navier–Stokes, and fluid continuity equations:

$$\frac{\partial \theta}{\partial t} + (\mathbf{w}\nabla)\theta = \frac{q_v}{c_{p,f}\rho_f} + \frac{1}{c_{p,f}\rho_f}\left\{ \nabla[\lambda_f\nabla\theta] + \frac{\partial p}{\partial x} + (\mathbf{w}\nabla)p \right\} \tag{7.19}$$

$$\frac{\partial \mathbf{w}}{\partial t} + (\mathbf{w}\nabla)\mathbf{w} = K - \frac{\nabla p}{\rho} + \mu_f\nabla^2\mathbf{w} + \frac{\mu_f}{3}\nabla(\mathbf{w}\nabla) \tag{7.20}$$

$$\frac{d\rho_f}{dt} + \rho_f\nabla\mathbf{w} = 0 \tag{7.21}$$

where the temperature and oil velocity are functions of space and time coordinates (x, y, z, t) and K is an external force operating on individual oil unit (e.g., gravitational force).

In the approach presented below, the following assumptions have been made:

- Stationary state
- Absence of an internal heat source $(q_v = 0)$
- Low oil speed (the nonlinear terms can be omitted)
- Incompressible fluid $\Delta\rho = 0$
- Laminar flow $[(\text{Re})_{f,d_h} < 1800]$, dominant forced convection over free convection, and, consequently, oil flow parallel to the pipe axis
- Cylindrical symmetry is assumed; that is, the cables are located in the center of the pipe
- The radiation heat transfer can be neglected in comparison with other heat transfer phenomena

7.3.3.1 Oil Velocity Profile.
The oil velocity profile can be Evaluated from Equations (7.20) and (7.21), assuming, additionally, invariable pressure drop per unit pipe length:

$$\frac{1}{r}\frac{\partial}{\partial r}\left(r\frac{\partial w(r)}{\partial r} \right) = \frac{1}{\mu_f}\frac{dp}{dx} \qquad \text{for } r \in (D_s^*/2, D_p/2); x > 0$$

$$w(r)|_{r=D_e^*/2} = 0; \qquad w(r)|_{r=D_p/2} = 0 \tag{7.22}$$

---

[3]The temperature of the pipe coating is measured, but in the analysis presented here the pipe temperature, also denoted by θ_p, is used. In the computer implementation, the pipe temperature is computed properly and is somewhat higher than the temperature of the coating.

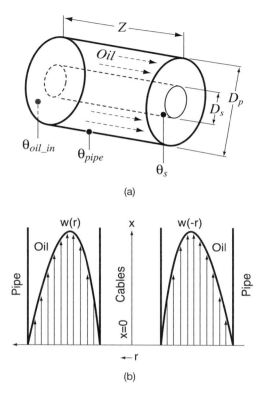

Figure 7-5 (a) Temperature profile in the pipe. (b) Oil velocity profile.

The physical interpretation of this equation is shown in Figure 7-5. The unknown part $(1/\mu_f) \cdot (dp/dx)$ in Equation (7.22) can be eliminated using the average mass velocity of the oil:

$$\overline{w} = \frac{1}{\|S\|} \iint_S w(r)dS = \frac{8}{D_p^2 - D_s^{*2}} \int_{D_s^*/2}^{D_p/2} w(r) \cdot r \, dr \qquad (7.23)$$

where S and $\|S\|$ are the oil filled cross section and its area, respectively, and $\overline{w} \equiv M/\{\pi \cdot [D_p^2 - (D_s^*)^2]/4\}$.

7.3.3.2 *Oil Temperature Distribution.*

The oil temperature distribution can be computed from Equation (7.19), which is quite complicated. Calculations can be considerably simplified by assuming constant pipe temperature[4] θ_p and given fluid velocity profile. As mentioned earlier, in real-time applications the boundary condi-

---

[4]The pipe temperature will change along the cable route, but for the relatively short distances for which the cooling analysis is performed, this assumption holds true.

tions are those imposed by the measured quantities; that is, the oil inlet and pipe temperatures. Therefore, we have

$$\frac{1}{r}\frac{\partial}{\partial r}\left(r\frac{\partial \theta_f}{\partial r}\right) + \frac{\partial^2 \theta_f}{\partial x^2} = \rho_f c_{p,f}\frac{w}{\lambda_f}\frac{\partial \theta_f}{\partial x} \qquad \text{for } r \times x \in (D_s^*/2, D_p/2) \times [0, +\infty)$$

(7.24)

$$\theta_f|_{x=0} = \theta_f^{in}; \qquad \theta_f|_{r=D_p} = \theta_p; \qquad -\lambda\frac{d\theta_f}{dr}\bigg|_{r=D_S^*} = \frac{W_c \cdot (1 + \lambda_1) + W_d}{\pi \cdot D_s^*}$$

where $W_o = n(W_c(1 + \lambda_1) + W_d)$ is the total power dissipated by the cables seen by oil.

The solution of Equation (7.24), obtained with the help of a numerical program (Zubert et al., 2000), is given by

$$\theta_f^\infty = \theta_p + 0.1535 \cdot \frac{\beta \cdot W_o}{2\pi \cdot \lambda_f(\theta_f)} \qquad (7.25)$$

$$\theta_f(Z) = \theta_f^{in} + (\theta_f^\infty - \theta_f^{in}) \cdot [1 - \exp(\alpha Z)] \qquad (7.26)$$

where

$$\alpha = \phi \cdot \frac{2 \cdot a_f(\theta_f)}{w \cdot D_p \cdot d_h} \qquad (7.27)$$

and denoting by $\phi_o = \dfrac{W_o}{2\pi \cdot \lambda_f(\theta_f) \cdot (\theta_f^{in} - \theta_p)}$, parameter ϕ can be approximated by

$$\phi = \begin{cases} -31.76 \cdot \exp(0.463 \cdot \phi_o) - 60.2 & \text{for } \phi_o \le 0 \\ 2.56 \cdot \phi_o - 88.7 & \text{for } 0 \le \phi_o \le 20 \\ 26.82 \cdot \exp(-0.140 \cdot (\phi_o - 20)) - 58.2 & \text{for } 20 \le \phi_o \end{cases} \qquad (7.28)$$

The above function was obtained numerically with the help of the program RES-CUER (Zubert et al., 2000). θ_f^∞ is a hypothetical temperature at $Z = \infty$, assuming that the hot-spot section has an infinite length. We can observe that Equation (7.26) is identical to Equation (7.11) with the exponent α taking the form given by Equation (7.27). However, there is a difference in the interpretation of θ_f^∞. In Buller's development, this is the oil temperature in the case in which there is no circulation. In Equation (7.25), this is the temperature at $Z = \infty$; that is, at the place where there is no longitudinal temperature gradient but under an assumption that oil is moving.

The rating Equation (7.15) is used again with the revised thermal resistances and new coefficient α.

7.3.3.3 *Thermal Resistances of the Oil Film.* The proposed thermal model will be similar to the one discussed in Section 7.3.1. The major difference is in calculation of the oil-film thermal resistances as described below.

A general expression of the thermal resistance of the oil in a pipe can be written as (Anders, 1997)

$$T_2 = \frac{1}{\pi D h_t} \tag{7.29}$$

where h_t is a heat transfer coefficient with the diameter D taking the value of D_s^* or D_p, depending on which thermal resistance we are computing. The heat transfer coefficient can be computed using the definition of a Nusselt number:

$$\text{Nu}_{f,d_h} = \frac{h_t d_h}{\lambda_f(\theta_f)} \tag{7.30}$$

Thus, from Equations (7.29) and (7.30), we obtain

$$T_{fs} = \frac{d_h}{\pi \cdot D_s^* \cdot \lambda_f(\theta_f) \cdot (\text{Nu})_{s,d_h}} \tag{7.31}$$

$$T_{fp} = \frac{d_h}{\pi \cdot D_p \cdot \lambda_f(\theta_f) \cdot (\text{Nu})_{p,d_h}} \tag{7.32}$$

Performing analysis using the RESCUER software (Zubert et al., 2000), the following expressions were obtained for the Nusselt numbers:

$$(Nu)_{s,D_p}^+ = \begin{cases} 123.9 & \text{for } Z \le 19.3 \cdot 10^{-8} \cdot \alpha_0^{-1} \\ 4.22 \cdot (\alpha_0 Z)^{-1/3} & \text{for } 19.3 \cdot 10^{-8} \cdot \alpha_0^{-1} \\ & \quad \le Z \le 0.001 \cdot \alpha_0^{-1} \\ 14.8 + \dfrac{65.074}{(\alpha_0 Z)^{0.03986} \exp(116.87 \cdot \alpha_0 Z)} & \text{for } 0.001 \cdot \alpha_0^{-1} \le Z \end{cases} \tag{7.33}$$

$$(Nu)_{p,D_p}^+ = \begin{cases} 48.9 & \text{for } Z \le 18.9 \cdot 10^{-8} \cdot \alpha_0^{-1} \\ 1.57 \cdot (\alpha_0 Z)^{-1/3} & \text{for } 18.9 \cdot 10^{-8} \cdot \alpha_0^{-1} \\ & \quad \le Z \le 0.001 \cdot \alpha_0^{-1} \\ 15.07 + \dfrac{0.9671}{(\alpha_0 Z)^{0.3049} \exp(402.27 \cdot \alpha_0 Z)} & \text{for } 0.001 \cdot \alpha_0^{-1} \le Z \end{cases} \tag{7.34}$$

where $(Nu)_{s,D_p}^+$ and $(Nu)_{p,D_p}^+$ are the Nusselt numbers for the cable and pipe surface, respectively, derived analytically with the characteristic diameter D_p, and

$$\alpha_0 = \frac{2}{(\text{Pe})_{f,D_p} \cdot D_p}$$

The average Peclet number is computed from

$$(\text{Pe})_{f,d_h} = (\text{Pr})_f \cdot (\text{Re})_{f,d_h} \tag{7.35}$$

and the average Reynolds number for the equivalent hydraulic diameter d_h is given by

$$(Re)_{f,d_h} = \frac{w \cdot d_h \cdot \rho_f}{\mu_f} \qquad (7.36)$$

The Nusselt numbers have been calculated for D_p characteristic dimension and the assumed value of $\beta = (D_s^*/D_p) = 0.7$. This can be extrapolated for others cases as follows (Pietruchow and Krasnosczekow, 1959):

$$(Nu)_{p,d_h} = (Nu)_{p,D_p}^+ \cdot \frac{d_h}{D_p} \cdot (\mu_f/\mu_p)^{1/8} \qquad (7.37)$$

$$(Nu)_{s,d_h} = (Nu)_{s,D_p}^+ \cdot \frac{d_h}{D_p} \cdot (\mu_f/\mu_s)^{1/8} \qquad (7.38)$$

where d_h denotes the hydraulic characteristic diameter $d_h = D_p - D_s^*$.

Temperature rise of the conductors is computed using a lump-sum representation of the oil thermal resistance as shown in Figure 7-6.

In order to evaluate various parameters of the oil, we need to first estimate its temperature. Based on numerical approximations with $\beta = 0.7$, the following expression was obtained:

$$\theta_f(x) \cong 0.451 \cdot \theta_s(x) + 0.549 \cdot \theta_p(x) \qquad (7.39)$$

Example 7.2

We will continue our analysis of the system examined in Example 7.1, but this time the thermal model described in this section will be used. We will show one iteration of the calculations for the case of slow oil circulation.

The initial conditions are assumed to be the same as in Example 7.1; however, we will need, additionally, the temperatures at the cable and pipe surfaces. Let us assume initially that these temperatures are the same as in the case with no circulation, as computed in Example 7.1 with 60°C bulk oil temperature. Thus, using Equation (7.39), we obtain

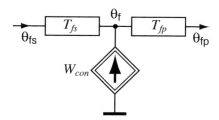

Figure 7-6   Equivalent model of heat transfer between cable, oil, and pipe.

$$\theta_f \cong 0.451 \cdot \theta_s + 0.549 \cdot \theta_p = 0.451 \cdot 60.9 + 0.549 \cdot 5.55 = 57.9°C$$

The properties of oil at this temperature are as follows (see Appendix D):

$$\mu_f = \mu_f(\theta_f) = 164.25 \cdot 57.9^{-2.0072} \cong 0.0475 \text{ N} \cdot \text{s/m}^2$$

$$\mu_s = \mu_f(\theta_s) = 164.25 \cdot 60.9^{-2.0072} \cong 0.0430 \text{ N} \cdot \text{s/m}^2$$

$$\mu_p = \mu_f(\theta_p) = 164.25 \cdot 55.5^{-2.0072} \cong 0.0518 \text{ N} \cdot \text{s/m}^2$$

$$\rho_f = \rho_f(\theta_f) = 903.6 - 0.566 \cdot 57.9 \cong 870.8 \text{ kg/m}^3$$

$$v_f(\theta_f) = \mu_f(\theta_f)/\rho_f(\theta_f) \cong 0.475/870.8 = 0.0000546 \text{ m}^2/\text{s}$$

$$a_f(\theta_f) = -1.696 \cdot 10^{-10} \cdot 57.9 + 1.67 \cdot 10^{-7} \cong 1.068 \cdot 10^{-7} \text{ m}^2/\text{s}$$

$$(\text{Pr})_f = v_f(\theta_f)/a_f(\theta_f) \cong 0.0000546/(1.068 \cdot 10^{-7}) \cong 510.9$$

$$\lambda_f(\theta_f) = -1.5855 \cdot 10^{-4} \cdot \theta_f + 0.2095 = -1.5855 \cdot 10^{-4} \cdot 57.9 + 0.2095$$
$$\cong 0.2003 \text{ W/(K} \cdot \text{m)}$$

We can now compute the oil temperatures using Equations (7.25) and (7.26):

$$\theta_f^\infty = \theta_p + 0.1535 \cdot \frac{\beta \cdot W_o}{2\pi \cdot \lambda_f(\theta_f)} = 55.5 + 0.1535 \cdot \frac{0.7 \cdot 82.5}{2\pi \cdot 0.2003} = 57.0°C$$

Since $\theta_f^{in} < \theta_p$, the first condition from Equation (7.28) applies and

$$\phi = -31.76 \cdot \exp\left(\frac{0.463 \cdot W_o}{2\pi \cdot \lambda_f(\theta_f) \cdot (\theta_f^{in} - \theta_p)}\right) - 60.20$$

$$= -31.76 \cdot \exp\left(\frac{0.463 \cdot 82.5}{2\pi \cdot 0.129 \cdot (50 - 55.5)}\right) - 60.2 \cong -60.3$$

$$\alpha = \phi \cdot \frac{2 \cdot a_f(\theta_f)}{w \cdot D_p \cdot d_h} = -60.3 \cdot \frac{2 \cdot 1.068 \cdot 10^{-7}}{0.0375 \cdot 0.2064 \cdot 0.0608} = -0.0274$$

$$\theta_f(Z) = \theta_f^{in} + (\theta_f^\infty - \theta_f^{in}) \cdot [1 - \exp(\alpha Z)]$$
$$\cong 50 + (57 - 50) \cdot [1 - \exp(-0.0274 \cdot 10)] = 51.7°C$$

Hence, the temperature at $Z = 10$ m is about two degrees warmer than at the oil inlet. This implies that, in the proposed approach, the cooling distance is shorter than the one computed by Buller's model, where, as we can recall from Example 7.1, the temperature at this distance was only 50.3°C.

In order to evaluate the oil film resistances, we will need the Nusselt numbers. We will start by computing the Reynolds and Peclet numbers from Equations (7.36) and (7.35), respectively:

$$(\text{Re})_{f,d_h} = \frac{w \cdot d_h \cdot \rho_f}{\mu_f} = \frac{0.0375 \cdot (0.2064 - 0.1456) \cdot 870.8}{0.0475} = 41.7$$

$$(\text{Pe})_{f,d_h} = (\text{Pr})_f \cdot (\text{Re})_{f,d_h} = 511.4 \cdot 41.7 = 21341$$

Since $0.001 \cdot (\text{Pe})_{f,d_h} \cdot D_p/2 = 2.20 < Z = 10$, from Equations (7.33), (7.34), (7.37), and (7.38), we have

$$(Nu)_{s,d_h} = \left(14.8 + \frac{65.074}{\left(\dfrac{2Z}{(\text{Pe})_{f,d_h} \cdot D_p}\right)^{0.03986} \exp\left(116.87 \cdot \dfrac{2Z}{(\text{Pe})_{f,d_h} \cdot D_p}\right)}\right) \cdot \frac{d_h}{D_p} \cdot \left(\frac{\mu_f}{\mu_s}\right)^{1/8}$$

$$= \left(14.8 + \frac{65.074}{\left(\dfrac{2 \cdot 10}{21341 \cdot 0.2064}\right)^{0.03986} \exp\left(116.87 \cdot \dfrac{2 \cdot 10}{21341 \cdot 0.2064}\right)}\right)$$

$$\cdot \frac{0.2064 - 0.1456}{0.2064} \cdot \left(\frac{0.0475}{0.0430}\right)^{1/8} \cong 18.6$$

$$(Nu)_{p,d_h} = \left(15.07 + \frac{0.9671}{\left(\dfrac{2Z}{(\text{Pe})_{f,D_p} \cdot D_p}\right)^{0.3049} \exp\left(402.27 \cdot \dfrac{2Z}{(\text{Pe})_{f,D_p} \cdot D_p}\right)}\right) \cdot \frac{d_h}{D_p} \cdot \left(\frac{\mu_f}{\mu_p}\right)^{1/8}$$

$$= \left(15.07 + \frac{0.9671}{\left(\dfrac{2 \cdot 10}{21341 \cdot 0.2064}\right)^{0.3049} \exp\left(402.27 \cdot \dfrac{2 \cdot 10}{21341 \cdot 0.2064}\right)}\right)$$

$$\cdot \frac{0.2064 - 0.1456}{0.2064} \cdot \left(\frac{0.0475}{0.0518}\right)^{1/8} \cong 4.63$$

The oil film thermal resistances are obtained from Equations (7.31) and (7.32):

$$T_{fs} = \frac{d_h}{\pi \cdot D_s^* \cdot \lambda_f(\theta_f) \cdot (Nu)_{s,d_h}} = \frac{0.2064 - 0.1456}{\pi \cdot 0.1456 \cdot 0.2003 \cdot 18.6} = 0.0357 \text{ K} \cdot \text{m/W}$$

$$T_{fp} = \frac{d_h}{\pi \cdot D_p \cdot \lambda_f(\theta_f) \cdot (Nu)_{p,d_h}} = \frac{0.2064 - 0.1456}{\pi \cdot 0.2064 \cdot 0.2003 \cdot 4.63} = 0.101 \text{ K} \cdot \text{m/W}$$

The thermal resistance between oil and ambient earth is equal to

$$T_{f-amb} = T_{fp} + T_4'' + T_4''' = 0.101 + 0.017 + 0.343 = 0.461 \text{ K} \cdot \text{m/W}$$

Similarly, for joule losses $T_{f-amb} = 0.407 \text{ K} \cdot \text{m/W}$.
Substituting these resistance in Equations (7.17) and (7.15), we obtain

$$\theta_d = W_d[\tfrac{1}{2}T_1 + 3T_{fs} + 3T_{f-amb}(1 - e^{\alpha Z})]$$
$$= 4.83 \cdot [0.5 \cdot 0.422 + 3 \cdot 0.0357 + 3 \cdot 0.461 \cdot (1 - e^{-0.274})] = 3.14°C$$

$$I = \left[\frac{\theta_c^{max} - \theta_{amb} - (\theta_f^{in} - \theta_{amb})e^{\alpha Z} - \theta_d}{R_{ac}[T_1 + 3(1 + \lambda_1)T_{fs} + 3[(1 + \lambda_1)T_{fp} + T_4^{eff}](1 - e^{\alpha Z})]} \right]^{0.5}$$

$$= \left[\frac{70 - 25 - (50 - 25)e^{-0.274} - 3.14}{0.0000246[0.422 + 3 \cdot 1.01 \cdot 0.0357 + 3 \cdot [1.01 \cdot 0.101 + 1.312 \cdot (0.017 + 0.289)](1 - e^{-0.274})]} \right]$$

$$= 1019 \text{ A}$$

Thus, the proposed method for real-time rating calculations gives, in this case, only slightly higher ampacity than Buller's model.

We will conclude this example by calculating the effective cooling distance from Equation (7.18):

$$Z_{cooling} = \frac{-3}{\alpha} = \frac{-3}{-0.0274} = 109 \text{ m}$$

The cooling distance is thus about 30% of the value computed with Buller's model. ∎

7.4 CONCLUDING REMARKS

This chapter presented detailed mathematical models for analysis of the temperature distribution along a pipe-type cable cooled by means of fluid circulation. The work in this area started over 50 years ago. A new rigorous model presented here was compared with some of the early developments. The major limitation of Buller's model is related to the calculation method of the thermal resistances of the oil film. These values clearly depend on the distance of the calculation point from the oil inlet point. In Buller's model, this relationship is neglected. However, remarkably close results were obtained with both models, suggesting that the original model by Buller (1952) can still safely be used.

The new model presented in this chapter is somewhat more complex than Buller's approach, and the main difference comes from the calculation of the oil-film resistances, with the proposed model using rigorously developed Nusselt numbers. These numbers depend not only on the oil velocity, but also on the distance from the oil inlet point.

7.5 CHAPTER SUMMARY

In the case of slow oil circulation, the cable rating is computed from

$$I = \left[\frac{\theta_c^{max} - \theta_{amb} - (\theta_f^{in} - \theta_{amb})e^{\alpha Z} - \theta_d}{R_{ac}[T_i + 3(1 + \lambda_1)T_{fs} + 3[(1 + \lambda_1)T_{fp} + T_{4\mu}^{eff}](1 - e^{\alpha Z})]} \right]^{0.5}$$

where

$$T_{f-amb} = T_{fp} + T_4'' + T_4'''$$

In Buller's model,

$$\alpha = \frac{-3}{M c_{p,f} \rho_f T_{f-amb}}$$

and in the model proposed in this chapter,

$$\alpha = \phi \cdot \frac{2 \cdot a_f(\theta_f)}{w \cdot D_p \cdot d_h}$$

with

$$\phi = \begin{cases} -31.76 \cdot \exp(0.463 \cdot \phi_o) - 60.2 & \text{for } \phi_o \leq 0 \\ 2.56 \cdot \phi_o - 88.7 & \text{for } 0 \leq \phi_o \leq 20 \\ 26.82 \cdot \exp(-0.140 \cdot (\phi_o - 20)) - 58.2 & \text{for } 20 \leq \phi_o \end{cases}$$

and

$$\phi_o = \frac{W_o}{2\pi \cdot \lambda_f(\theta_f) \cdot (\theta_f^{in} - \theta_p)}$$

The temperature rise caused by dielectric losses is obtained as

$$\theta_d = W_d [\tfrac{1}{2} T_1 + 3 T_{fs} + 3(T_{fp} + T)(1 - e^{\alpha Z})]$$

The variables are defined at the beginning of this chapter.

The cable-oil and oil-pipe film resistance can be computed in several different ways. Buller proposed the following expressions:

$$T_{fp} = \frac{9.512 \cdot 10^{-3} \cdot (\mu_f)^{0.4}}{(D_s^* \cdot w)^{0.8}} \cdot \left[1 - \left(\frac{D_s^*}{D_p} \right)^2 \right]^{0.2}$$

$$T_{fs} = \frac{9.512 \cdot 10^{-3} \cdot (\mu_f)^{0.4}}{(D_s^* \cdot w)^{0.8}} \cdot \left[\left(\frac{D_p}{D_s^*} \right)^2 - 1 \right]^{0.2}$$

For a laminar flow, Buller argued that there is little to be gained by using the relatively complex expressions available, and it is better to neglect flow altogether and use the technique based on natural convection and conduction developed by Buller and Neher (1950). For approximate work, the following simple equations can be used:

$$T_{fs} = \frac{0.0057}{D_s^*} \qquad T_{ps} = \frac{0.0057}{D_p}$$

In the model developed for real-time applications, the following expressions are proposed for the calculation of the film resistances:

$$T_{fs} = \frac{d_h}{\pi \cdot D_s^* \cdot \lambda_f(\theta_f) \cdot (Nu)_{s,d_h}}$$

$$T_{fp} = \frac{d_h}{\pi \cdot D_p \cdot \lambda_f(\theta_f) \cdot (Nu)_{p,d_h}}$$

with the Nusselt numbers defined by

$$(Nu)_{s,D_p}^+ = \begin{cases} 123.9 & \text{for } Z \le 19.3 \cdot 10^{-8} \cdot \alpha_0^{-1} \\ 4.22 \cdot (\alpha_0 Z)^{-1/3} & \text{for } 19.3 \cdot 10^{-8} \cdot \alpha_0^{-1} \\ & \quad \le Z \le 0.001 \cdot \alpha_0^{-1} \\ 14.8 + \dfrac{65.074}{(\alpha_0 Z)^{0.03986} \exp(116.87 \cdot \alpha_0 Z)} & \text{for } 0.001 \cdot \alpha_o^{-1} \le Z \end{cases}$$

$$(Nu)_{p,D_p}^+ = \begin{cases} 48.9 & \text{for } Z \le 18.9 \cdot 10^{-8} \cdot \alpha_0^{-1} \\ 1.57 \cdot (\alpha_0 Z)^{-1/3} & \text{for } 18.9 \cdot 10^{-8} \cdot \alpha_0^{-1} \\ & \quad \le Z \le 0.001 \cdot \alpha_0^{-1} \\ 15.07 + \dfrac{0.9671}{(\alpha_0 Z)^{0.3049} \exp(402.27 \cdot \alpha_0 Z)} & \text{for } 0.001 \cdot \alpha_o^{-1} \le Z \end{cases}$$

where $(Nu)_{s,D_p}^+$, $(Nu)_{p,D_p}^+$ are the Nusselt numbers for the cable and pipe surface, respectively, derived analytically with the characteristic diameter D_p and

$$\alpha_0 = \frac{2}{(Pe)_{f,D_p} \cdot D_p}$$

The average Peclet number is computed from

$$(Pe)_{f,d_h} = (Pr)_f \cdot (Re)_{f,d_h}$$

and the average Reynolds number for the equivalent hydraulic diameter d_h is given by

$$(Re)_{f,d_h} = \frac{w \cdot d_h \cdot \rho_f}{\mu_f}$$

The Nusselt numbers have been calculated for D_p characteristic dimension and the assumed value of $\beta = (D_s^*/D_p) = 0.7$. This can be extrapolated for others cases as follows:

$$(Nu)_{p,d_h} = (Nu)_{p,D_p}^+ \cdot \frac{d_h}{D_p} \cdot (\mu_f/\mu_p)^{1/8}$$

$$(Nu)_{s,d_h} = (Nu)_{s,D_p}^+ \cdot \frac{d_h}{D_p} \cdot (\mu_f/\mu_s)^{1/8}$$

where d_h denotes the hydraulic characteristic diameter $d_h = D_p - D_s^*$.

REFERENCES

Buller, F.H, .Neher, J.H., (1950) "The Thermal Resistance Between Oil and A Surrounding Pipe Duct Wall", *AIEE Transaction 69,* Part I, pp. 342–349.

Buller, F.H., Falcone, A.J., and Roberts, W.J. (1951), "Forced-air Cooling of Station Cables," *AIEE Transactions on Power Apparatus and Systems,* Vol. 70, pp. 798–803.

Buller, F.H. (1952), "Artificial Cooling of Power Cables," *AIEE Transactions on Power Apparatus and Systems,* Vol. 84, No. 9, pp. 795–806.

Burrell, R.W. (1965), "Application of Oil Cooling in High-Pressure Oil-Filled Pipe-Cable Circuits," *AIEE Transactions on Power Apparatus and Systems,* Vol. 71, pp. 634–631.

Chato, J. (1977), "Free and Forced Convective Cooling of Pipe-Type Electric Cable Systems," *EPRI EL-147, Project 7821, Final Report,* August 1977.

Chato, J., Crowley, J.M., et al. (1980), "Free and Forced Convective Cooling of Pipe-Type Electric Cables," *EPRI EL-1872, Vol. 1, Project 7583-1.*

CIGRE, (1979), "The Calculation of Continuous Ratings for Forced-Cooled Cable," Report of Study Committee 21, WG 21-08, *Electra,* No. 66, pp. 59–84.

CIGRE, (1985), "The Calculation of Continuous Ratings for Forced-Cooled High-Pressure Oil-Filled Pipe-Type Cables," Report of Study Committee 21, WG 21-08, *Electra,* No. 113, pp. 97–121.

Giaro, J.A. (1956), "The Efficiency of Forced Cooling of Power Cables Compared with Systems Laying with Natural Cooling," CIGRE Report No. 207, Paris, France.

Glicksman, L.R., Sanders, J.V., Robsenow, W.M., and Buckweitz, M.D. (1978), "Heat Conduction in the Cable Insulation of Forced-Cooled Underground Transmission Lines," *IEEE Trans. on Power Apparatus and Systems,* Vol. PAS-97, No. 1, pp. 134–139.

Kitogawa, K. (1964), "Forced Cooling of Power Cables in Japan. Its Studies and Performance," CIGRE Report 213, Paris, France.

McAdams, W.H.M. (1942), *Heat Transmission,* 2nd ed., McGraw-Hill, New York.

Oudin, J.M., et al. (1960), "French Research and Development in the Field of High and Extra High Voltage Cables," CIGRE Report No. 230, Paris, France.

Pietruchow B.S., Krasnosczekow E.A. (1959), "Tieplootdacza priwiazkostnom tieczenji zidkosti w trubach w uslowjach suszczestwiennogo izmienienija wiszkosti," *Tieploperedacza i tieplowoje modelirowanje* [in Russian], published by. ANSSSR, 1959, pp. 165.

Purnhagen, D.W. (1984), "Designer's Handbook for Forced-Cooled High-Pressure Oil-Filled Pipe-Type Cable Systems," *EPRI EL-3624, Project 7801-5,* Final Report, July 1984.

Ralston, P., and West, O.H. (1960), "The Artificial Cooling by Water of Underground Cables," CIGRE Report No. 215, Paris, France.

Zubert, M., Napieralska, M., and Napieralski, A. (2000), "RESCUER—The New Solution in Multidomain Simulations," *Microelectronics Journal, 31*(11–12) pp. 945–954.

Model Cables

Six model cables are described in this appendix. The cables are used throughout the book to illustrate various concepts as they are developed. Design and computed parameters for the model cables are summarized in Table A1.

The installation conditions of the model cables are also described below. However, the installation information may vary in the examples to show the sensitivity of cable rating to variations in various laying parameters.

MODEL CABLE NO. 1

This is a 10 kV single-conductor XLPE cable. Conductor resistance at 20°C is taken from IEC 60228 (1978). The cable has copper screen wires with a given initial electrical resistance (at 20°C) and a PVC jacket. All thermal and electrical parameters are as specified in IEC 60287 (1994), and these values are given in the various tables in (Anders, 1997). The cable cross section is shown in Figure A1.

The laying conditions are assumed as follows. Cables are located 1 m below the ground in a flat configuration. Uniform soil properties are assumed throughout. Spacing between cables is equal to one cable diameter (spacing between centers is equal to two cable diameters). Ambient soil temperature is 15°C. The thermal resistivity of the soil is equal to 1.0 K·m/W. The cables are solidly bonded and not transposed.

MODEL CABLE NO. 2

This is a 145 kV, three-core, XLPE cable. The cable has a lead sheath, steel wire armor, jute bedding, and a PVC serving with the thermal resistivity of 4 K · m/W. The dielectric losses are very small but included in the analysis. The cable cross section is shown in Figure A2.

The installation conditions are assumed as follows: the cable is located underground in the soil with thermal resistivity 0.6 K · m/W: the ambient air temperature is 25°C.

Rating of Electric Power Cables in Unfavorable Thermal Environment. By George J. Anders **303**
ISBN 0-471-67909-7 © 2004 the Institute of Electrical and Electronics Engineers.

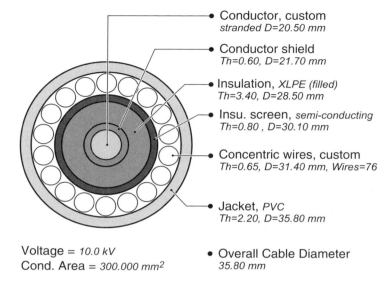

Conductor, custom
stranded D=20.50 mm

Conductor shield
Th=0.60, D=21.70 mm

Insulation, *XLPE (filled)*
Th=3.40, D=28.50 mm

Insu. screen, *semi-conducting*
Th=0.80 , D=30.10 mm

Concentric wires, custom
Th=0.65, D=31.40 mm, Wires=76

Jacket, *PVC*
Th=2.20, D=35.80 mm

Voltage = *10.0 kV*
Cond. Area = *300.000 mm²*

Overall Cable Diameter
35.80 mm

Figure A1 Cross section of the model cable No. 1.

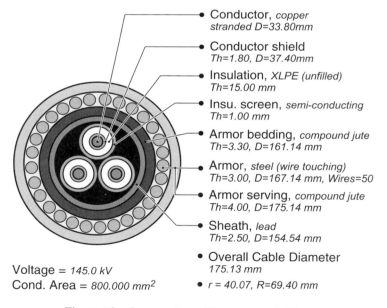

Conductor, *copper*
stranded D=33.80mm

Conductor shield
Th=1.80, D=37.40mm

Insulation, *XLPE (unfilled)*
Th=15.00 mm

Insu. screen, *semi-conducting*
Th=1.00 mm

Armor bedding, *compound jute*
Th=3.30, D=161.14 mm

Armor, *steel (wire touching)*
Th=3.00, D=167.14 mm, Wires=50

Armor serving, *compound jute*
Th=4.00, D=175.14 mm

Sheath, *lead*
Th=2.50, D=154.54 mm

Overall Cable Diameter
175.13 mm

Voltage = *145.0 kV*
Cond. Area = *800.000 mm²*

r = *40.07, R=69.40 mm*

Figure A2 Cross section of the model cable No. 2.

MODEL CABLE NO. 3

This is a 138 kV, high-pressure, liquid-filled (HPLF) cable. All parameters are the same as in the Neher and McGrath (1957) paper (see Table A1). The cable shield consists of an intercalated 7/8 (0.003) in bronze tape, one 1 in a lay; and a single 0.1 (0.2) in, D-shaped, bronze skid wire, 1.5 in a lay. The cables lie in a cradle configuration and operate at 85% load-loss factor. Several parameters are different from those used in IEC 60287 (1994). The ones that are different are given as follows:

Thermal resistivity of insulation	5.5 K ·m/W
thermal resistivity of pipe coating	1.0 K · m/W
dielectric constant	3.5
tan δ	0.005

The cable cross section is shown in Figure A3.

The laying conditions are assumed as follows. The cables are located in a steel pipe of 8.625 in outside diameter. The pipe is covered with an asphalt mastic covering 0.5 in thick. The center of the pipe is located 3 ft below the ground. The soil is uniform throughout. The ambient soil temperature is 25°C. The thermal resistivity of the soil is equal to 0.8 K · m/W.

MODEL CABLE NO. 4

This is a 230 kV, low-pressure, oil-filled cable with 1250 kcmil (633 mm^2) copper, hollow-core conductor. The cable has paper insulation with an insulation screen composed of four layers of carbon tapes. On top of insulation screen is a lead sheath reinforced with three layers of copper tapes with 50% overlap. The tapes are 25 mm wide and 1.3 mm thick. The lay of tapes is equal to 115.2 mm. Armor is composed of 51, #4 AWG copper wires with a lay length of 121.8 mm. The armor bedding is saturated jute and a layer of polyethylene with the equivalent thermal resistivity 4.27 K · m/W. The armor serving is a saturated jute. All remaining thermal and electrical parameters are as specified in IEC 60287 (1994).

The cable cross section is shown in Figure A4. Three cables are laid in a thermal backfill as shown in Figure A5.

Soil ambient temperature is 15°C. Thermal resistivities of the soil and backfill are shown in Figure A5. Cable sheaths are two-point bonded.

MODEL CABLE NO. 5

This is a 400 kV, paper–polypropylene–paper cable with 2000 mm^2 copper segmental conductor and aluminum corrugated sheath. The outer covering is a PE jacket. The cable cross section is shown in Figure A6.

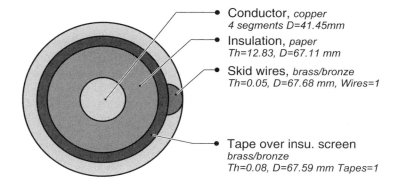

Conductor, *copper*
4 segments D=41.45mm

Insulation, *paper*
Th=12.83, D=67.11 mm

Skid wires, *brass/bronze*
Th=0.05, D=67.68 mm, Wires=1

Tape over insu. screen
brass/bronze
Th=0.08, D=67.59 mm Tapes=1

Voltage = *138.0 kV*   Cond. Area = *800.000 mm²*

Figure A3   Cross section of the model cable No. 3.

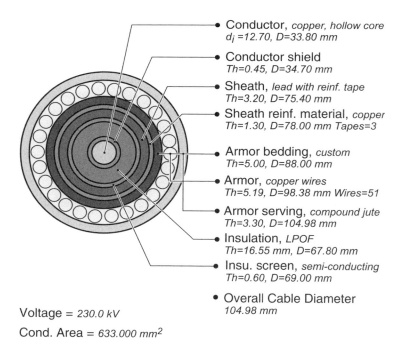

Conductor, *copper, hollow core*
d_i =12.70, D=33.80 mm

Conductor shield
Th=0.45, D=34.70 mm

Sheath, *lead with reinf. tape*
Th=3.20, D=75.40 mm

Sheath reinf. material, *copper*
Th=1.30, D=78.00 mm Tapes=3

Armor bedding, *custom*
Th=5.00, D=88.00 mm

Armor, *copper wires*
Th=5.19, D=98.38 mm Wires=51

Armor serving, *compound jute*
Th=3.30, D=104.98 mm

Insulation, *LPOF*
Th=16.55 mm, D=67.80 mm

Insu. screen, *semi-conducting*
Th=0.60, D=69.00 mm

Overall Cable Diameter
104.98 mm

Voltage = *230.0 kV*

Cond. Area = *633.000 mm²*

Figure A4   Cross section of the model cable No. 4.

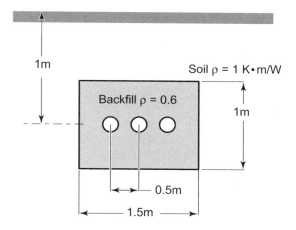

Figure A5   Laying conditions for model cable No. 4.

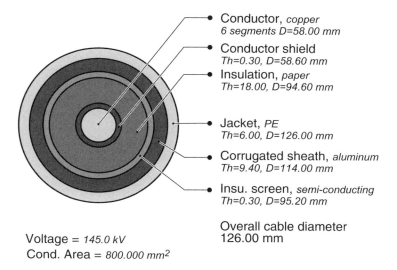

Conductor, *copper*
6 segments D=58.00 mm

Conductor shield
Th=0.30, D=58.60 mm

Insulation, *paper*
Th=18.00, D=94.60 mm

Jacket, *PE*
Th=6.00, D=126.00 mm

Corrugated sheath, *aluminum*
Th=9.40, D=114.00 mm

Insu. screen, *semi-conducting*
Th=0.30, D=95.20 mm

Overall cable diameter
126.00 mm

Voltage = *145.0 kV*
Cond. Area = *800.000 mm²*

Figure A6   Cross section of the model cable No. 5.

The cables are laid in a flat formation without transposition, directly in the soil, with thermal resistivity of 1 K · m/W and ambient temperature equal to 25°C. The sheaths are cross bonded with unknown minor section lengths. The centers of the cables are 1.8 m below the ground and phases are 0.5 m apart.

CABLE MODEL NO. 6

This is a 110 kV cable with a Milliken copper conductor of 1000 mm², an insulation thickness of 18 mm, and a copper screen of 100 mm² cross section under the cable jacket. The outer diameter of the cable core is 90 mm. The copper screens are single-point bonded, so that there are no losses caused by circulating currents. The cable cross section is shown in Figure A7.

The circuit is laid 1 m underground with an axial distance of 0.4 m. For a thermal resistivity of moist soil of 1.0 K · m/W and for dry soil of 2.5 K · m/W, respectively, as well as an ambient soil temperature of 20°C, the cable has a rating of 1098 A with 100% load factor.

Laying the cables in a bundled formation yields a rating of 932 A, if the copper screens are single-point bonded; and 791 A, if the copper screens are solidly bonded. If the cables are bundled and laid into a nonmetallic pipe with an inner diameter of 300 mm, the rating becomes 978 A, whereas the losses of a magnetic pipe with the same dimensions will reduce the rating to 835 A.

The symbols used in this table and the quantities they represent are given in the following list. Following the practice adopted in the IEC Standard 60287, all cable component diameters are given in millimeters.

D_a' = external diameter of armor, mm
D_d = internal diameter of a duct or pipe, mm
D_e = external diameter of cable, or equivalent diameter of a group of cores in pipe-type cable, mm
D_i = diameter over insulation, mm
D_s = external diameter of metal sheath, mm
D_{oc} = the diameter of the imaginary coaxial cylinder that just touches the crests of a corrugated sheath, mm
D_{is} = diameter over core screen, mm
D_{it} = the diameter of the imaginary cylinder that just touches the inside surface of the troughs of a corrugated sheath, mm
D_o = external diameter of duct or pipe, mm
D_s = external diameter of metal sheath, mm
I = current in one conductor (r.m.s. value), A
R = alternating current resistance of conductor at its maximum operating temperature, Ω/m
R = distance between conductor centers of a three-core cable, mm
R_{s0} = resistance of sheath at 20°C, Ω/m
R_0 = dc resistance of conductor at 20°C, Ω/m

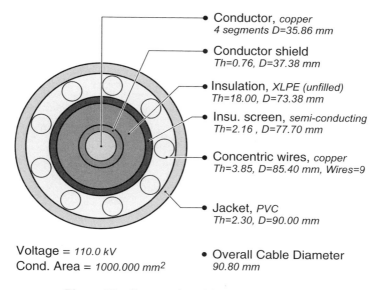

Conductor, *copper*
4 segments D=35.86 mm

Conductor shield
Th=0.76, D=37.38 mm

Insulation, *XLPE (unfilled)*
Th=18.00, D=73.38 mm

Insu. screen, *semi-conducting*
Th=2.16 , D=77.70 mm

Concentric wires, *copper*
Th=3.85, D=85.40 mm, Wires=9

Jacket, *PVC*
Th=2.30, D=90.00 mm

Voltage = *110.0 kV*
Cond. Area = *1000.000 mm²*

Overall Cable Diameter
90.80 mm

Figure A7　Cross section of the model cable No. 6.

Table A1　Parameters of model cables

	Cable No. 1	Cable No. 2	Cable No. 3	Cable No. 4	Cable No. 5	Cable No. 6
			Construction			
	Cu stranded	Cu stranded	Cu segmental	Cu hollow core	Cu segmental	Cu segmental
Conductor						
S (mm²)	300	800	1010	633	2000	1000
d_c (mm)	20.5	33.8	41.45	33.8	58.0	35.86
d_{cs} (mm)	21.7	37.4	41.45	34.7	58.6	37.48
d_i (mm)				17.5		
c (mm)		40.07				
s (mm)		69.4	67.59			
Insulation	XLPE	XLPE	Paper	Paper	PPLP	XLPE
D_i (mm)	28.5	67.4	67.11	67.8	94.6	73.38
D_{is} (mm)	30.1	69.4	67.26	69.0	95.2	77.70
t_i (mm)		15.0				
t (mm)		32.0				
t_1 (mm)		16.0				
Screen/sheath	Cu wires	Lead	Tape and 1 skid wire	Lead with 3 Cu tapes	Corrugated aluminum	9 Cu wires
D_s (mm)	31.4	154.4	67.59	75.4		85.4
D_{oc} (mm)					102.0	

Table A1   *Continued*

	Cable No. 1	Cable No. 2	Cable No. 3	Cable No. 4	Cable No. 5	Cable No. 6
			Given cable parameters			
	Cu stranded	Cu stranded	Cu segmental	Cu hollow core	Cu segmental	Cu segmental
D_{it} (mm)					113.0	
d (mm)		152.04		73.8		
d_2 (mm)				76.0 (78 over tape)		
t_s (mm)		2.5		3.2	6.0	3.85
Armor		Steel wires touching— 50 wires		Copper		
D'_a (mm)		167.14		98.4		
δ_0 (mm)		3.0		5.189		
Jacket–bedding	PVC	Jute	Somastic pipe coating	Jute– polyethylene– jute	PE	PVC
t_2 (mm)		3.3		5.0		
t_3 (mm)		4.0		3.3		
t_i (mm)	2.2				6.0	2.30
D_e (mm)	35.8	175.13	244.48	105	125.2	90.0
Duct or pipe			Steel			
D_d (mm)			206.38			
D_0 (mm)			219.08			
			Given cable parameters			
θ (°C)	90	90	70	85	85	90
f (Hz)	50	50	60	50	60	50
ε			3.5	3.5	2.8	2.5
$\tan \delta$			0.005	0.0033	0.001	0.001
R_0 (ohm/km)	0.0601					
R_{s0} (ohm/km)	0.759					
			Computed parameters (hottest cable)			
I (A)	629	942	902	771	1365	1098
R (ohm/km)	0.0781	0.033	0.0245	0.0356	0.0126	0.02312
y_s	0.014	0.097	0.050	0.031	0.132	0.03
y_p	0.0047	0.075	0.054	0.001	0.005	0.001
X (ohm/km)	0.096		0.052	0.154	0.168	0.141
X_1 (ohm/km)	0.102			0.170	0.186	
X_m (ohm/km)	0.044			0.044	0.052	0.044
λ_1	0.090	0.123	0.010	0.229	0.150	0
λ_2		0.210	0.311	0.990		0
T_1 (K · m /W)	0.214	0.754	0.422	0.568	0.579	0.431
T_2 (K · m /W)		0.040	0.082	0.082		0.05

Table A1 *Continued*

	Cable No. 1	Cable No. 2	Cable No. 3	Cable No. 4	Cable No. 5	Cable No. 6
			Given cable parameters			
	Cu stranded	Cu stranded	Cu segmental	Cu hollow core	Cu segmental	Cu segmental
T_3 (K · m /W)	0.104	0.047	0.017	0.066	0.056	0
T_4 (K · m/W)	1.933	0.288	0.343 (0.289)[1]	0.814	1.276	2.806[2]
W_c (W/m)	30.85	29.16	19.93	22.11	23.40	27.90
W_d (W/m)	0	0.58	4.83	5.46	6.53	0.26
W_I (W/m)	33.59	38.86	26.32	49.08	26.92	27.90
W_t (W/m)	33.59	118.32	31.15	54.53	33.45	28.16

[1]The second T_4 value corresponds to the Neher/McGrath method with load-loss factor $\mu = 0.85$.
[2]The external thermal resistance of this cable includes the mutual heating effect.

S = cross-sectional area of conductor, mm^2

T_1 = thermal resistance per core between conductor and sheath, K · m/W

T_2 = thermal resistance between sheet and armor, K · m/W

T_3 = thermal resistance of external serving, K · m/W

T_4 = thermal resistance of surrounding medium (ratio of cable surface temperature rise above ambient to the losses per unit length), K · m/W

W_c = losses in conductor per unit length, W/m

W_I = total I^2R power loss of each cable, W/m

W_d = dielectric losses per unit length per phase, W/m

W_t = total power dissipated in the cable per unit length, W/m

X = reactance of sheath (two-core cables and three-core cables in trefoil arrangement), W/m

X_1 = reactance of sheath (cables in flat formation), Ω/m

X_m = mutual reactance between the sheath of one cable and the conductors of the other two when cables are in flat formation, Ω/m

c = distance between the axes of conductors and the axis of the cable for three-core cables (= $0.55\, r_1 + 0.29\, t$ for sector-shaped conductors), mm

d = mean diameter of sheath or screen, mm

d_2 = mean diameter of reinforcement, mm

d_c = external diameter of conductor, mm

d_{cs} = external diameter of conductor shield, mm

d_i = internal diameter of hollow conductor, mm

f = system frequency, Hz

r = distance between a conductor center and the cable center in a three-core cable, mm

s = axial separation of conductors, mm

t = insulation thickness between conductors, mm

t_1 = insulation thickness between conductors and sheath, mm

t_2 = thickness of the bedding, mm

t_3 = thickness of the serving, mm

t_i = thickness of core insulation, including screening tapes plus half the thickness of any nonmetallic tapes over the laid-up cores, mm

t_J = thickness of the jacket, mm

t_s = thickness of the sheath, mm

y_p = proximity-effect factor

y_s = skin-effect factor

δ_0 = equivalent thickness of armor or reinforcement, mm

$\tan \delta$ = loss factor of insulation

ε = relative permittivity of insulation

θ = maximum operating temperature of conductor, °C

λ_1, λ_2 = ratio of the total losses in metallic sheaths and armor, respectively, to the total conductor losses (or losses in one sheath or armor to the losses in one conductor)

REFERENCES

Anders, G.J. (1997), *Rating of Electric Power Cables—Ampacity Calculations for Transmission, Distribution and Industrial Applications,* IEEE Press, New York, McGraw-Hill, (1998).

IEC Standard 228 (1978) "Conductors of Insulated Cables," Second edition. First Supplement (1982).

IEC Standard 60287 (1969, 1982, 1994), "*Calculation of the Continuous Current Rating of Cables (100% load factor),*" 1st edition 1969, 2nd edition 1982, 3rd edition 1994–1995.

IEC Standard 60287, Part 2-1 (1994), "Calculation of thermal resistances."

Neher, J.H., and McGrath, M.H. (1957), "Calculation of the Temperature Rise and Load Capability of Cable Systems," *AIEE Trans.,* Vol. 76, Part 3, pp. 755–772.

Computations of the Mean Moisture Contents in Media Surrounding Underground Power Cables

INTRODUCTION

The soils used in underground cable installations are composed of granulated minerals, water, air, and organic materials. The thermal resistivities of these components range from 0.11 K·m/W for quartz to 0.165 K·m/W for water and 40.0 K·/W for air. To obtain the minimum thermal resistivity, the soil should contain the maximum amount of rock particles and the minimum amount of air.

As it is impossible to pack the solid particles without interstices, it is better to have them filled with water than with air. Water also acts as a binder helping to compact the soil.

In general, the factors appearing to affect the thermal resistivity of most soils are:

1. Soil density
2. Moisture content
3. The volume of air voids in relation to the volume of water
4. Particle surface area per unit volume
5. Mineral composition of solids

Moisture content, in particular, is not constant but varies according to climatic conditions, and may vary because of the thermal gradients caused by the heat dissipated by buried power cables. The method of evaluating moisture content in the soil discussed here deals with moisture fluctuations in the soil caused by climatic conditions only.

Formulas developed by Thornthwaite for agricultural purposes are adopted in the case of cable backfills and, in the Toronto area, climate (Vassallo, 1969; Anders and Radhakrishna, 1988).

SOIL–WATER BALANCE

Water precipitates in the form of snow, rain, or dew. Part of it is lost as surface runoff, part of it percolates through the soil layers, and the remainder is retained in

Rating of Electric Power Cables in Unfavorable Thermal Environment. By George J. Anders
ISBN 0-471-67909-7 © 2004 the Institute of Electrical and Electronics Engineers.

the soil. The precipitation fraction that is retained depends on the type of soil and its previous moisture content. The maximum moisture content that a soil can sustain for any extended period is defined as the field capacity, usually expressed as a percentage by weight of dry soil or as "inches of water" (per foot of depth of soil). Water returns to the atmosphere by evaporation from ground surfaces and by transpiration from vegetation.

The interest here is limited to evaporation from soil surfaces and transpiration from plants; these two effects combined are called "evapotranspiration." There is a distinction between the amount of water that actually evapotranspirates and the amount that would transpire and evaporate if it were available.

If enough water is available, evapotranspiration rises to a maximum that depends only on the climate. This may be called "potential evapotranspiration." Thus, the potential evapotranspiration does not represent the actual transfer of water to the atmosphere but the transfer that would take place under ideal conditions of moisture content and vegetation. Nevertheless, the research reported Thornthwaite (1948) and Vassello (1969) indicate that the results obtained considering the potential evapotranspiration as a basic element are consistent with those obtained with independent investigations.

DETERMINATION OF POTENTIAL EVAPOTRANSPIRATION

In each area, the relation between mean weekly temperature and potential evapotranspiration can be expressed by an equation of the form

$$PE = 0.145 \cdot [50 \cdot (TW - 32)/(9 \cdot I)]^A \tag{B1}$$

where:
PE = weekly potential evapotranspiration (inches of water)
TW = mean temperature for the past week in degrees F
I = heat index for the current year
A = a constant function of I

The constant I is given as a sum of 12 monthly heat indexes, each given by the equation

$$i = [(TM - 32)/9]^{1.514} \tag{B2}$$

where TM is the mean monthly temperature.
Thus,

$$I = \sum_{i=1}^{12} i \tag{B3}$$

The relationship between I and A is closely approximated by the expression

$$A = 675 \cdot 10^{-9} \cdot I^3 - 771 \cdot 10^{-7} \cdot I^2 + 10^{-5} \cdot I + 0.49239 \tag{B4}$$

Equation (B1) gives unadjusted values of potential evapotranspiration, as this equation was first developed for a standard month of 30 days, each day having 12 hours of possible sunshine. It is necessary to apply a correction factor because the number of sunlight hours, when evapotranspiration is higher, varies with the season and latitude, and the month length ranges from 28 to 31 days.

The correction factors C, according to Thornthwaite (1948), for the Toronto area are 0.81, 0.82, 1.02, 1.12, 1.26, 1.28, 1.29, 1.20, 1.04, 10.95, 0.81, and 0.77 for January to December. The adjusted potential evapotranspiration, PEC, is now given by

$$PEC = PE \cdot C \tag{B5}$$

WEEKLY SOIL–WATER BALANCE CALCULATIONS

The computations of weekly water balance in the soil start in the month in which frost leaves the ground. It is assumed that at this point the soil is saturated with water; that is, its moisture content is at field capacity. The starting and ending dates in a year for water balance computation have been established by examining the mean soil temperatures provided by the Weather Service Bureau, in this case, by Environment Canada for the 66-year period.

The total moisture content in a given week is obtained by summing the soil moisture content of the previous week and the weekly total precipitation. The actual evapotranspiration, AE, is assumed to be proportional to potential evapotranspiration, PEC. In fact, when the soil moisture content falls below the field capacity, the rate of evapotranspiration falls too; thus

$$AE = \frac{\text{Previous balance}}{\text{Field capacity}} \cdot PEC \tag{B6}$$

The water balance is obtained by subtracting the actual evapotranspiration from the total moisture content.

REFERENCES

Anders, G.J., and Radhakrishna, H.J. (1988), "Computation on the Temperature Field and Moisture Content in the Vicinity of Current Carrying Underground Power Cables," *IEE Proceedings,* Part C, Vol. 153, No.1, pp. 51–62.

Thornthwaite, C.W. (1948), "An Approach Toward a Rational Classification of Climate," *Geographical Review,* Vol. 38.

Vassallo, D.G. (1969), "The Influence of Climatic Conditions on the Moisture Content of Cable Backfills," Ontario Hydro Research Division Report, No. E69-9-K, February 1969.

Estimation of Backfill Thermal Resistivity

For the purpose of Monte Carlo analysis discussed in Chapter 4, the following procedure has been developed for modification of the thermal resistance of the backfill to take into account the moisture migration problem due to thermal gradients.

First, we need to create the curves that relate the effect of the cable heat rate on the moisture content in the soil surrounding the cables. Examples of such curves for the cable system described in Chapter 4 have been established for several representative initial moisture conditions φ_i in the soil and are shown in Figure C1.

The information contained in the curves in Figure C1 is used in the analysis in the following way. At the beginning of each Monte Carlo simulation, values of W and φ_i are selected at random from the respective histograms. Next, the initial moisture content φ_i is used to select the appropriate "graph" and the value of W is used to select the appropriate "curve" in the chosen graph. The moisture content at the cable surface read from this curve is recorded and the thermal resistivity computed form the curves in Figure 4-13. This resistivity is used for the entire backfill zone. We also need to establish whether drying out can possibly occur. It is reasonable to assume that if it occurs, the entire backfill will be affected. The detailed analysis of the moisture migration in the vicinity of the cables considered in Chapter 4 has shown that low moisture content prevails in the entire backfill zone (see Figure C2) and the moisture content increases quite rapidly above this zone (see Figure C1).

One way to create the curves shown in Figure C2 would be to conduct laboratory experiments for the sand under consideration as described, for instance, in Groeneveld et al. (1984). Another possibility is to use available analytical tools to determine the steady-state moisture distribution around the cable for different values of W.

In the example described in Chapter 4, the heat-rate values, W, range from 9.6 W/m to 25.6 W/m with all circuits in service and from 15.0 W/m to 33.0 W/m with one circuit out of service. These losses were established for the study years based on the given total peak demand at stations A and B. Curves similar to those shown in Figure C1 could be developed for values of losses varying by 1 W/m in the defined range.

Analysis of the results of moisture content computations in downtown Toronto, performed with the aid of the Thornthwaite method for the 66-year period, revealed

Rating of Electric Power Cables in Unfavorable Thermal Environment. By George J. Anders   **317**
ISBN 0-471-67909-7 © 2004 the Institute of Electrical and Electronics Engineers.

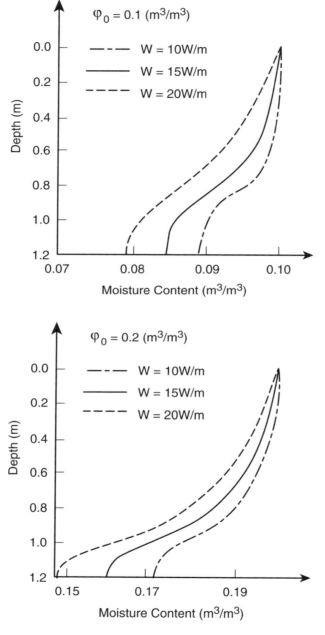

Figure C1   Effect of the heat rate W on the moisture distribution in the sand.

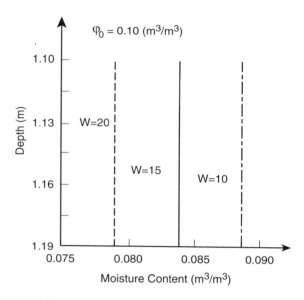

Figure C2   Moisture content profile in the immediate vicinity of the cables.

that the moisture content in the sandy backfill varies between 32 and 95% of field capacity in the month of August. Ten families of curves shown in Figure C1 have been developed for the initial moisture content in the range of 32–95% of field capacity.

The following considerations were taken into account when computer runs were performed to obtain the graphs shown in Figure C1. Since the Thornthwaite method already takes into account the loss of moisture in the soil due to evapotranspiration, the computer results should reflect the migration of the moisture due to the thermal gradients alone. This has been achieved by specifying constant moisture boundaries at the ground surface and at the bottom of the study region (5 m), with the moisture values equal to the initial moisture content. Another boundary consideration that should be taken into account is the value of the ambient air and soil temperatures. In the computer runs performed here, these values were kept constant at 20°C and 15°C, respectively. Sensitivity studies that were performed have shown that the variations in the boundary temperatures have a very small effect on the computed moisture content.

REFERENCE

Groeneveld, G.J., Snijders, A.L., Koopmans, G., and Vermeer, J. (1984), "Improved Method to Calculate the Critical Conditions for Drying Out Sandy Soils Around Power Cables," *IEE Proceedings,* Vol. 131, Part C, No. 2, pp. 42–53.

Equations for Dielectric Fluid Parameters

The following equations represent the dependence of oil parameters on temperature. These relations were used in sample calculations in Chapter 7. All equations are valid for $\theta \in \langle 0°C, 110°C \rangle$ with an exception of the kinematic viscosity, which is valid for $\theta \in \langle 20°C, 110°C \rangle$.

Thermal diffusivity:

$$a_f(\theta) = -1.696 \cdot 10^{-10} \cdot \theta + 1.67 \cdot 10^{-7} \ [m^2/s]$$

Kinematic viscosity:

$$\mu_f(\theta) = 164.25 \cdot \theta^{-2.0072} \ [N \cdot s/m^2]$$

Density:

$$\rho_f(\theta) = 903.6 - 0.566 \cdot \theta \ [kg/m^3]$$

Thermal conductivity:

$$\lambda_f(\theta) = -1.5855 \cdot 10^{-4} \cdot \theta + 0.2095 \ [W/(K \cdot m)]$$

Rating of Electric Power Cables in Unfavorable Thermal Environment. By George J. Anders
ISBN 0-471-67909-7 © 2004 the Institute of Electrical and Electronics Engineers.

▰▰▰ Index

absorption coefficient, 24
absorptivity, 4
air convection in pipe, 102
air thermal conductivity, 104, 252
air viscosity, 253, 321
armor loss factor, 20, 44, 49
   magnetic armor, 50
   nonmagnetic wire, 50
   three-conductor cables, 51
artificial cooling, 280
   forced circulation, 281
   slow circulation, 281
attainment factor, 26, 31, 149, 151
average losses
   daily, 232
   weekly, 232

backfill thermal resistivity. *See* thermal
   resistivity
bonding arrangements, 272
Buller's model, 282, 287

cable bundle
   air filling factor, 255
   conductor filling factor, 255
   heat conduction, 254
   stranding factor, 255
   unequal loading, 262
cable crossing
   cyclic rating, 147
   derating factor, 124, 162
   heating by a steam pipe, 131
   important factors, 122
   influence of the crossing angle, 138
   influence of the heat source distance, 138
   multiple heat sources, 134
   mutual thermal resistance, 163
   remedial actions, 123

   screen longitudinal heat flow, 140
   simulatneous heating effects, 138
   soil dryout, 156
   transient rating, 145
   utility practices, 122
cable earthing, 272
cable material properties, 53
cable reactance, 181
cables in air, 24
   external thermal resistance, 64
   multi-layer installation, 270
   on floors or trays, 267
   on ladders, 269
   on ventilated trays, 268
   several cables per phase, 270
   under ceiling, 268
cables in pipes, 102
characteristic diameter, 226, 232, 238
   arbitrary load, 234
   sinusoidal load, 233
   rectilinear load, 234
circulating current losses, 41
conductor electrical resistance, 38
conductor joule losses, 82
conductor losses, 214
convection coefficient, 250, 252
critical current, 217, 221
critical isotherm, 23
critical temperature, 217
cross-bonded systems, 42
cycle length, 226
cyclic load, 147
cyclic rating, 90
cyclic rating factor, 32, 90

daily load cycle, 224
   Neher-McGrath approach, 225
daily load factor, 60

Rating of Electric Power Cables in Unfavorable Thermal Environment. By George J. Anders   **323**
ISBN 0-471-67909-7 © 2004 the Institute of Electrical and Electronics Engineers.

daily loss factor, 226
decision analysis, 201
decision tree. *See* decision analysis
deeply buried cables, 237
derating factors, 84, 131
   cable crossing, 124
   groups of cables in air, 266
dielectric constant, 40
dielectric fluid
   density, 288
   kinematic viscosity, 288
dielectric fluid circulation
   thermal effects, 283
dielectric loss factor, 41
dielectric losses, 10, 40, 220
directly drilled tunnels, 212
dry zone, 22, 99
dynamic feeder rating. *See* real time rating

eddy current losses, 41, 46
effective cooling distance, 287
effective thermal resistivity, 255
electrical and thermal networks analogy, 9
electrical capacitance of insulation, 41
emissivity, 3, 251
energy balance equation, 6, 127, 141, 250
energy losses, 217
expected monetary value. *See* decision
   analysis
exponential integral, 30
external heat source, 125
external thermal resistance, 60
   between oil and ambient, 283
   buried cables, 60
   cables in air, 64, 251
   cables in backfill, 62
   cables in ducts, 63
   cables in tunnel, 240
   correction factor, 172
   groups of cables in air, 67
   Neher-McGrath approach, 171
   unequally loaded cables, 64
   with load variations, 225

fictitious diameter. *See* characteristic
   diameter
field capacity. *See* soil moisture content
fluid continuity equation, 292
fluidized thermal backfill, 167

forced convection coefficient, 254
forced cooling, 290
Fourier-Kirchoff equation, 292
Fourier's law, 2, 81, 127
friction coefficient of the air flow, 104

geometric factor, 172
Grashoff number, 252, 284
groups of cables, 30

heat conduction, 2
heat convection, 3, 252
   forced, 3
   heat transfer coefficient, 3
   natural, 3
heat isolating plate, 123
heat transfer coefficient, 251
heat transfer equations, 2, 5
heat transfer rate, 4
hydraulic diameter, 284

insulation thermal resistance. *See* thermal
   resistance
   SL-type cables, 58
   three-core shielded cables, 55
internal thermal capacitance, 11
internal thermal resistance, 11, *See* thermal
   resistance

Kennelly's principle, 29, 126
kinematic viscosity of air, 104

ladder network, 10
   long-duration transients, 11
   short-duration transients, 13
   two-loop equivalent, 15
law of conservation of energy, 4
line heat source, 166
load factor, 227
load-loss factor, 34, 60, 213, 224, 244
longitudinal heat flow, 80, 106
longitudinal thermal resistance, 78, 126
loss factor. *See also* load loss factor
   armor, 20, 44
   daily, 228, 231
   sheath, 20, 44
   weekly, 231
   yearly, 237
loss factor approximations, 226

lump parameter network. *See* ladder network

magnetic field reduction, 113
magnetic induction, 115
moisture migration, 21, 99
   two-zone model, 22
Monte Carlo simulation, 188, 194, 196
multicore cables, 247
multicore cables, 247
multiple cables per phase, 270
mutual thermal resistance, 131, 134, 158,
   See also thermal resistance

natural convection, 252
Navier-Stokes equation, 292
Newton's law, 3, 9
Nusselt number, 252, 284, 295

parametric analysis, 170
Peclet number, 284, 295
pipe loss factor, 52
potential evapotranspiration. *See* soil
   moisture content
Prandtl number, 252
principle of superposition, 61
probabilistic thermal analysis, 188
probability distributions
   ambient temperature, 190
   load, 190
   soil and backfill, 190
properties of insulating oil, 297
proximity effect, 39
proximity effect factor, 39

radial heat flow, 80, 127
radiation
   radiative heat transfer rate, 4
radiation heat transfer coefficient, 9, 252
   Black's approximation, 252, 257
   Neher-McGrath approximation, 252
rated current, 215
rating equations
   steady-state, 19
real time rating, 211, 223
reinforcing tape
   length of lay, 50
remote terminal unit, 211

Reynolds number, 104, 254, 284, 289, 290, 296
RTU. *See* remote terminal unit

screening factor, 56
sector-shaped conductors
   proximity effect factor, 39
   skin effect factor, 39
sheath bonding arrangements, 42
sheath loss factor, 20, 41, 44, 181
   bonded both ends-flat configuration, 45
   bonded both ends-triangular
      configuration, 45
   cross bonded system, 47
   large segmental conductors, 46
   pipe-type cables, 49
   single-point bonded system, 46
   three-conductor cables, 48
sheath reactance, 44
   flat transposed formation, 45
   trefoil formation, 44
sheath resistance, 44
single-point-bonded system, 42
skin effect, 39
slow oil circulation
   real time ratings, 291
soil diffusivity, 29, 225
soil dryout, 99, 156, 213
soil moisture content, 192, 226
   field capacity, 314
   potential evapotranspiration, 314
   weekly soil-water balance, 315
solar absorption coefficient, 67
solar radiation, 24
solidly-bonded systems, 42
specific heat of cable materials, 53
Stefan-Boltzmann constant, 250
Stefan-Boltzmann law, 3
stochastic optimization, 202
   assumption variables, 203
   decision variables, 202
   forecast variables, 203
   logistic distribution, 205
stranding factor, 255
strength probability distribution, 199
superposition principle, 8

telecommunication cables, 248
thermal backfill, 62

thermal backfill *(continued)*
  geometric factor, 172
  Neher-McGrath approach, 62
thermal backfill design
  constraints, 173
  parameters, 169
  sensitivity measures, 179
  tornado chart, 169
thermal capacitance, 8, 10, 69, 223
  armor, 72
  concentric layers, 72
  conductor, 71
  insulation, 71
  oil in pipe-type cables, 72
  oil in the conductor, 69
  reinforcing tape, 72
thermal conductivity of air, 104
thermal probe technique, 167
thermal property analyzer, 167
thermal radiation, 3
thermal resistance, 8
  between sheath and armor, 58
  cylindrical layer, 8
  equivalent, 21
  external, 20, 60
  internal, 91
  jacket, 249
  longitudinal, 78
  multicore cables, 264
  mutual, 125, 134, 158
  of insulation, 19, 54
  of oil film, 287, 295
  of outer covering, 19, 58
  pipe-type cables, 59
  three-conductor cables, 54

three-core cables with fillers, 55
  total, 125, 214, 244
thermal resistances, 52
thermal resistivities of cable materials, 53
thermal resistivity, 3
  measurements, 166
  of backfill, 317
  of soil, 192
thermal sands, 167
Thornthwaite model, 193, 313
three-core cable
  equivalent representation, 12
tornado chart, 169, 170
total heat transfer coefficient, 9
transfer function, 26
  poles and zeros, 27

unfavorable region
  cyclic rating, 90
  derating factor, 84

Van Wormer theory, 10
  coefficient, 12, 71
  coefficient for dielectric losses, 13
  coefficient for short duration transients, 13
ventilated pipes
  forced convection, 108
  magnetic field reduction, 113
ventilated routes, 101
volumetric specific heat, 10

weekly load cycle, 231
withstand probability distribution, 199